Contents

Preface to the Instructor IR-xi
Preface to the Student IR-xv

CHAPTER 1 — Integrated Review: Getting Ready for Organizing and Summarizing Data IR-1

Note: The material in this chapter should be covered in conjunction with Chapter 1 Data Collection and prior to Chapter 2 Organizing and Summarizing Data.

1.IR1 Fundamentals of Fractions IR-1
    1.IR1.1 Identify Terms Used with Fractions IR-1
    1.IR1.2 Find Equivalent Fractions IR-4
    1.IR1.3 Write a Fraction in Lowest Terms IR-5

1.IR2 Fundamentals of Decimals IR-9
    1.IR2.1 Understand the Place Value of a Decimal Number IR-9
    1.IR2.2 Read and Write Decimals in Words IR-10
    1.IR2.3 Write Decimals as Fractions IR-12
    1.IR2.4 Round Decimals to a Given Place Value IR-13

1.IR3 The Real Number Line IR-16
    1.IR3.1 Classify Numbers IR-16
    1.IR3.2 Plot Points on a Real Number Line IR-19
    1.IR3.3 Use Inequalities to Order Real Numbers IR-20
    1.IR3.4 Compute the Absolute Value of a Real Number IR-23

1.IR4 Multiplying and Dividing Fractions IR-24
    1.IR4.1 Multiply Fractions IR-24
    1.IR4.2 Divide Fractions IR-27

1.IR5 Adding and Subtracting Fractions IR-31
    1.IR5.1 Add or Subtract Fractions with Like Denominators IR-31
    1.IR5.2 Find the Least Common Denominator and Equivalent Fractions IR-34
    1.IR5.3 Add or Subtract Fractions with Unlike Denominators IR-36

1.IR6 Operations on Decimals IR-42
    1.IR6.1 Add or Subtract Decimals IR-42
    1.IR6.2 Multiply Decimals IR-46
    1.IR6.3 Divide a Decimal by a Whole Number IR-50
    1.IR6.4 Divide a Decimal by a Decimal IR-53
    1.IR6.5 Divide Decimals by Powers of 10 IR-54
    1.IR6.6 Convert a Fraction to a Decimal IR-55

1.IR7 Fundamentals of Percent Notation IR-58
    1.IR7.1 Define Percent IR-58
    1.IR7.2 Convert Percents to Decimals and Decimals to Percents IR-59
    1.IR7.3 Convert Percents to Fractions and Fractions to Percents IR-61

1.IR8 Language Used in Modeling IR-65
    1.IR8.1 Use Models That Involve Addition and Subtraction IR-65
    1.IR8.2 Use Models That Involve Multiplication IR-66
    1.IR8.3 Use Models That Involve Division IR-67

CHAPTER 2 — Integrated Review: Getting Ready for Numerically Summarizing Data IR-70

Note: The material in this chapter should be covered in conjunction with Chapter 2 Organizing and Summarizing Data, and prior to Chapter 3 Numerically Summarizing Data.

2.IR1 Exponents and the Order of Operations IR-70
    2.IR1.1 Evaluate Exponential Expressions IR-70
    2.IR1.2 Apply the Rules for Order of Operations IR-72

2.IR2 Square Roots IR-78
    2.IR2.1 Evaluate Square Roots IR-78
    2.IR2.2 Determine Whether a Square Root Is Rational, Irrational, or Not a Real Number IR-80

2.IR3 Simplifying Algebraic Expressions; Summation Notation IR-83
    2.IR3.1 Evaluate Algebraic Expressions IR-83
    2.IR3.2 Identify Like Terms and Unlike Terms IR-84
    2.IR3.3 Use the Distributive Property IR-86
    2.IR3.4 Simplify Algebraic Expressions by Combining Like Terms IR-87
    2.IR3.5 Use Summation Notation IR-88

2.IR4 Solving Linear Equations IR-91
    2.IR4.1 Determine Whether a Number Is a Solution to an Equation IR-91
    2.IR4.2 Solve Linear Equations IR-93
    2.IR4.3 Determine Whether an Equation Is a Conditional Equation, an Identity, or a Contradiction IR-97

2.IR5 Using Linear Equations to Solve Problems IR-101
    2.IR5.1 Translate English Sentences into Mathematical Statements IR-101
    2.IR5.2 Model and Solve Direct Translation Problems IR-102
    2.IR5.3 Solve for a Variable in a Formula IR-106

CHAPTER 3 — Integrated Review: Getting Ready for Least-Squares Regression IR-111

Note: The material in this chapter should be covered in conjunction with Chapter 3 Numerically Summarizing Data, and prior to Chapter 4 Describing the Relation between Two Variables.

3.IR1 The Rectangular Coordinate System and Equations in Two Variables IR-111
    3.IR1.1 Plot Points in the Rectangular Coordinate System IR-111
    3.IR1.2 Determine Whether an Ordered Pair Satisfies an Equation IR-115
    3.IR1.3 Create a Table of Values That Satisfy an Equation IR-117

3.IR2 Graphing Equations in Two Variables IR-124
    3.IR2.1 Graph a Line by Plotting Points IR-124
    3.IR2.2 Graph a Line Using Intercepts IR-129
    3.IR2.3 Graph Vertical and Horizontal Lines IR-132

3.IR3 Slope IR-137
    3.IR3.1 Find the Slope of a Line Given Two Points IR-137
    3.IR3.2 Find the Slope of Vertical and Horizontal Lines IR-140

INTEGRATED REVIEW MATERIALS
to accompany
Sullivan Statistics

INTEGRATED REVIEW MATERIALS
to accompany
Sullivan Statistics

Michael Sullivan, III
Joliet Junior College

Photo Credits: p. 111 Science History Images/Alamy Stock Photo.
Cover Credits:
Interactive Statistics: Billion Photos/Shutterstock, Peter Kotoff/Shutterstock, Big Pants Production/Shutterstock
Sullivan Statistics: Africa Studio/Shutterstock
Texas Screenshots: All Texas Instruments Graphing Calculator graphs © copyright by Texas Instruments Education Technology

The author and publisher of this book have used their best efforts in preparing this book. These efforts include the development, research, and testing of the theories and programs to determine their effectiveness. The author and publisher make no warranty of any kind, expressed or implied, with regard to these programs or the documentation contained in this book. The author and publisher shall not be liable in any event for incidental or consequential damages in connection with, or arising out of, the furnishing, performance, or use of these programs.

Copyright © 2021, 2018, 2015 Pearson Education, Inc.
Publishing as Pearson, 330 Hudson Street, NY NY 10013

PEARSON, ALWAYS LEARNING, MYLAB(TM) STATISTICS are exclusive trademarks owned by Pearson Education, Inc. or its affiliates in the United States and/or other countries.

All rights reserved. No part of this publication may be reproduced, stored in a retrieval system, or transmitted, in any form or by any means, electronic, mechanical, photocopying, recording, or otherwise, without the prior written permission of the publisher. Printed in the United States of America.

5 2024

ISBN-13: 978-0-13-662530-8
ISBN-10: 0-13-662530-4

Contents **IR-vii**

      3.IR3.3 Graph a Line Using Its Slope and a Point on the Line IR-142

      3.IR3.4 Work with Applications of Slope IR-143

3.IR4 Slope-Intercept Form of a Line IR-147

      3.IR4.1 Use the Slope-Intercept Form to Identify the Slope and y-Intercept of a Line IR-147

      3.IR4.2 Graph a Line Whose Equation Is in Slope-Intercept Form IR-148

      3.IR4.3 Graph a Line Whose Equation Is in the Form $Ax + By = C$ IR-150

      3.IR4.4 Find the Equation of a Line Given Its Slope and y-Intercept IR-150

      3.IR4.5 Work with Linear Models in Slope-Intercept Form IR-151

3.IR5 Point-Slope Form of a Line IR-156

      3.IR5.1 Find the Equation of a Line Given a Point and a Slope IR-156

      3.IR5.2 Find the Equation of a Line Given Two Points IR-158

      3.IR5.3 Build Linear Models from Data IR-160

CHAPTER 4 Integrated Review: Getting Ready for Probability IR-167

Note: The material in this chapter should be covered in conjunction with Chapter 4 Describing the Relation between Two Variables, and prior to Chapter 5 Probability.

4.IR1 Scientific Notation IR-167

      4.IR1.1 Convert Decimal Notation to Scientific Notation IR-167

      4.IR1.2 Convert Scientific Notation to Decimal Notation IR-168

      4.IR1.3 Use Scientific Notation to Multiply and Divide IR-169

4.IR2 Interval Notation; Intersection and Union of Sets IR-173

      4.IR2.1 Represent Inequalities Using the Real Number Line and Interval Notation IR-173

      4.IR2.2 Determine the Intersection or Union of Two Sets IR-176

4.IR3 Linear Inequalities IR-179

      4.IR3.1 Solve Linear Inequalities Using Properties of Inequalities IR-179

      4.IR3.2 Model Inequality Problems IR-185

Note: There is no integrated review material to be covered in conjunction with Chapter 5 Probability, and prior to Chapter 6 Discrete Probability Distributions.

CHAPTER 6 Integrated Review: Getting Ready for the Normal Probability Distribution IR-189

Note: The material in this chapter should be covered in conjunction with Chapter 6 Discrete Probability Distributions, and prior to Chapter 7 The Normal Probability Distribution.

6.IR1 Perimeter and Area of Polygons and Circles IR-189

      6.IR1.1 Find the Perimeter and Area of a Rectangle and a Square IR-190

      6.IR1.2 Find the Perimeter and Area of a Parallelogram and a Trapezoid IR-196

      6.IR1.3 Find the Perimeter and Area of a Triangle IR-198

      6.IR1.4 Find the Circumference and Area of a Circle IR-198

Note: There is no integrated review material to be covered in conjunction with Chapter 7 The Normal Probability Distribution, and prior to Chapter 8 Sampling Distributions.

CHAPTER 8 — Integrated Review: Getting Ready for Confidence Intervals IR-205

Note: The material in this chapter should be covered in conjunction with Chapter 8 Sampling Distributions, and prior to Chapter 9 Estimating the Value of a Parameter.

8.IR1 Compound Inequalities IR-205
- **8.IR1.1** Solve Compound Inequalities Involving "and" IR-205
- **8.IR1.2** Solve Compound Inequalities Involving "or" IR-208
- **8.IR1.3** Solve Problems Using Compound Inequalities IR-209

8.IR2 Absolute Value Equations and Inequalities IR-213
- **8.IR2.1** Solve Absolute Value Equations IR-213
- **8.IR2.2** Solve Absolute Value Inequalities Involving $<$ or $\leq$ IR-216
- **8.IR2.3** Solve Absolute Value Inequalities Involving $>$ or $\geq$ IR-218
- **8.IR2.4** Solve Applied Problems Involving Absolute Value Inequalities IR-220

APPENDIX A — Functions, Exponential Functions, Logarithmic Functions IR-225

Note: The material in this chapter is optional and may be covered as part of an integrated review course.

A.IR1 Relations IR-225
- **A.IR1.1** Understand Relations IR-225
- **A.IR1.2** Find the Domain and the Range of a Relation IR-226
- **A.IR1.3** Graph a Relation Defined by an Equation IR-229

A.IR2 An Introduction to Functions IR-232
- **A.IR2.1** Determine Whether a Relation Expressed as a Map or Ordered Pairs Represents a Function IR-232
- **A.IR2.2** Determine Whether a Relation Expressed as an Equation Represents a Function IR-234
- **A.IR2.3** Determine Whether a Relation Expressed as a Graph Represents a Function IR-235
- **A.IR2.4** Find the Value of a Function IR-236
- **A.IR2.5** Find the Domain of a Function IR-238
- **A.IR2.6** Work with Applications of Functions IR-239

A.IR3 Functions and Their Graphs IR-243
- **A.IR3.1** Graph a Function IR-243
- **A.IR3.2** Obtain Information from the Graph of a Function IR-244
- **A.IR3.3** Know Properties and Graphs of Basic Functions IR-248
- **A.IR3.4** Interpret Graphs of Functions IR-249

A.IR4 Linear Functions and Models IR-253
- **A.IR4.1** Graph Linear Functions IR-253
- **A.IR4.2** Find the Zero of a Linear Function IR-254
- **A.IR4.3** Build Linear Models from Verbal Descriptions IR-256

Getting Ready for Exponential and Logarithmic Functions IR-263
- **1** Simplify Exponential Expressions Using the Product Rule IR-263
- **2** Simplify Exponential Expressions Using the Quotient Rule IR-264

     3 Evaluate Exponential Expressions with a Zero or Negative Exponent IR-264
     4 Simplify Exponential Expressions Using the Power Rule IR-267
     5 Simplify Exponential Expressions Containing Products or Quotients IR-268
A.IR5 Exponential Functions IR-271
  A.IR5.1 Evaluate Exponential Expressions IR-272
  A.IR5.2 Graph Exponential Functions IR-273
  A.IR5.3 Define the Number *e* IR-276
  A.IR5.4 Solve Exponential Equations IR-277
  A.IR5.5 Use Exponential Models That Describe Our World IR-278
A.IR6 Logarithmic Functions IR-285
  A.IR6.1 Change Exponential Equations to Logarithmic Equations IR-285
  A.IR6.2 Change Logarithmic Equations to Exponential Equations IR-286
  A.IR6.3 Evaluate Logarithmic Functions IR-286
  A.IR6.4 Determine the Domain of a Logarithmic Function IR-288
  A.IR6.5 Graph Logarithmic Functions IR-288
  A.IR6.6 Work with Natural and Common Logarithms IR-291
  A.IR6.7 Solve Logarithmic Equations IR-291
  A.IR6.8 Use Logarithmic Models That Describe Our World IR-293
A.IR7 Properties of Logarithms IR-296
  A.IR7.1 Understand the Properties of Logarithms IR-296
  A.IR7.2 Write a Logarithmic Expression as a Sum or Difference of Logarithms IR-297
  A.IR7.3 Write a Logarithmic Expression as a Single Logarithm IR-300
  A.IR7.4 Evaluate a Logarithm Whose Base Is Neither 10 Nor *e* IR-301
A.IR8 Exponential Equations IR-304
  A.IR8.1 Solve Exponential Equations IR-304
  A.IR8.2 Solve Equations Involving Exponential Models IR-306

Answers to Selected Exercises IR-AN-1

Preface to Instructors for Integrated Review

In an effort to increase completion rates at community colleges, four-year colleges, and universities, there has been a push to change the traditional mathematics course sequence for students who place into developmental mathematics courses. The traditional developmental mathematics course sequence has been a gatekeeper for many to achieve a college degree. One solution to eliminate the barrier to collegiate mathematics is the corequisite class, in which students are exposed to material typically taught in prerequisite courses (such as Intermediate Algebra) on a just-in-time basis in the corresponding college level course.

Integrated Review Materials to Accompany Sullivan Statistics is a solution for a corequisite statistics course. We carefully considered content presented in traditional developmental mathematics courses and developed a learning path for presenting material that students will need to be successful in Statistics. We also wrote new material that represents content that is normally not covered in a traditional developmental sequence, but would be of benefit to students. Of particular importance in selecting and writing material was to develop the mathematical thinking and maturity that comes with successful completion of a developmental mathematics curriculum. Therefore, this course includes material on study skills as well as developing mathematical skills, mathematical thinking, and statistical thinking. The material in the *Integrated Review* is available as a print component and also in the *Integrated Review* course in MyLab Statistics.

The logic behind developing material was to expose students to the mathematics they will need prior to working on the corresponding statistics material. As an example, Chapter 1 of Statistics is heavy on data collection and the language of statistics. There is not a lot of "number crunching." While students work on Chapter 1 of Statistics, the corresponding corequisite material is meant to develop the skills students will need to be successful in Chapter 2 of Statistics. Below is a summary of the organization.

- While studying Chapter 1 of Statistics, work on the Chapter 1 corequisite material to get ready for Chapter 2 of Statistics.
- While studying Chapter 2 of Statistics, work on the Chapter 2 corequisite material to get ready for Chapter 3 of Statistics.
- And so on . . .

It is important to understand that the material written for the corequisite portion of the course is based on the Sullivan Developmental Mathematics series. Therefore, the writing style and pedagogy are consistent with that found in the Sullivan Statistics text. In addition, the material has been well-vetted by the marketplace.

All of the Integrated Review material in MyLab Statistics is hidden from student view and unassigned by default. To make content visible to the student, click Manage Course > Edit Course Menu. Click eText Contents, then the chapters for which you want to make content viewable. Uncheck the box "Hide".

The Learning Path

As part of the Integrated Review material, we have developed a learning path that may be used as part of your course. This is simply one potential learning path. Utilize the materials available as you desire for the benefit of your students.

In the learning path provided below, the assumption is that some students will already have mastered some of the objectives covered within a given Integrated

Review chapter. We recognize that it would be frustrating for students to work through homework problems on content they already understand. To address this scenario, we have developed chapter-level quizzes they can take prior to working on a particular Integrated Review chapter. This allows the identification of the skills each student needs to remediate.

- **Step 1—Chapter-Level Skills Check:** Each Integrated Review chapter begins with a chapter-level quiz. Any student who scores a certain level of mastery (established by the instructor) on the quiz is considered to have mastered the material within the Integrated Review chapter and is allowed to proceed to the next chapter.
- **Step 2—Section-Level Mastery Homework:** If a student does not demonstrate mastery on the chapter-level quiz, then the student must complete section-level personalized homework for those objectives not yet mastered within the section. The instructor establishes the score required to achieve mastery. However, it is recommended that the level of mastery be set at 90%. If the student scores a certain level of mastery within a particular section on the chapter-level quiz, then the student may move to the next section's personalized homework.

This learning path is unassigned by default in MyLab Statistics. To edit or assign the assignments associated with this learning path, go to the Assignment Manager within MyLab Statistics.

The learning path that we provide as part of the course is simply one potential approach to presenting the Integrated Review material. The Integrated Review content may be presented using traditional lecture (utilizing the complete text available), emporium (lab-based), hybrid (blended), or online approaches. It is our intent to provide the resources necessary to provide each instructor with the utmost in flexibility.

Resources Available to Students

The corequisite course offers a variety of remediation tools for the student. The two that students are most likely to use are the objective-level videos and the text. As mentioned earlier, the corequisite material is based on the Sullivan Developmental Mathematics series.

- **Video** The videos that students may use for remediation were created by the Developmental Mathematics author team with many of the videos done by Michael Sullivan. The videos are created at the objective level so that students will not waste time watching video on material they have demonstrated mastery on. Each video includes a complete development of the topic along with one or more examples illustrating the mathematics. When the amount of material within a particular objective warrants it, more than one video per objective may be available. This is done to keep video length down to a manageable level. Students will be alerted to the availability of a video with the ▶ icon. The videos are available in MyLab Statistics and are captioned in English and Spanish.
- **Text** Each section of the corequisite material is based on material originally written for the Developmental Mathematics series. The text presents explanations, definition, examples, and homework problems for the student. In addition, the answers to all Quick Check exercises and odd numbered end-of-section exercises are available to the student.

The Structure of the Integrated Review Text

Students who enroll in a corequisite class likely have hectic lives coupled with anxiety and trepidation as it pertains to quantitative disciplines. Therefore, any text must provide pedagogical support that makes the text and supplements valuable to students as they study and do assignments. Pedagogy must be presented within a framework that teaches students how to study mathematics and statistics.

Integrated Review has a set of pedagogical features that help students develop good study skills, garner an understanding of the connections between topics, and work smarter in the process. The pedagogy is based upon the more than 25 years of classroom teaching experience the author brings to this text.

Examples are often the determining factor in how valuable a textbook is to a student. Students look to examples to provide them with guidance and instruction when they need it most—the times when they are away from the instructor and the classroom. *Integrated Review* has several example formats in an attempt to provide superior guidance and instruction for students. The formats include:

Innovative Examples

Examples have a two-column format in which annotations are provided to the **left** of the arithmetic or algebra, rather than the right, as is the practice in most texts. Because we read from **left to right,** placing the annotation on the left will make more sense to the student. It becomes clear that the annotation describes what we are about to do instead of what was just done. The annotations may be thought of as the teacher's voice offering guidance and clarification immediately before writing the next step in the solution on the board. Consider the following:

EXAMPLE 3 **Solving a Linear Equation by Combining Like Terms**

Solve the linear equation: $3y - 2 + 5y = 2y + 5 + 4y + 3$

Solution

$$3y - 2 + 5y = 2y + 5 + 4y + 3$$

Combine like terms: $\quad 8y - 2 = 6y + 8$

Subtract 6y from both sides: $\quad 8y - 2 - 6y = 6y + 8 - 6y$

$$2y - 2 = 8$$

Add 2 to both sides: $\quad 2y - 2 + 2 = 8 + 2$

$$2y = 10$$

Divide both sides by 2: $\quad \dfrac{2y}{2} = \dfrac{10}{2}$

$$y = 5$$

Check $\quad 3y - 2 + 5y = 2y + 5 + 4y + 3$

Let $y = 5$ in the original equation: $\quad 3(5) - 2 + 5(5) \stackrel{?}{=} 2(5) + 5 + 4(5) + 3$

$$15 - 2 + 25 \stackrel{?}{=} 10 + 5 + 20 + 3$$

$$38 = 38 \quad \text{True}$$

The solution set is $\{5\}$.

Quick ✓

In Problems 11–13, solve each linear equation and verify your solution.

11. $2x + 3 + 5x + 1 = 4x + 10$

12. $4b + 3 - b - 8 - 5b = 2b - 1 - b - 1$

13. $2w + 8 - 7w + 1 = 3w - 1 + 2w - 5$

Showcase Examples

Showcase Examples are used strategically to introduce key topics or important problem-solving techniques. These examples provide "how-to" instruction by offering a guided, step-by-step approach to solving a problem. Students can then immediately see how each of the steps is employed. All showcase examples have the words "How to" in the example title and have a three-column format in which the column on the left describes a step, the middle column provides an annotation (in red) to explain the step, and the right column presents the arithmetic/algebra. With this format, students can see each step in the problem-solving process in context so that the steps make more sense.

EXAMPLE 7   How to Round a Decimal

Round 0.387 to the nearest hundredth.

Step-by-Step Solution

Step 1: Identify and underline the digit you want to round to.	We wish to round to the nearest hundredth: 0.3$\underline{8}$7
Step 2: If the digit to the right of the underlined digit is a 5 or more, add 1 to the underlined digit. If the digit to the right of the underlined digit is 4 or less, leave the underlined digit as is.	The digit to the right of 8 is 7. Because 7 is greater than 5, add 1 to 8: $1 + 8 = 9$
Step 3: Drop the digits to the right of the underlined digit.	0.387 rounded to the nearest hundredth is 0.39.

Quick Check Exercises

Placed at the conclusion of most examples, the *Quick Check* exercises provide students with an opportunity for immediate reinforcement. By working the problems that mirror the example just presented, students get instant feedback and gain confidence in their understanding of the concept. All *Quick Check* exercises are available in MyLab Statistics. In addition, if the student selects View the Textbook in the Learning Aids he/she will be directed to the portion of the text that illustrates the concepts being assessed in the Quick Check exercise. Finally, all *Quick Check* exercise answers are provided with the Student Answers.

Preface to Students for Integrated Review

Welcome to *Integrated Review Materials to Accompany Sullivan Statistics*. A question you may be asking yourself is, "What is integrated review?" In an effort to streamline the curriculum, many schools have decided to offer courses that simultaneously review skills first presented in high school with a corresponding college-level course, rather than require students to take one or more semesters of prerequisite courses before the college-level course. The logic behind this new model is that most students have seen a majority of the content in the prerequisite courses at some point in their high school careers. Rather than reteach the same material, we have developed a new course that reviews the material needed to be successful in statistics on a just-in-time basis.

Below, we provide some helpful tips that will allow you to be successful in this course and your other courses.

What to Do the First Week of the Semester

The first week of the semester gives you the opportunity to prepare for a successful course. Here are the things you should do:

1. **Pick a good seat.** Choose a seat that gives you a good view of the room. Sit close enough to the front so you can easily see the board and hear the professor.

2. **Read the syllabus to learn about your instructor and the course.** Take note of your instructor's name, office location, e-mail address, telephone number, and office hours. Pay attention to any additional help such as tutoring centers, videos, software, and online tutorials. Be sure you fully understand all of the instructor's policies for the class, including those on absences, missed exams or quizzes, and homework. Know important dates and put them in your planner, tablet, computer, or phone.

3. **Learn the names of some of your classmates and exchange contact information.** One of the best ways to learn math (or any subject) is through group study sessions. Try to create time each week to study with your classmates. Knowing how to get in contact with classmates is also useful, because you can obtain the assignment for the day if you ever miss class.

4. **Budget your time.** Most students have a tendency to "bite off more than they can chew." To help with time management, consider the following general rule: Plan on studying at least two hours outside of class for each hour in class. So, if you enrolled in a four-hour math class, you should set aside at least eight hours each week to study for the course. You will also need to set aside time for other courses. Consider your work schedule and personal life when creating your time budget. A time chart is provided on the following page for you to use to manage your time. Fill it in with all your obligations. Do you have enough time set aside to study?

5. **Organization** Get a three-ring binder that is dedicated to only this class. We like three-ring binders because it is easy to add paper when needed and you can add hand-outs provided by the instructor into the binder. Don't use this binder for other classes.

Work Smart: Study Skills
Plan on studying at least two hours outside of class for each hour in class every week.

	Monday	Tuesday	Wednesday	Thursday	Friday	Saturday	Sunday
7 am							
8 am							
9 am							
10 am							
11 am							
Noon							
1 pm							
2 pm							
3 pm							
4 pm							
5 pm							
6 pm							
7 pm							
8 pm							
9 pm							

What to Do Before, During, and After Class

Now that the semester is under way, we present the following ideas for what to do before, during, and after each class meeting. These suggestions may sound overwhelming, but we guarantee that by following them, you will be successful in mathematics (and other courses). Also, you will find that following this plan will make studying for exams much easier.

Before Class

1. Read the section or sections that will be covered in the upcoming class meeting. Watch the video lectures that accompany the text.

2. Based on your reading, write down a list of questions. Your questions will probably be answered through the lecture, but if not, you can then ask any questions that are not answered completely. Also, write down any important new vocabulary and formulas in your notebook.

3. Arrive early and make sure you are mentally prepared for class. Your mind should be alert and ready to concentrate for the entire class. Get a good night of sleep and eat healthy meals to maintain your energy and concentration.

During Class

1. Stay alert. Do not doze off or daydream during class. Understanding the lecture will be very difficult when you "return to class."

2. Avoid distractions such as cell phones and websites that do not pertain to the class.

3. Take thorough notes. It is normal not to understand certain topics the first time you hear them in a lecture. However, this does not mean you should throw your hands up in despair. Rather, continue to take class notes. When you have more time to think about the material, your notes will make sense.

4. Do not be afraid to ask questions. In fact, instructors love questions for two reasons. First, if one student has a question, other students are likely to have the same question. Second, by asking questions, you teach the teacher what topics cause difficulty.

After Class

1. Reread (and possibly rewrite) your class notes. You may be amazed at how often your confusion during class disappears after studying your in-class notes after class.

2. Reread the section. This is an especially important step. Once you have heard the lecture, the section will make more sense and you will understand much more.

3. Do your homework as soon as possible. Do not procrastinate. Homework is not optional. There is an old Chinese proverb that says,

> I hear . . . and I forget
> I see . . . and I remember
> I do . . . and I understand

This proverb applies to any situation in life in which you want to succeed. Would a pianist expect to be the best if she didn't practice? The only way you are going to learn algebra and statistics is by doing algebra and statistics.

4. When you get a problem wrong, try to figure out why you got the problem wrong. Once you figure out why, rework a similar problem. If you can't discover your error, be sure to ask for help.

5. If you have questions, visit your professor during office hours. Try to connect with others in your class to ask and answer questions about the homework. You can also go to the tutoring center on campus, if available.

How to Use the Resources

The *Integrated Review* course offers a variety of tools for you. The two that you are most likely to use are the objective-level videos and the text.

- **Video** Every objective has one or more classroom lecture video of the authors teaching their students marked by an ▶ icon. These "live" classroom lectures can be used to supplement your instructor's presentations and your reading of the text. They can be found in the Multimedia Library of MyLab Statistics.

- **Text** The text presents explanations, definition, examples, and homework problems. In addition, the answers to all Quick Check exercises and odd numbered end-of-section exercises are available to you.

How to Prepare for an Exam

The following steps are time-tested suggestions to help you prepare for an exam.

Step 1: Revisit your homework and the chapter review problems About one week before your exam, start to redo your homework assignments. If you don't understand a topic, seek out help. Work the problems in the chapter review as well. If you get a problem wrong, look back in the text and your notes to try to figure out the solution. If you can't find a solution, seek out help from your classmates or instructor.

Step 2: Test yourself A day or two before the exam, take the chapter test under test conditions. Be sure to check your answers in the back of the book. If you get any problems wrong, determine why and remedy the situation. A great source are the complete video solutions to all problems on the chapter test. Do not gamble on what may or may not be on the exam—master everything.

Step 3: Exam day Be sure to arrive early at the location of the exam. Prepare your mind for the exam. Be sure you are well-rested and well-nourished. Don't try to pull "all-nighters." All-nighters do not work because your brain can only process a certain amount of material at a time.

CHAPTER 1
Integrated Review: Getting Ready for Organizing and Summarizing Data

Outline

1.IR1 Fundamentals of Fractions
1.IR2 Fundamentals of Decimals
1.IR3 The Real Number Line
1.IR4 Multiplying and Dividing Fractions
1.IR5 Adding and Subtracting Fractions
1.IR6 Operations on Decimals
1.IR7 Fundamentals of Percent Notation
1.IR8 Language Used in Modeling

1.IR1 Fundamentals of Fractions

Objectives

① Identify Terms Used with Fractions
② Find Equivalent Fractions
③ Write a Fraction in Lowest Terms

① Identify Terms Used with Fractions

Fractions are used to represent a part of a whole, such as part of a pie.

The visual representation of one whole can take many forms. For example, we can use a pie to represent one whole and divide it into 8 *equal parts*. In a fraction, the **denominator** represents the number of equal parts the whole is divided into.

  is a representation for $\dfrac{\square}{8}$ ← denominator

Another common representation of one whole is a rectangle. The rectangle below has been divided into 7 equal parts.

 is a representation for $\dfrac{\square}{7}$ ← denominator

The number of parts we wish to consider is shaded, and this number, which is called the **numerator**, is written in the top position of the fraction. If 5 of the 8 pieces of a pie are shaded, we represent the shaded part of the pie as $\dfrac{5}{8}$. So,

  is represented as $\dfrac{5}{8}$ ← numerator (number of shaded parts)
← denominator (number of equal parts in the whole)

A fraction bar also indicates division. Thus $\dfrac{5}{8}$ means "five divided by eight" and can be written $8\overline{)5}$.

IR-1

EXAMPLE 1 Writing Fractions

Write a fraction to represent the shaded area. Identify the numerator and the denominator of each fraction.

(a)

(b)

(c) (image of rectangle divided into 4 parts with 1 shaded)

(d) (image of a ruler from 0 in. to 1, with markings)

Solution

(a) There are 3 out of 6 regions shaded. The fraction that represents the shaded region is $\frac{3}{6}$. The numerator is 3 (number of shaded sections) and the denominator is 6 (number of equal parts).

(b) There are 2 out of 3 regions shaded. The fraction that represents the shaded region is $\frac{2}{3}$. The numerator is 2 and the denominator is 3.

(c) The fraction that represents the shaded region is $\frac{1}{4}$, where 1 is the numerator and 4 is the denominator.

(d) Even though the line segments that divide the distance from 0 to 1 are of different lengths, there are 16 equal divisions, or 16 equal parts, from 0 to 1. The orange line includes 10 segments, so the fraction $\frac{10}{16}$ represents its length, where 10 is the numerator and 16 is the denominator.

We can also use fractions to represent a portion of a group of like items.

EXAMPLE 2 Writing Fractions

(a) The figure below is a flock of 7 sheep. What fraction of the flock represents the number of black sheep?

(b) In a survey of 365 adult Americans, 58 stated that deep dish was their favorite pizza. What fraction of adult Americans report deep dish as their favorite pizza? *Source:* Harris Interactive

Solution

(a) Because 3 sheep are black and there are a total of 7 sheep, the fraction of the flock that is black is

$\frac{3}{7}$ ← numerator (number of black sheep)
      ← denominator (total sheep in the flock)

(b) The poll reports that 58 out of 365 adult Americans stated that deep dish was their favorite pizza. This fraction is written

$\frac{58}{365}$ ← numerator (number who prefer deep dish)
        ← denominator (number surveyed)

Each of the fractions written in Examples 1 and 2 is a proper fraction. In a **proper fraction,** the numerator is less than the denominator, so the quantity represented is less than one whole.

If you cut a pie into four equal pieces, and all four pieces were eaten, then $\frac{4}{4}$ pieces or 1 whole pie was eaten. See Figure 1(a). If you have two pies, each cut into four equal pieces, and five pieces were eaten, then $\frac{4}{4}$ or 1 whole pie was eaten plus $\frac{1}{4}$ of the other pie. So, $\frac{5}{4}$ pies were eaten. See Figure 1(b).

Figure 1

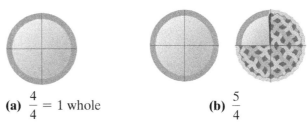

(a) $\frac{4}{4} = 1$ whole (b) $\frac{5}{4}$

The fractions $\frac{4}{4}$ and $\frac{5}{4}$ are called **improper fractions** because the numerator is greater than or equal to the denominator.

EXAMPLE 3   **Writing Fractions**

Write a fraction to represent the shaded region. Identify the numerator and the denominator of each fraction.

(a) 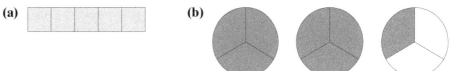 (b)

Solution

Work Smart

When the numerator and denominator of a fraction are the same nonzero value the fraction equals 1.

$\frac{a}{a} = 1$ if $a \neq 0$

(a) There are 5 out of 5 parts shaded. The fraction is

$\frac{5}{5}$ ← numerator (number of shaded parts)
    ← denominator (total number of equal parts)

The numerator is 5 and the denominator is 5, so this is an improper fraction. When the numerator and denominator are the same, the fraction represents 1, because 5 out of 5 equal pieces is one whole.

(b) There are 7 shaded regions and each pie is divided into 3 equal parts.

The fraction is $\frac{7}{3}$. ← numerator (number of shaded regions)
         ← denominator (number of equal parts in one whole)

The numerator is 7 and the the denominator is 3, so this is an improper fraction.

We can see in the figure from Example 3(b) that $\frac{7}{3}$ pies is the same as 2 whole pies and $\frac{1}{3}$ of another pie. We can write this as $2\frac{1}{3}$ pies. We call $2\frac{1}{3}$ a **mixed number** because it mixes an integer and a proper fraction.

For any nonzero whole number a $(a \neq 0)$,

- $\frac{0}{a} = 0$

- $\frac{a}{0}$ is undefined

This means the denominator of a fraction may not equal 0. However, the numerator of a fraction may equal zero. For example, if a pie is divided into five pieces, but nobody takes a piece, then $\frac{0}{5}$ or 0 pieces were taken from the pie.

Quick ✓

1. *True or False* In the fraction $\frac{4}{6}$, a whole was divided into 4 equal parts.
2. *True or False* In a fraction, the numerator can be equal to, greater than, or less than the nonzero denominator.
3. In the fraction $\frac{4}{9}$, a whole is divided into _____ equal parts.
4. In the fraction $\frac{8}{6}$, the numerator is _____ and the denominator is _____.
5. If the numerator is less than the denominator in a fraction, the fraction is called _____.
6. If the numerator is greater than or equal to the denominator, the fraction is called _____.
7. Write the fraction that represents the shaded region.
8. In a class of 30 statistics students, 9 students received an A. What fraction of the class received an A?
9. Determine which of the following are proper fractions:

$$\frac{12}{5}, \frac{4}{6}, \frac{9}{1}, \frac{1}{2}, \frac{8}{8}$$

10. Determine which of the following are improper fractions:

$$\frac{3}{15}, \frac{100}{100}, \frac{2}{1}, \frac{16}{3}, \frac{999}{1000}$$

▶ ❷ Find Equivalent Fractions

Definition

Equivalent fractions are fractions that represent the same part of a whole.

Figure 2

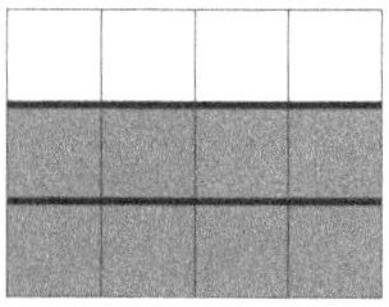

For example, $\frac{2}{3}$ and $\frac{8}{12}$ are equivalent fractions. To understand why, consider Figure 2. If we divide the whole into 12 equal parts and shade 8 of these parts, it is clear that the shaded region represents $\frac{8}{12}$ of the rectangle. If we then consider only the 3 equal parts separated by the thick black lines, we can see that 2 parts are shaded, forming the fraction $\frac{2}{3}$. In each case, the same portion of the rectangle is shaded, so $\frac{2}{3}$ and $\frac{8}{12}$ are equivalent fractions.

How do we obtain equivalent fractions? The answer lies in the following property.

In Words

We can obtain an equivalent fraction by multiplying the numerator and denominator of the fraction by the same nonzero number.

If $a, b,$ and c are integers, then

$$\frac{a}{b} = \frac{a \cdot c}{b \cdot c} \quad \text{if } b \neq 0, c \neq 0$$

Why does this work? Remember the fraction bar represents division. So, $\frac{5}{5}$ means $5 \div 5$, which equals 1. In fact, for any nonzero integer, a, $a \div a = \frac{a}{a} = 1$. Also, any integer multiplied by 1 results in the integer. That is, $1 \cdot a = a \cdot 1 = a$. In the expression $\frac{a}{b} = \frac{a \cdot c}{b \cdot c}, c \neq 0$ we are multiplying $\frac{a}{b}$ by 1, where the 1 is in the form $\frac{c}{c}$.

EXAMPLE 4 Writing an Equivalent Fraction

Write the fraction $\frac{3}{5}$ as an equivalent fraction with a denominator of 20.

Solution

We want to find the number of parts out of 20 that is the same as 3 parts out of 5. Using fraction notation, we want to know $\frac{3}{5} = \frac{?}{20}$. What do we need to multiply 5 by to obtain 20? Because $5 \times 4 = 20$, form an equivalent fraction by multiplying the numerator and denominator of $\frac{3}{5}$ by 4.

$$\frac{3}{5} = \frac{3 \cdot 4}{5 \cdot 4}$$
$$= \frac{12}{20}$$

We can write any integer as an equivalent fraction by first using the fact that $a = \frac{a}{1}$ and then multiplying by 1, written in the form $\frac{c}{c}$. For example, to write 5 as an equivalent fraction with denominator 3, write $5 = \frac{5}{1} = \frac{5 \cdot 3}{1 \cdot 3} = \frac{15}{3}$.

Quick ✓

In Problems 11–14, rewrite each fraction with the denominator indicated.

11. $\frac{1}{2}$ with denominator 6

12. $\frac{1}{4}$ with denominator 20

13. $\frac{7}{5}$ with denominator 35

14. $\frac{4}{13}$ with denominator 39

❸ Write a Fraction in Lowest Terms

Definition

A fraction is written in **lowest terms** when the numerator and the denominator share no common factor other than 1.

In Words

To write a fraction in lowest terms, find any common factors between the numerator and denominator, and divide out the common factors.

Write fractions in lowest terms by using the fact that

$$\frac{a \cdot c}{b \cdot c} = \frac{a}{b}$$

Thus, to write a fraction in lowest terms, write the numerator and the denominator as a product of primes and then divide out common factors.

EXAMPLE 5 Writing a Fraction in Lowest Terms

Write $\frac{15}{18}$ in lowest terms.

Solution

Write the numerator and the denominator as the product of primes, and divide out the common factors.

Work Smart

It may help to think of "divide out common factor" as

$$\frac{15}{18} = \frac{15 \div 3}{18 \div 3} = \frac{5}{6}$$

$$\frac{15}{18} = \frac{3 \cdot 5}{3 \cdot 3 \cdot 2}$$

Divide out common factors: $= \frac{3 \cdot 5}{3 \cdot 3 \cdot 2}$

$$= \frac{5}{3 \cdot 2}$$

Multiply: $= \frac{5}{6}$

EXAMPLE 6 **Writing a Fraction in Lowest Terms**

Write $\frac{36}{48}$ in lowest terms.

Solution
Write the numerator and the denominator as the product of primes, and divide out the common factors.

$$\frac{36}{48} = \frac{2 \cdot 2 \cdot 3 \cdot 3}{2 \cdot 2 \cdot 2 \cdot 2 \cdot 3}$$

Divide out common factors: $= \frac{2 \cdot 2 \cdot 3 \cdot 3}{2 \cdot 2 \cdot 2 \cdot 2 \cdot 3}$

$$= \frac{3}{2 \cdot 2}$$

Multiply: $= \frac{3}{4}$

Work Smart

Eventually, you may learn to divide out common factors that are not prime.

$$\frac{36}{48} = \frac{3 \cdot 12}{4 \cdot 12}$$

$$= \frac{3}{4}$$

As suggested by the Work Smart, we may also divide out common factors that are not prime.

EXAMPLE 7 **Writing a Fraction in Lowest Terms**

Write $\frac{28}{45}$ in lowest terms.

Solution

$$\frac{28}{45} = \frac{2 \cdot 2 \cdot 7}{3 \cdot 3 \cdot 5}$$

Because the numerator and denominator share no common prime factors, $\frac{28}{45}$ is in lowest terms.

Quick ✓

In Problems 15–20, write each fraction in lowest terms.

15. $\frac{21}{14}$ **16.** $\frac{27}{30}$ **17.** $\frac{18}{24}$ **18.** $\frac{96}{144}$ **19.** $\frac{76}{105}$ **20.** $\frac{152}{380}$

Determine Whether Two Fractions Are Equivalent

We have learned that two fractions are equivalent if they represent the same part of the whole. We can determine if two fractions are equivalent by writing each fraction in lowest terms to see if they simplify to the same fraction.

EXAMPLE 8 Determining Whether Two Fractions Are Equivalent

Determine whether $\frac{12}{40}$ and $\frac{9}{30}$ are equivalent fractions.

Solution

To determine whether two fractions are equivalent, write each fraction in lowest terms.

Work Smart

$$\frac{12}{40} = \frac{3 \cdot 4}{10 \cdot 4}$$
$$= \frac{3}{10}$$

$$\frac{12}{40} = \frac{2 \cdot 2 \cdot 3}{2 \cdot 2 \cdot 2 \cdot 5}$$

Divide out common factors:
$$= \frac{2 \cdot 2 \cdot 3}{2 \cdot 2 \cdot 2 \cdot 5}$$
$$= \frac{3}{10}$$

$$\frac{9}{30} = \frac{3 \cdot 3}{2 \cdot 3 \cdot 5}$$

Divide out common factors:
$$= \frac{3 \cdot 3}{2 \cdot 3 \cdot 5}$$
$$= \frac{3}{10}$$

Because both fractions simplify to $\frac{3}{10}$, we know that $\frac{12}{40} = \frac{9}{30}$.

From Example 8, we make the following observation. When we multiply the numbers that lie across the equal sign from each other diagonally, we are finding **cross products**. If the cross products are equal, then the fractions are equivalent. From Example 8, we know that $\frac{12}{40}$ and $\frac{9}{30}$ are equivalent fractions. To find the cross products, set the fractions equal to each other and multiply on the diagonal as shown below.

In Words

When we write $\stackrel{?}{=}$, we are asking whether or not the quantities are equal. The symbol $\neq$ means "is not equal to."

$$\frac{12}{40} \stackrel{?}{=} \frac{9}{30}$$

$40 \cdot 9 = 360$

$12 \cdot 30 = 360$

Because the cross products are equal (both are 360), the fractions $\frac{12}{40}$ and $\frac{9}{30}$ are equivalent.

Checking for Equality of Fractions

If $a \cdot d = b \cdot c$ and $b \neq 0, d \neq 0$, then $\frac{a}{b} = \frac{c}{d}$.

EXAMPLE 9 Determining Whether Two Fractions Are Equivalent

Use the cross products to determine whether the fractions are equivalent.

(a) $\frac{6}{10}$ and $\frac{15}{25}$

(b) $\frac{8}{40}$ and $\frac{6}{14}$

Solution

Find the cross products.

(a)

$$\frac{6}{10} \stackrel{?}{=} \frac{15}{25}$$

$\leftarrow 10 \cdot 15 = 150$

$\leftarrow 6 \cdot 25 = 150$

Because the cross products are equal (both equal 150), the fractions are equivalent. So,

$$\frac{6}{10} = \frac{15}{25}$$

(b)

$$\frac{8}{40} \stackrel{?}{=} \frac{6}{14}$$

$\leftarrow 40 \cdot 6 = 240$

$\leftarrow 8 \cdot 14 = 112$

Because the cross products are not equal, the fractions are not equivalent. So,

$$\frac{8}{40} \neq \frac{6}{14}$$

Quick ✓

21. Two fractions that represent the same portion of a whole are called _____ _____.

22. *True or False* We can find a fraction equivalent to $\frac{3}{4}$ by computing the following:
$$\frac{3 \cdot 2}{4 \cdot 3} = \frac{6}{12}.$$

In Problems 23 and 24, determine whether the fractions are equivalent.

23. $\frac{2}{3}$ and $\frac{10}{15}$

24. $\frac{32}{45}$ and $\frac{8}{9}$

1.IR1 Exercises — MyLab Statistics
Underlined exercises have complete video solutions in MyLab.

Problems 1–24 are the Quick ✓s that follow the EXAMPLES.

Building Skills

In Problems 25–30, write a fraction to represent the shaded region. Identify the numerator and the denominator of each fraction. See Objective 1.

25.

26.

27.

28.

29.

30.

In Problems 31 and 32, determine which fractions are proper. See Objective 1.

31. $\frac{7}{8}, \frac{16}{1}, \frac{11}{10}, \frac{3}{4}, \frac{7}{7}$

32. $\frac{3}{5}, \frac{1}{6}, \frac{5}{4}, \frac{100}{100}, \frac{89}{90}$

In Problems 33 and 34, determine which fractions are improper. See Objective 1.

33. $\frac{1}{3}, \frac{7}{10}, \frac{9}{8}, \frac{12}{12}, \frac{50}{20}$

34. $\frac{7}{3}, \frac{8}{6}, \frac{2}{5}, \frac{56}{54}, \frac{1}{1}$

In Problems 35–42, rewrite the fraction with the denominator indicated. See Objective 2.

35. Write $\frac{1}{3}$ with a denominator of 15.

36. Write $\frac{1}{5}$ with a denominator of 20.

37. Write $\frac{4}{3}$ with a denominator of 12.

38. Write $\frac{7}{4}$ with a denominator of 28.

39. Write $\frac{13}{15}$ with a denominator of 45.

40. Write $\frac{9}{13}$ with a denominator of 39.

41. Write 5 with a denominator of 2.

42. Write 4 with a denominator of 12.

In Problems 43–50, write the fraction in lowest terms. See Objective 3.

43. $\frac{18}{27}$

44. $\frac{32}{44}$

45. $\frac{15}{46}$

46. $\frac{40}{21}$

47. $\frac{54}{63}$

48. $\frac{96}{56}$

49. $\frac{630}{495}$

50. $\frac{585}{546}$

In Problems 51–54, determine whether the fractions are equivalent. See Objective 3.

51. $\frac{6}{4}$ and $\frac{36}{24}$

52. $\frac{20}{25}$ and $\frac{6}{8}$

53. $\frac{49}{25}$ and $\frac{28}{20}$

54. $\frac{30}{55}$ and $\frac{12}{22}$

Applying the Concepts

55. **Video Game Violence** In a survey of 2278 adult Americans, 1321 believed there is a link between video games and teenagers showing violent behavior. Write the fraction of adult Americans who believe there is a link between video games and teenage violence. *Source:* Harris Interactive

56. **Pet Owners** In a survey of 1585 pet owners, 1442 considered their pet to be a part of the family. Write the fraction of pet owners who believe their pet is part of the family. *Source:* Harris Interactive

57. Saving for Retirement In a survey of 1000 adult Americans, 380 indicated that they were not confident that they will have enough saved to last through retirement. Write the fraction, in lowest terms, of adult Americans who are not confident they will have enough saved to last through retirement. *Source:* Pew Research

58. The Rich Are Different In a survey of 2500 adult Americans, 850 considered the rich less likely to be honest than the average person. Write the fraction, in lowest terms, of adult Americans who believe the rich less likely to be honest than the average person. *Source:* Pew Research

1.IR2 Fundamentals of Decimals

Objectives

1. Understand the Place Value of a Decimal Number
2. Read and Write Decimals in Words
3. Write Decimals as Fractions
4. Round Decimals to a Given Place Value

1 Understand the Place Value of a Decimal Number

Like fractional notation, decimal notation is used to describe parts of a whole. Numbers written in decimal notation are called **decimal fractions** or just **decimals.**

Decimals are used to describe a whole that is divided into 10 equal parts, 100 equal parts, 1000 equal parts, and so on. In other words, decimals are fractions with a denominator that is a power of 10.

The rectangle in Figure 3(a) is divided into 10 equal parts, one of which is shaded. Written as a fraction, each part is $\frac{1}{10}$ of the rectangle. Written as a decimal, each part is 0.1 of the rectangle. We read both the fraction $\frac{1}{10}$ and the decimal 0.1 as "one tenth."

The dot in the number 0.1 is called the **decimal point.**

The rectangle in Figure 3(b) is divided into 10 equal parts, and 3 of the parts are shaded. We represent 3 parts out of 10 as $\frac{3}{10}$ (fraction) or 0.3 (decimal). We read both the fraction $\frac{3}{10}$ and the decimal 0.3 as "three tenths."

Figure 3

(a) $\frac{1}{10} = 0.1$

(b) $\frac{3}{10} = 0.3$

The square in Figure 4(a) is divided into 100 equal parts. Written as a fraction, each part is $\frac{1}{100}$ of the square. Written as a decimal, each part is 0.01 of the square. We read both the fraction $\frac{1}{100}$ and the decimal 0.01 as "one hundredth."

To show the decimal $0.83 = \frac{83}{100}$, shade 83 of the 100 boxes, as shown in Figure 4(b).

Work Smart

Did you notice that we wrote the decimal number for $\frac{83}{100}$ as 0.83? We include the zero in front of the decimal point to emphasize that there is no whole number involved.

Figure 4

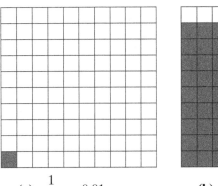

(a) $\frac{1}{100} = 0.01$

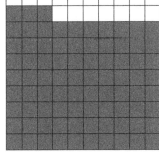

(b) $\frac{83}{100} = 0.83$

When discussing decimals, we must consider place value. See Figure 5. The number 9186.347 is read "nine thousand, one hundred eighty-six *and* three hundred forty-seven thousandths." The word *and* marks the location of the decimal point.

Work Smart

Note that the value of each place is 10 times the value of the place to its right. For example, "hundreds" is ten times as much as "tens."

Figure 5

EXAMPLE 1 Determining Place Value for Decimal Numbers

Determine the place value of the following digits in the number 801.4956. Indicate the fractional value that each digit represents.

(a) 4 (b) 8 (c) 9 (d) 6

Solution

(a) The 4 is in the tenths place. This represents $\frac{4}{10}$, or 0.4.

(b) The 8 is in the hundreds place. This represents 8×100, or 800.

(c) The 9 is in the hundredths place. This represents $\frac{9}{100}$, or 0.09.

(d) The 6 is in the ten-thousandths place. This represents $\frac{6}{10,000}$, or 0.0006.

Quick ✓

Tell the place value of each digit in the number 401.2657. Indicate the fractional value that each digit represents.

1. (a) 2 (b) 5 (c) 0
    (d) 6 (e) 7

In Problems 2–5, tell what digit appears in the indicated place.

2. 39.245; the thousandths place **3.** 5786.29; the ones place

4. 3.14; the hundredths place **5.** 745.963; the tenths place

In Problems 6–9, tell the place value of the indicated digit in the given number. Indicate the fractional value that each digit represents.

6. 271.35; the 3 **7.** 65,071.28; the 8

8. 0.196; the 6 **9.** 40.769; the 4

▶ ❷ Read and Write Decimals in Words

Table 1 shows several numbers written as decimals and in words.

Table 1

Decimal	Read as
0.7	Seven tenths
−0.31	Negative thirty-one hundredths
0.009	Nine thousandths
0.073	Seventy-three thousandths
−0.897	Negative eight hundred ninety-seven thousandths
0.0513	Five hundred thirteen ten-thousandths

Section 1.IR2 Fundamentals of Decimals   IR-11

Did you notice that all of the place value names for numbers to the right of the decimal point end in the suffix *ths*? Place value names that end in *ths* help distinguish decimal place values from whole number place values. For the number 587.42, the number 5 is in the *hundreds* place and the number 2 is in the *hundredths* place.

EXAMPLE 2   **Reading and Writing Decimal Numbers in Words**

Write each of the following numbers in words.

  (a) 0.6   (b) 0.84   (c) 0.019

Solution

Number	Words
(a) 0.6	Six tenths because $0.6 = \dfrac{6}{10}$
(b) 0.84	Eighty-four hundredths because $0.84 = \dfrac{84}{100}$
(c) 0.019	Nineteen thousandths because $0.019 = \dfrac{19}{1000}$

EXAMPLE 3   **Reading and Writing Word Names for Decimal Numbers**

Write each of the following numbers in words.

  (a) 5.4   (b) 7.05   (c) 5609.348   (d) 2.6305

Solution

Number	Words
(a) 5.4	Five and four tenths because $5.4 = 5\dfrac{4}{10}$
(b) 7.05	Seven and five hundredths because $7.05 = 7\dfrac{5}{100}$
(c) 5609.348	Five thousand, six hundred nine and three hundred forty-eight thousandths because $5609.348 = 5609\dfrac{348}{1000}$
(d) 2.6305	Two and six thousand, three hundred five ten-thousandths because $2.6305 = 2\dfrac{6305}{10,000}$

Work Smart

The word *and* is used to denote the location of the decimal point in a number.

Writing Decimal Numbers in Words

Step 1: Read the number to the left of the decimal point as you normally would.
Step 2: Read the decimal point as the word *and*.
Step 3: Read the part of the number to the right of the decimal point as if it were a whole number.
Step 4: Use the place value of the digit farthest right in the number. The numbers to the right of the decimal ends in *ths*.

EXAMPLE 4   **Writing Decimals in Words**

Write each number in words.

  (a) 0.105   (b) 100.005

Solution

(a) 0.105 is written one hundred five thousandths because $0.105 = \frac{105}{1000}$.

(b) 100.005 is written one hundred and five thousandths because $100.005 = 100\frac{5}{1000}$.

Do you see the importance of the word *and* when writing a decimal name?

Decimal notation is also used with money. Checks are usually written in the following format.

ARDYS JOHNSON — 581
Phone: 555-2737
4250 West 18th Avenue
Chicago, IL 60601-2180
Date: July 1, 20 __ — 2-74/710

PAY TO THE ORDER OF Food Mart $ 36.12

Thirty-six and 12/100 ———— Dollars

FOR CLASSROOM USE ONLY

SCB SKY CENTRAL BANK, CHICAGO, ILLINOIS

MEMO groceries   Ardys Johnson

⑁:071000744⑁: 0581 00867⑁ 40856⑁"

EXAMPLE 5 **Writing Word Names for Decimals**

Write the word name for a purchase of $84.23 at Target, as though you were writing a check.

Solution

The word name for $84.23 is eighty-four and $\frac{23}{100}$ dollars.

Quick ✓

10. When writing a decimal number, the word ___ is used to represent the decimal point.

In Problems 11–16, write the word name for each number.

11. 0.23
12. 31.4
13. 0.018
14. 0.4521
15. 200.05
16. 0.205

17. Write the word name for $95.23, your monthly electric bill, as though you were writing a check.

▶ ❸ Write Decimals as Fractions

Writing a decimal as a fraction is an extension of reading and writing decimals.

EXAMPLE 6 **Writing Decimals as Fractions**

Convert each decimal to a fraction or a mixed number in lowest terms, if possible.

(a) 0.33   (b) 0.6   (c) 0.125   (d) 11.78

Solution

(a) 0.33 is equivalent to 33 hundredths, or $\frac{33}{100}$.

(b) 0.6 is equivalent to 6 tenths, or $\frac{6}{10}$. Because $\frac{6}{10} = \frac{2 \cdot 3}{2 \cdot 5} = \frac{2 \cdot 3}{2 \cdot 5} = \frac{3}{5}$, we write 0.6 as $\frac{3}{5}$.

(c) 0.125 is equivalent to one hundred twenty-five thousandths, or $\frac{125}{1000}$. However, $\frac{125}{1000} = \frac{1 \cdot 125}{8 \cdot 125} = \frac{1 \cdot \cancel{125}}{8 \cdot \cancel{125}} = \frac{1}{8}$. So $0.125 = \frac{1}{8}$.

(d) 11.78 is equivalent to $11\frac{78}{100} = 11\frac{2 \cdot 39}{2 \cdot 50} = 11\frac{\cancel{2} \cdot 39}{\cancel{2} \cdot 50} = 11\frac{39}{50}$.

If the number to be converted to a fraction contains an integer, such as Example 6(d), keep the integer and change the decimal to a fraction. The answer will be a mixed number.

To convert a decimal to a fraction, use the following steps.

Converting a Decimal to a Fraction

Step 1: Identify the place value of the last digit in the decimal. If there is a number to the left of the decimal point, it is the integer part and is left unchanged.

Step 2: Write the decimal as a fraction using the place value of the last digit as the denominator, and write in lowest terms.

Quick ✓

In Problems 18–22, convert each decimal to a fraction and write in lowest terms, if possible.

18. 0.1   **19.** 0.01   **20.** 0.003   **21.** 3.75   **22.** 0.625

▶ ❹ Round Decimals to a Given Place Value

EXAMPLE 7 **How to Round a Decimal**

Round 0.387 to the nearest hundredth.

Step-by-Step Solution

Step 1: Identify and underline the digit you want to round to.

We wish to round to the nearest hundredth: 0.3$\underline{8}$7

Step 2: If the digit to the right of the underlined digit is a 5 or more, add 1 to the underlined digit. If the digit to the right of the underlined digit is 4 or less, leave the underlined digit as is.

The digit to the right of 8 is 7. Because 7 is greater than 5, add 1 to 8: $1 + 8 = 9$

Step 3: Drop the digits to the right of the underlined digit.

0.387 rounded to the nearest hundredth is 0.39.

To round decimals to a specified place value, use the following steps.

> **Rounding a Decimal to a Specified Place Value**
>
> **Step 1:** Identify and underline the digit you want to round to.
>
> **Step 2:** If the digit to the right is 5 or more, add 1 to the underlined digit. If the digit to the right of the underlined digit is 4 or less, leave the underlined digit as is.
>
> **Step 3:** Drop the digits to the right of the underlined digit.

EXAMPLE 8 Rounding a Decimal Number

Round 0.36924 to the nearest thousandth.

Work Smart

We can also say "round to three decimal places" instead of "round to the nearest thousandth."

Solution

To round 0.36924 to the nearest thousandth, we see that the number 9 is in the thousandths place: 0.36924. The number to the right of 9 is 2. Because 2 is less than 5, round 0.36924 to 0.369.

EXAMPLE 9 Rounding a Decimal Number

Round 41.59673 to the nearest hundredth.

Work Smart

The hundredths place is two decimal places to the right of the decimal point, so we must write the 0 in the hundredths place. Writing 41.59673 to the nearest hundredth as 41.6 is incorrect.

Solution

To round 41.59673 to the nearest hundredth, we see that the number 9 is in the hundredths place: 41.59673. The number to the right of 9 is 6, and 6 is greater than 5. Add 1 to 9, and obtain 10. So 41.59673 is rounded to 41.60.

Quick ✓

In Problems 23–30, round each number to the given decimal place.

23. 0.12 to the nearest tenth
24. 0.981 to the nearest hundredth
25. 3.679 to the nearest hundredth
26. 0.2973 to the nearest thousandth
27. 24.003 to the nearest hundredth
28. 1.89 to the nearest tenth
29. 0.0399 to the nearest thousandth
30. 73.98 to the nearest tenth

EXAMPLE 10 Rounding Money

When you bought a new lamp for $29.99, the sales tax was computed to be $1.85938. Round $1.85938 to the nearest cent.

Solution

Rounding to the nearest cent means rounding to the nearest hundredth. To round $1.85938, we see that the number 5 is in the hundredths (cents) place: $1.85938. The number to the right of 5 is 9. So $1.85938 rounded to the nearest cent is $1.86.

Quick ✓

In Problems 31 and 32, round each amount to the nearest cent.

31. $1.9118625
32. $3.9852

33. Three cans of soup cost $2.80, so one can of soup costs $0.933333. To the nearest cent, what does one can of soup cost?

34. Two boxes of cereal cost $5.89, so one box costs $2.945. To the nearest cent, what does one box of cereal cost?

1.IR2 Exercises — MyLab Statistics
Underlined exercises have complete video solutions in MyLab.

Problems 1–34 are the Quick ✓s that follow the EXAMPLES.

Building Skills

In Problems 35–40, tell the digit in the indicated place. See Objective 1.

35. 57.091; the hundredths place

36. 156.2; the ones place

37. 38,549.067; the thousandths place

38. 4275.36; the tenths place

39. 0.26397; the ten-thousandths place

40. 3.67241; the hundred-thousandths place

In Problems 41–46, tell the place value of the indicated digit in the given number. See Objective 1.

41. 451.2; the 2   **42.** 1296.78; the 8

43. 26,781.095; the 9   **44.** 0.12345; the 3

45. 2.9386; the 2   **46.** 74.08091; the 9

In Problems 47–54, write each decimal in words. See Objective 2.

47. 0.21   **48.** 0.237   **49.** 841.6   **50.** 1.73

51. 0.306   **52.** 0.214   **53.** 0.006   **54.** 0.014

In Problems 55–58, write each dollar amount in words. See Objective 2.

55. Stock Price In August, Facebook stock ended the day at a price of $125.48.

56. Stock Price Each share of GAP stock sold recently for $25.61.

57. Cable TV A local cable TV provider is offering a special on cable TV, Internet, and phone service for $89.99 per month.

58. Cable TV A local cable TV provider is offering a special on cable TV and Internet, for $49.99 per month.

In Problems 59–68, write each decimal as a fraction or mixed number in lowest terms. See Objective 3.

59. 0.3   **60.** 0.9   **61.** 0.25   **62.** 0.75   **63.** −0.4

64. −0.8   **65.** 0.28   **66.** 0.76   **67.** 0.05   **68.** 0.01

In Problems 69–74, round each number to the nearest tenth. See Objective 4.

69. 0.43   **70.** 0.22   **71.** −0.456

72. −0.678   **73.** 12.782   **74.** 65.991

In Problems 75–80, round each number to the nearest hundredth. See Objective 4.

75. 4.982   **76.** 11.463

77. −19.299   **78.** −31.395

79. 0.7763   **80.** 0.6852

In Problems 81–86, round each number to the nearest thousandth. See Objective 4.

81. 17.2865   **82.** 845.4396

83. 0.0024   **84.** 0.0063

85. 68.9995   **86.** 45.9996

In Problems 87–90, round each dollar amount as indicated. See Objective 4.

87. $123.568 to the nearest cent

88. $294.783 to the nearest cent

89. $17.42 to the nearest dollar

90. $1599.99 to the nearest dollar

Applying the Concepts

91. NCAA Swimming At the 2016 NCAA National Swimming Championships, Ryan Murphy, nephew of the author, set the NCAA record in the 200 backstroke with a time of 1.5955 minutes. Round this number to the nearest thousandth. Round this number to the nearest hundredth.

92. Olympic Swimming At the 2016 Summer Olympics, American swimmer Ryan Murphy swam the 100-meter backstroke race in a time of 51.85 seconds. Round this number to the nearest tenth. (*Source:* 2016 Olympics website)

93. Cell Phone The HTC One™ X cell phone is thin, only 0.9144 cm. Round this number to the nearest hundredth. (*Source:* AT&T)

94. Currency One U.S. dollar was worth 0.8969 euro recently. Round this number to the nearest hundredth. (*Source:* Google Finance)

95. Drought The 2012 drought left Missouri about 6.575 cm below normal rainfall. Round this number to the nearest tenth. (*Source:* University of Missouri Climate Center)

96. Drought Rainfall in Oklahoma was 17.8662 cm below normal during the summer of 2012. Round this number to the nearest tenth. (*Source:* Oklahoma Climatological Survey)

1.IR3 The Real Number Line

Objectives

1. Classify Numbers
2. Plot Points on a Real Number Line
3. Use Inequalities to Order Real Numbers
4. Compute the Absolute Value of a Real Number

This section discusses the *real number system*. In short, real numbers are numbers that we use to count or measure things, such as 25 students in your class, 18.4 miles per gallon, or a $130 debt.

Numbers are organized in *sets*. A **set** is a well-defined collection of objects. For example, we can identify the students enrolled in Elementary Statistics at your college as a set. The collection of numbers 0, 1, 2, 3, 4, 5, 6, 7, 8, and 9 may also be identified as a set. If we let A represent this set of numbers, then we can write

$$A = \{0, 1, 2, 3, 4, 5, 6, 7, 8, 9\}$$

In this notation, braces $\{\ \}$ are used to enclose the objects, or **elements,** in the set. A set with no elements in it is called an **empty set.** Empty sets are denoted by the symbol $\varnothing$ or $\{\ \}$.

> **Work Smart**
> The use of the word "real" to describe numbers leads us to question, "Are there 'nonreal' numbers?" The answer is yes. We use the word "imaginary" to describe nonreal numbers. Imaginary does not mean that these numbers are made up, however. We will not discuss imaginary numbers in the text.

EXAMPLE 1 Writing a Set

Write the set that represents the vowels.

Solution
The vowels are *a, e, i, o,* and *u*. If we let V represent this set, then

$$V = \{a, e, i, o, u\}$$

Quick ✓

1. Write the set that represents the first four positive, odd numbers.
2. Write the set that represents the states in the United States with names that begin with the letter A.
3. Write the set that represents the states in the United States with names that begin with the letter Z.

❶ Classify Numbers

We will develop the real number system by looking at the history of numbers. The first types of numbers that humans worked with are the *natural numbers*, or *counting numbers*.

Definition
The **natural numbers**, or **counting numbers**, are the numbers in the set $\{1, 2, 3, \ldots\}$.

The three dots in this definition are called an *ellipsis* and indicate that the pattern continues indefinitely.

The natural numbers are often used to count things. For example, we can count the number of cars waiting at a Wendy's drive-thru. We can represent the counting numbers graphically using a number line. See Figure 6. The arrow on the right indicates the direction in which the numbers increase.

Figure 6
The natural numbers

Since we do not count the number of cars waiting in the drive-thru by saying, "zero, one, two, three ...," zero is not a natural, or counting, number. When we add the number 0 to the set of counting numbers, we get the set of *whole numbers*.

Definition

The **whole numbers** are the numbers in the set $\{0, 1, 2, 3, \ldots\}$.

Figure 7
The whole numbers.

Figure 7 represents the whole numbers on the number line. Notice that the set of natural numbers is included in the set of whole numbers.

By expanding the numbers to the left of zero on the number line, we obtain the set called *integers*.

Definition

The **integers** are the numbers in the set $\{\ldots, -3, -2, -1, 0, 1, 2, 3, \ldots\}$.

Figure 8
The integers.

Figure 8 represents the integers on the number line. Notice that the whole numbers and natural numbers are included in the set of integers.

Integers are useful in many situations. For example, we could not discuss temperatures above 0°F (positive counting numbers) or below 0°F (negative counting numbers) without integers. A debt of 300 dollars can be represented as an integer by -300 dollars.

How can we represent a part of a whole, such as a part of last night's leftover pizza or part of a dollar? To address this problem, we enlarge our number system to include *rational numbers*.

> **In Words**
> A rational number is a number that can be expressed as a fraction where the numerator is any integer and the denominator is any nonzero integer.

Definition

A **rational number** is a number that can be written in the form $\dfrac{p}{q}$, where p and q are integers. However, q cannot equal zero.

Examples of rational numbers are $\dfrac{2}{5}, \dfrac{5}{2}, \dfrac{0}{8}, -\dfrac{7}{9}$, and $\dfrac{31}{4}$. Because $\dfrac{p}{1} = p$ for any integer p, it follows that all integers are also rational numbers. For example, 7 is an integer, but it is also a rational number because it can be written as $\dfrac{7}{1}$. We illustrate this idea below.

Work Smart
Remember, all integers are also rational numbers. For example, $42 = \dfrac{42}{1}$.

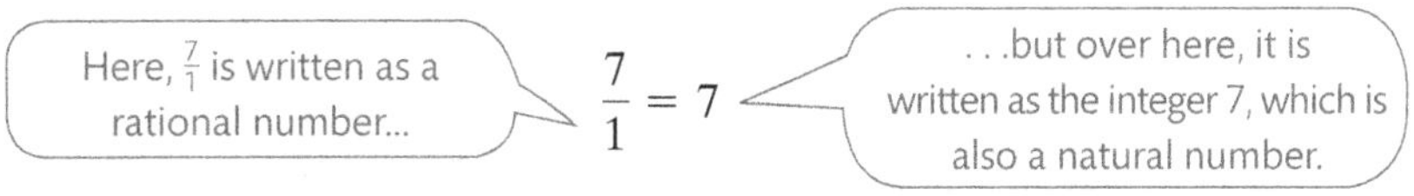

> Here, $\dfrac{7}{1}$ is written as a rational number...
> $\dfrac{7}{1} = 7$
> ...but over here, it is written as the integer 7, which is also a natural number.

In addition to representing rational numbers as fractions, we can also represent rational numbers in decimal form as either repeating decimals or terminating decimals. Table 2 shows various rational numbers in fraction form and decimal form.

The repeating decimal $0.\overline{3}$ and the terminating decimal -0.375 are rational numbers because they represent fractions.

Decimals that neither terminate nor repeat are called *irrational numbers*.

> **In Words**
> Numbers that cannot be written as the ratio of two integers are irrational.

Definition

An **irrational number** is a number that has a decimal representation that neither terminates nor repeats. Therefore, irrational numbers cannot be written as the quotient (ratio) of two integers.

Table 2		
Fraction Form of Rational Number	Decimal Form of Rational Number	Terminating or Repeating Decimal
$\frac{7}{2}$	3.5	Terminating
$\frac{1}{3}$	$0.333\ldots = 0.\overline{3}$	Repeating
$-\frac{3}{8}$	-0.375	Terminating
$-\frac{15}{11}$	$-1.3636\ldots = -1.\overline{36}$	Repeating

An example of an irrational number is 1.343343334... because the decimal neither terminates nor repeats. Other examples of irrational numbers are the symbols $\sqrt{2}$, π, and e. Because irrational numbers have decimals that neither terminate, nor repeat, we typically represent their decimal values rounded to a specified number of places. For example, $\sqrt{2}$ has a decimal value of 1.4142 rounded to four decimal places, π has a decimal value of 3.142 rounded to three decimal places, and e has a decimal approximation of 2.7183 rounded to four decimal places.

Now we are ready for a formal definition of the set of *real numbers*.

Definition

The set of rational numbers combined with the set of irrational numbers is called the set of **real numbers.**

Figure 9 shows the relationships among the various types of numbers. Note that the oval that represents the whole numbers surrounds the oval that represents the natural numbers. This means the set of whole numbers includes all the natural numbers.

Figure 9

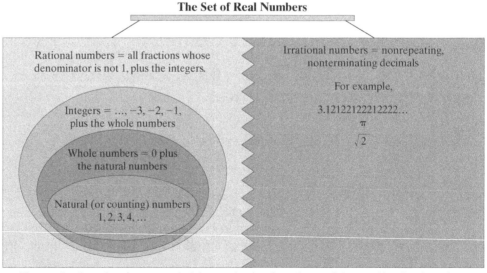

The set of real numbers is composed of the set of rational numbers and the set of irrational numbers

EXAMPLE 2 **Classifying Numbers in a Set**

List the numbers in the set

$$\left\{9, -\frac{2}{7}, -4, 0, -4.010010001\ldots, 3.\overline{632}, 18.3737\ldots\right\}$$

that are

(a) Natural numbers (b) Whole numbers
(c) Integers (d) Rational numbers
(e) Irrational numbers (f) Real numbers

Solution

(a) 9 is the only natural number.

(b) 0 and 9 are the whole numbers.

(c) 9, −4, and 0 are the integers.

(d) 9, $-\frac{2}{7}$, −4, 0, $3.\overline{632}$, and 18.3737... are the rational numbers.

(e) −4.0100.10001... is the only irrational number because the decimal does not repeat, nor does it terminate.

(f) All the numbers listed are real numbers. Real numbers consist of rational numbers together with irrational numbers.

Quick ✓

4. *True or False* Every integer is a rational number.

5. Real numbers that can be represented with a terminating or repeating decimal are called _____ numbers.

In Problems 6–11, list the numbers in the set $\left\{\frac{11}{5}, -5, 12, 2.\overline{76}, 0, 2.737737773..., \frac{18}{4}\right\}$

that are

6. Natural numbers 7. Whole numbers
8. Integers 9. Rational numbers
10. Irrational numbers 11. Real numbers

▶ ❷ Plot Points on a Real Number Line

Look back at Figure 8 on page IR-17. Notice the gaps between the integers plotted on the number line. These gaps are filled in with the real numbers that are not integers.

To construct a **real number line,** pick a point on a line somewhere in the center, and label it 0. This point is called the **origin.** The point 1 unit to the right of 0 corresponds to the real number 1. The distance between 0 and 1 determines the **scale** of the number line. For example, the point representing 2 is twice as far from 0 as 1 is. See Figure 10.

Figure 10
The real number line.

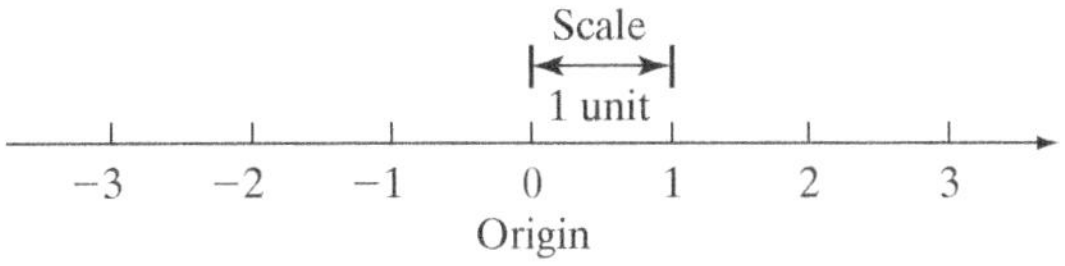

Notice that an arrowhead on the right end of the line indicates the direction in which the numbers increase. Points to the left of 0 correspond to the real numbers −1, −2, and so on.

Definition
The real number associated with a point *P* is called the **coordinate** of *P*.

EXAMPLE 3 Plotting Points on a Real Number Line

On a real number line, label the points with coordinates $0, 6, -2, 2.5, -\frac{1}{2}$.

Solution
Draw a real number line and then plot the points. See Figure 11. Notice that 2.5 is midway between 2 and 3. Also notice that $-\frac{1}{2}$ is midway between -1 and 0.

Figure 11

Quick ✓

12. The point on the real number line whose coordinate is 0 is called the _____.

13. On a real number line, label the points with coordinates $0, 3, -2, \frac{1}{2}$, and 3.5.

The real number line consists of three classes (or categories) of real numbers, as shown in Figure 12.

Figure 12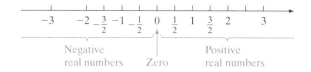

- The **negative real numbers** are the coordinates of points to the left of 0.
- The real number 0 is the coordinate of the origin.
- The **positive real numbers** are the coordinates of points to the right 0.

The **sign** of a number refers to whether the number is a positive or a negative real number. For example, the sign of -4 is negative and the sign of 100 is positive.

③ Use Inequalities to Order Real Numbers

Given two numbers (points) a and b, a must be to the left of b (denoted $a < b$) or the same as b (denoted $a = b$) or to the right of b (denoted $a > b$). See Figure 13.

If a is less than or equal to b, write $a \leq b$. Similarly, $a \geq b$ means that a is greater than or equal to b. Collectively, the symbols $<, >, \leq,$ and $\geq$ are called **inequality symbols.** The "arrowhead" in an inequality always points to the smaller number. For $3 < 5$, the "arrowhead" points to 3.

Note that $a < b$ and $b > a$ mean the same thing. For example, $2 < 3$ and $3 > 2$ mean the same thing. Do you see why?

Figure 13

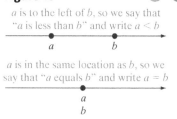

a is to the left of b, so we say that "a is less than b" and write $a < b$

a is in the same location as b, so we say that "a equals b" and write $a = b$

a is to the right of b, so we say that "a is greater than b" and write $a > b$

EXAMPLE 4 Using Inequality Symbols

(a) We know that $3 is less than $7 and that 3 apples is fewer than 7 apples. Using the real number line, we say $3 < 7$ because the point whose coordinate is 3 lies to the left of the point whose coordinate is 7 on a real number line.

(b) Being $2 in debt is not as bad as being $5 in debt, so $-2 > -5$. Using the real number line, $-2 > -5$ because the point whose coordinate is -2 lies to the right of the point whose coordinate is -5 on a real number line.

(c) $2.7 > \frac{5}{2}$ because $\frac{5}{2} = 2.5$ and $2.7 > 2.5$.

(d) $\frac{5}{6} > \frac{4}{5}$ because $\frac{5}{6} = \frac{25}{30}$ and $\frac{4}{5} = \frac{24}{30}$, and 25 out of 30 parts is more than 24 out of 30 parts. We could also write $\frac{5}{6} = 0.8\overline{3}$ and $\frac{4}{5} = 0.8$. Because $0.8\overline{3}$ is greater than 0.80, $\frac{5}{6} > \frac{4}{5}$.

Work Smart
Write fractions with a common denominator or change fractions to decimals to compare the location of the numbers on the number line.

Quick ✓

14. The symbols $<, >, \leq, \geq$ are called _____ symbols.

In Problems 15–20, insert the symbol $<, >$, or $=$, to make a true statement.

15. $2 \underline{} 9$

16. $-5 \underline{} -3$

17. $\frac{4}{5} \underline{} \frac{1}{2}$

18. $\frac{4}{7} \underline{} 0.5$

19. $\frac{4}{3} \underline{} \frac{20}{15}$

20. $-\frac{4}{3} \underline{} -\frac{5}{4}$

Based upon the discussion so far, we conclude that

$a > 0$ is equivalent to a is positive
$a < 0$ is equivalent to a is negative

We sometimes read $a > 0$ as "a is positive." If $a \geq 0$, then $a > 0$ or $a = 0$, so we may read this as "a is nonnegative" or "a is greater than or equal to zero."

▶ Compare Decimals

In statistics, it is very important that you are able to compare decimals. One way to compare decimals is to compare digits in corresponding places. To see why this works, consider 0.8 and 0.76. We know that $0.8 = 0.80 = \frac{80}{100}$, and $0.76 = \frac{76}{100}$. Because 80 parts out of 100 is greater than 76 parts out of 100, $0.80 > 0.76$.

Comparing Two Positive Decimals

Step 1: Start at the left and compare digits in corresponding decimal places, left to right.

Step 2: When two digits in the same decimal place are not equal, the number with the larger digit is the larger of the two numbers. It may be necessary to add zeros to the right of the last decimal place.

Work Smart
Writing a 0 to the right of any decimal does not change its value.

EXAMPLE 5 **Comparing Two Positive Decimals**

Which is larger: 2.04 or 2.038?

Solution

Write the decimals so that the decimal points are aligned. Add a zero to the right of the 2.04 so that both decimals have a number in the thousandths place.

The numbers in the ones place are the same: $\begin{cases} 2.040 \\ 2.038 \end{cases}$

The numbers in the tenths place are the same: $\begin{cases} 2.040 \\ 2.038 \end{cases}$

The numbers in the hundredths place are different: $\begin{cases} 2.040 \\ 2.038 \end{cases}$

Because 4 is greater than 3 in the hundredths place, 2.04 is larger than 2.038.

We can also use the real number line to visualize order. Note from the real number line in Figure 14 that 2.04 is to the right of 2.038, so 2.04 is greater than 2.038. Notice in the real number in Figure 14, the scale is 0.001.

Figure 14

EXAMPLE 6 Comparing Two Positive Decimals

Insert the symbol $<$, $>$, or $=$ to make a true statement.

0.06 _____ 0.59

Solution

5 is larger than 0 in the tenths place: $\begin{cases} 0.06 \\ 0.59 \end{cases}$

Because 5 is greater than 0 in the tenths place, $0.06 < 0.59$.

EXAMPLE 7 Comparing Two Positive Decimals

Insert the symbol $<$, $>$, or $=$ to make a true statement.

0.04 _____ 0.043

Solution
Add a zero to the right of the 4 in 0.04 so that both decimals have a number in the thousandths place.

The numbers in the tenths place are the same: $\begin{cases} 0.040 \\ 0.043 \end{cases}$

The numbers in the hundredths place are the same: $\begin{cases} 0.040 \\ 0.043 \end{cases}$

The numbers in the thousandths place are different: $\begin{cases} 0.040 \\ 0.043 \end{cases}$

Because 0 is less than 3 in the thousandths place, $0.04 < 0.043$.

Work Smart
You can also compare decimals by writing them as fractions. For Example 7,

$0.040 = \dfrac{40}{1000}$ and $0.043 = \dfrac{43}{1000}$.

$\dfrac{40}{1000} < \dfrac{43}{1000}$, so $0.04 < 0.043$.

EXAMPLE 8 Comparing Two Positive Decimals

Insert the symbol $<$, $>$, or $=$ to make a true statement:

1 _____ 0.999

Solution
Add the decimal point and three zeros to the right of 1.

1 is larger than 0 in the ones place: $\begin{cases} 1.000 \\ 0.999 \end{cases}$

Because 1 is greater than 0 in the ones place, $1 > 0.999$.

Quick ✓

In Problems 21–26, insert $<$, $>$, or $=$ to make a true statement.

21. 153.978 _____ 153.897

22. 0.345 _____ 0.34500

23. 0.14 _____ 0.1

24. 0.097 _____ 0.098

25. 0.025 _____ 0.05

26. 0.008 _____ 0.01

4 Compute the Absolute Value of a Real Number

The real number line can be used to describe the concept of *absolute value*.

> **In Words**
> Think of absolute value as the number of units you must count to get from 0 to a number. The absolute value of a number can never be negative because it represents a distance.

Definition

The **absolute value** of a number a, written $|a|$, is the distance from 0 to a on a real number line.

For example, because the distance from 0 to 3 on a real number line is 3, the absolute value of 3, $|3|$, is 3. Because the distance from 0 to -3 on a real number line is 3, $|-3| = 3$. See Figure 15.

Figure 15

EXAMPLE 9   **Computing Absolute Value**

Evaluate each of the following:

(a) $|6|$ (b) $|-7|$ (c) $|0|$ (d) $-|-1.5|$

Solution

(a) $|6| = 6$ because the distance from 0 to 6 on a real number line is 6.

(b) $|-7| = 7$ because the distance from 0 to -7 on a real number line is 7.

(c) $|0| = 0$ because the distance from 0 to 0 on a real number line is 0.

(d) $-|-1.5| = -1.5$

Quick ✓

27. The distance from zero to a point on a real number line whose coordinate is a is called the _____ _____ of a.

In Problems 28–30, evaluate each expression.

28. $|-15|$ 29. $\left|\dfrac{3}{4}\right|$ 30. $-|-4|$

1.IR3 Exercises   MyLab Statistics

Underlined exercises have complete video solutions in MyLab.

Problems 1–30 are the Quick ✓'s that follow the EXAMPLES.

Building Skills

In Problems 31–36, write each set. See Objective 1.

31. A is the set of *whole* numbers less than 5.

32. B is the set of *natural* numbers less than 25.

33. D is the set of *natural* numbers less than 5.

34. C is the set of *integers* between -6 and 4, not including -6 or 4.

35. E is the set of even *natural* numbers less than 1.

36. F is the set of *whole* numbers less than 0.

In Problems 37–42, list the elements in the set

$$\left\{-4, 3, -\dfrac{13}{2}, 0, 2.303003000\ldots\right\}$$ *that are described.*

See Objective 1.

37. natural numbers

38. whole numbers

39. integers

40. rational numbers

41. irrational numbers

42. real numbers

In Problems 43–48, list the elements in the set $\left\{-4.2, 3.\overline{5}, \pi, \dfrac{5}{5}\right\}$ *that are described. See Objective 1.*

43. real numbers
44. rational numbers
45. irrational numbers
46. integers
47. whole numbers
48. natural numbers

In Problems 49 and 50, plot the points in each set on a real number line. See Objective 2.

49. $\left\{0, \dfrac{3}{3}, -1.5, -2, \dfrac{4}{3}\right\}$

50. $\left\{\dfrac{3}{4}, \dfrac{0}{2}, -\dfrac{5}{4}, -0.5, 1.5\right\}$

In Problems 51–66, insert >, <, = *to make a true statement. See Objective 3.*

51. $-1 \_\_ 0$
52. $-8 \_\_ -8.5$
53. $\dfrac{5}{8} \_\_ \dfrac{6}{11}$
54. $\dfrac{5}{12} \_\_ \dfrac{2}{3}$
55. $\dfrac{2}{9} \_\_ 0.22$
56. $\dfrac{5}{11} \_\_ 0.\overline{45}$
57. $\dfrac{42}{6} \_\_ 7$
58. $\dfrac{3}{4} \_\_ \dfrac{3}{5}$
59. $0.58 \_\_ 0.59$
60. $0.25 \_\_ 0.242$
61. $0.08 \_\_ 0.008$
62. $0.042 \_\_ 0.044$
63. $0.23 \_\_ 0.228$
64. $0.671 \_\_ 0.67$
65. $0.07 \_\_ \dfrac{7}{100}$
66. $0.19 \_\_ \dfrac{19}{100}$

In Problems 67–74, evaluate each expression. See Objective 4.

67. $|-12|$
68. $|-8|$
69. $|4|$
70. $|7|$
71. $\left|-\dfrac{3}{8}\right|$
72. $\left|-\dfrac{13}{9}\right|$
73. $-|-2.1|$
74. $-|-3.2|$

In Problems 75 and 76, (a) plot the points on a real number line, and (b) write the numbers in ascending order.

75. $\left\{\dfrac{3}{5}, -1, -\dfrac{1}{2}, 1, 3.5, |-7|, -4.5\right\}$

76. $\left\{8, -2, |-4|, -1.5, -\dfrac{4}{3}, 0, -\dfrac{15}{3}\right\}$

77. **Flight Time** The following data represent the flight time (in minutes) of seven flights from Las Vegas, Nevada, to Newark, New Jersey, on United Airlines. Arrange the data in ascending order.

    282, 270, 260, 266, 257, 261, 267

78. **Exam Time** The following data represent the amount of time (in minutes) that nine students took to complete an online portion of an exam in Sullivan's Statistics course. Arrange the data in ascending order.

    61, 128, 85, 122, 79, 95, 86, 90

79. **pH in Water** The acidity or alkalinity of a solution is measured using pH. A pH less than 7 is acidic; a pH greater than 7 is alkaline. The following data represent the pH of twelve bottles of water. Arrange the data in ascending order.

    5.15, 5.09, 5.26, 5.20, 5.02, 5.23, 5.28, 5.26, 5.13, 5.25, 5.21, 5.24

80. **M&Ms** The following data represent the weights (in grams) of ten plain M&M candies. Arrange the data in ascending order.

    0.87, 0.82, 0.90, 0.87, 0.95, 0.92, 0.88, 0.84, 0.81, 0.93

1.IR4 Multiplying and Dividing Fractions

Objectives

1. Multiply Fractions
2. Divide Fractions

▶ ① Multiply Fractions

Figure 16(a) illustrates $\dfrac{3}{5}$ because 3 parts out of 5 equal parts are shaded. Now, suppose we wanted to take one-half of $\dfrac{3}{5}$. This could be done by dividing the shaded region that represents $\dfrac{3}{5}$ into two halves and shading one-half darker as shown in Figure 16(b). Now, the darker part is $\dfrac{3}{10}$. From this, we might conclude that one-half of $\dfrac{3}{5}$ is $\dfrac{3}{10}$, or $\dfrac{1}{2} \cdot \dfrac{3}{5} = \dfrac{3}{10}$, because the word "of" suggests multiplication.

Section 1.IR4 Multiplying and Dividing Fractions IR-25

Figure 16

(a) $\dfrac{3}{5}$ (b) $\dfrac{1}{2}$ of $\dfrac{3}{5}$ is $\dfrac{3}{10}$

In Words
To find the product of two or more fractions, multiply the numerators together. Then multiply the denominators together. Write the fraction in lowest terms, if necessary.

This would be correct and leads to the following property.

Multiplying Fractions

$$\dfrac{a}{b} \cdot \dfrac{c}{d} = \dfrac{a \cdot c}{b \cdot d}, \text{ where } b \text{ and } d \neq 0$$

EXAMPLE 1 **Multiplying Fractions**

Find the product: $\dfrac{4}{9} \cdot \dfrac{21}{8}$

Solution
Use the property just stated by multiplying the numerators and multiplying the denominators.

$$\dfrac{4}{9} \cdot \dfrac{21}{8} = \dfrac{4 \cdot 21}{9 \cdot 8}$$

Write the numerator and denominator as products of prime factors:
$$= \dfrac{2 \cdot 2 \cdot 3 \cdot 7}{3 \cdot 3 \cdot 2 \cdot 2 \cdot 2}$$

Divide out common factors:
$$= \dfrac{2 \cdot 2 \cdot 3 \cdot 7}{3 \cdot 3 \cdot 2 \cdot 2 \cdot 2}$$

Perform the multiplication:
$$= \dfrac{7}{6}$$

EXAMPLE 2 **Multiplying Fractions**

Find the product: $\dfrac{4}{15} \cdot \dfrac{25}{24} \cdot \dfrac{6}{5}$

Solution

$$\dfrac{4}{15} \cdot \dfrac{25}{24} \cdot \dfrac{6}{5} = \dfrac{4 \cdot 25 \cdot 6}{15 \cdot 24 \cdot 5}$$

Write the numerator and denominator as products of prime factors:
$$= \dfrac{2 \cdot 2 \cdot 5 \cdot 5 \cdot 2 \cdot 3}{3 \cdot 5 \cdot 2 \cdot 2 \cdot 2 \cdot 3 \cdot 5}$$

Divide out common factors:
$$= \dfrac{2 \cdot 2 \cdot 5 \cdot 5 \cdot 2 \cdot 3}{3 \cdot 5 \cdot 2 \cdot 2 \cdot 2 \cdot 3 \cdot 5}$$

If all of the factors divide out, be sure to use 1 as a factor:
$$= \dfrac{1}{3}$$

Work Smart
Instead of using prime factorization, we could do the following:

$$\dfrac{4}{15} \cdot \dfrac{25}{24} \cdot \dfrac{6}{5} = \dfrac{\overset{1}{4} \cdot \overset{5}{25} \cdot \overset{1}{6}}{\underset{3}{15} \cdot \underset{4}{24} \cdot \underset{1}{5}}$$
$$= \dfrac{1}{3}$$

Quick ✓

In Problems 1–6, find each product, and write the answer in lowest terms.

1. $\dfrac{1}{2} \cdot \dfrac{1}{2}$
2. $\dfrac{3}{7} \cdot \dfrac{35}{36}$
3. $\dfrac{5}{8} \cdot \dfrac{2}{15}$
4. $\dfrac{14}{25} \cdot \dfrac{15}{24} \cdot \dfrac{45}{28}$

EXAMPLE 3 **Moral Values in the United States**

In a recent survey conducted by Gallup it was found that $\dfrac{6}{11}$ of 880 adult Americans feel the state of moral values in the United States is getting better. How many of the adult Americans surveyed feel the state of moral values in the United States is getting better?

Solution

The word "of" in mathematics usually means multiplication. So, $\frac{6}{11}$ of 880 means $\frac{6}{11} \cdot 880$.

$$\frac{6}{11} \cdot 880 = \frac{6}{11} \cdot \frac{880}{1}$$
$$= \frac{6 \cdot 880}{11 \cdot 1}$$
$$= \frac{6 \cdot 11 \cdot 80}{11 \cdot 1}$$
$$= \frac{6 \cdot 80}{1}$$
$$= 480$$

Work Smart
Notice we did not use the prime factorization in Example 3. Do you see why?

Of the 880 adult Americans surveyed, 480 feel the state of moral values in the United States is getting better.

Quick ✓

5. What is $\frac{3}{5}$ of 100?

6. Sonia saves $\frac{2}{15}$ of her monthly income for retirement. If Sonia earned $4500 last month, how much did she save for retirement?

The following rules of signs apply to fractions.

- The product of two positive numbers is positive;
- The product of a positive number and a negative number is negative;
- The product of two negative numbers is positive.

EXAMPLE 4 Multiplying Fractions

Find the product: $\left(-\frac{9}{11}\right)\left(\frac{22}{15}\right)$

Solution
The product of a negative number and a positive number is negative. Attach the negative sign to the product and then multiply.

$$\left(-\frac{9}{11}\right)\left(\frac{22}{15}\right) = -\frac{9 \cdot 22}{11 \cdot 15}$$

Write the numerator and denominator as products of prime factors:
$$= -\frac{3 \cdot 3 \cdot 2 \cdot 11}{11 \cdot 3 \cdot 5}$$

Divide out common factors:
$$= -\frac{3 \cdot 3 \cdot 2 \cdot 11}{11 \cdot 3 \cdot 5}$$

Perform the multiplication:
$$= -\frac{6}{5}$$

Work Smart
Be careful! It is easy to forget to write the negative sign in your answer.

EXAMPLE 5 Multiplying Fractions

Find the product: $-\frac{56}{27} \cdot \left(-\frac{15}{14}\right)$

Section 1.IR4 Multiplying and Dividing Fractions   IR-27

Work Smart

Multiplying fractions does not require that the factors be written as a product of primes. Consider

$$\frac{56 \cdot 15}{27 \cdot 14} = \frac{4 \cdot 14 \cdot 3 \cdot 5}{9 \cdot 3 \cdot 14 \cdot 1}$$
$$= \frac{20}{9}$$

Another possible notation is

$$\frac{56 \cdot 15}{27 \cdot 14} = \frac{\overset{4}{\cancel{56}} \cdot \overset{5}{\cancel{15}}}{\underset{9}{\cancel{27}} \cdot \underset{1}{\cancel{14}}}$$
$$= \frac{20}{9}$$

Solution

Because the product of two negative numbers is positive, the answer is positive.

$$-\frac{56}{27} \cdot \left(-\frac{15}{14}\right) = \frac{56 \cdot 15}{27 \cdot 14}$$

Write the numerator and denominator as products of prime factors:
$$= \frac{2 \cdot 2 \cdot 2 \cdot 7 \cdot 3 \cdot 5}{3 \cdot 3 \cdot 3 \cdot 2 \cdot 7}$$

Divide out common factors:
$$= \frac{2 \cdot 2 \cdot 2 \cdot \cancel{7} \cdot \cancel{3} \cdot 5}{3 \cdot 3 \cdot \cancel{3} \cdot 2 \cdot \cancel{7}}$$

Perform the multiplication:
$$= \frac{20}{9}$$

EXAMPLE 6   **Multiplying Fractions**

Find the product: $\dfrac{39}{42} \cdot \left(-\dfrac{21}{13}\right)$

Solution

Because the product of a positive number and a negative number is negative, the answer is negative.

$$\frac{39}{42} \cdot \left(-\frac{21}{13}\right) = -\frac{39 \cdot 21}{42 \cdot 13}$$

Write the numerator and denominator as products of prime factors:
$$= -\frac{3 \cdot 13 \cdot 3 \cdot 7}{2 \cdot 3 \cdot 7 \cdot 13}$$

Divide out common factors:
$$= -\frac{\cancel{3} \cdot \cancel{13} \cdot 3 \cdot \cancel{7}}{2 \cdot \cancel{3} \cdot \cancel{7} \cdot \cancel{13}}$$

$$= -\frac{3}{2}$$

Quick ✓

In Problems 7–12, find each product, and write the answer in lowest terms.

7. $\left(-\dfrac{2}{3}\right)\left(-\dfrac{7}{15}\right)$ 8. $\dfrac{35}{32} \cdot \left(-\dfrac{16}{49}\right)$ 9. $-\dfrac{18}{35} \cdot \dfrac{63}{54}$

10. $\left(-\dfrac{9}{12}\right)\left(-\dfrac{8}{27}\right)\left(-\dfrac{9}{24}\right)$ 11. $-6 \cdot \left(-\dfrac{2}{15}\right)$ 12. $\dfrac{9}{2} \cdot 24 \cdot \left(-\dfrac{3}{4}\right)$

▶ ❷ Divide Fractions

To discuss division of fractions, we need to introduce the idea of a *reciprocal*, or *multiplicative inverse*.

In Words

Any two numbers whose product is 1 are called reciprocals or multiplicative inverses of each other.

Definition

For each nonzero rational number $\dfrac{a}{b}$, there is a rational number $\dfrac{b}{a}$, called the **reciprocal**, or the **multiplicative inverse**, that has the following property:

$$\frac{a}{b} \cdot \frac{b}{a} = 1, \text{ where } a \neq 0 \text{ and } b \neq 0$$

For example, $\frac{2}{3}$ and $\frac{3}{2}$ are reciprocals because $\frac{2}{3} \cdot \frac{3}{2} = \frac{6}{6} = 1$. To find the reciprocal of 5, first write 5 as $\frac{5}{1}$. The reciprocal of 5 is $\frac{1}{5}$ because $\frac{5}{1} \cdot \frac{1}{5} = \frac{5}{5} = 1$.

What is the reciprocal of -8? If a number is negative, its reciprocal must also be negative so that the product is positive. The reciprocal of -8 is $-\frac{1}{8}$ because $\left(-\frac{8}{1}\right)\left(-\frac{1}{8}\right) = \frac{8}{8} = 1$.

EXAMPLE 7 **Finding the Reciprocal of a Number**

Find the reciprocal of each of the following numbers.

(a) $\frac{9}{7}$ (b) $-\frac{4}{5}$ (c) -2 (d) $\frac{1}{15}$

Solution

(a) $\frac{7}{9}$ (b) $-\frac{5}{4}$ (c) $-\frac{1}{2}$ (d) 15

Quick ✓

In Problems 13 and 14, find the reciprocal of each number.

13. $-\frac{15}{4}$ 14. $\frac{1}{6}$

In Words

To divide two fractions, multiply by the reciprocal of the divisor. Put another way, "flip and multiply."

Divide two fractions by rewriting the division as an equivalent multiplication problem, using to the following procedure.

Dividing Fractions

$$\frac{a}{b} \div \frac{c}{d} = \frac{a}{b} \cdot \frac{d}{c} = \frac{a \cdot d}{b \cdot c}, \text{ where } b, c, d \neq 0$$

EXAMPLE 8 **How to Divide Fractions**

Find the quotient: $\frac{3}{10} \div \frac{12}{25}$

Step-by-Step Solution

Step 1: Write the equivalent multiplication problem.

$$\frac{3}{10} \div \frac{12}{25} = \frac{3}{10} \cdot \frac{25}{12}$$

Step 2: Multiply, write the product in factored form, and divide out common factors.

$$= \frac{3 \cdot 25}{10 \cdot 12}$$

Factor: $= \frac{3 \cdot 5 \cdot 5}{2 \cdot 5 \cdot 3 \cdot 4}$

Divide out common factors: $= \frac{3 \cdot 5 \cdot 5}{2 \cdot 5 \cdot 3 \cdot 4}$

$$= \frac{5}{2 \cdot 4}$$

Step 3: Multiply the remaining factors.

$$= \frac{5}{8}$$

Quick ✓

In Problems 15–18, find each quotient, and write the answer in lowest terms.

15. $\dfrac{16}{21} \div \dfrac{4}{7}$ **16.** $\dfrac{2}{3} \div \dfrac{1}{15}$ **17.** $15 \div \dfrac{25}{8}$ **18.** $\dfrac{14}{12} \div \dfrac{56}{6}$

EXAMPLE 9 Dividing Fractions

Find the quotient:

(a) $-\dfrac{44}{63} \div \left(-\dfrac{11}{21}\right)$ (b) $\dfrac{2}{15} \div (-10)$ (c) $-3 \div \dfrac{1}{6}$

Solution

(a) Because the quotient of two negative numbers is positive, the quotient is positive.

Write the equivalent multiplication problem: $-\dfrac{44}{63} \div \left(-\dfrac{11}{21}\right) = \dfrac{44}{63} \cdot \dfrac{21}{11}$

$$= \dfrac{44 \cdot 21}{63 \cdot 11}$$

Write the products in factored form: $= \dfrac{2 \cdot 2 \cdot 11 \cdot 3 \cdot 7}{3 \cdot 3 \cdot 7 \cdot 11}$

Divide out common factors: $= \dfrac{2 \cdot 2 \cdot \cancel{11} \cdot 3 \cdot \cancel{7}}{3 \cdot 3 \cdot \cancel{7} \cdot \cancel{11}}$

Multiply the remaining factors: $= \dfrac{4}{3}$

Work Smart

In Example 9(a), we could work smart by factoring as follows:
$$\dfrac{44 \cdot 21}{63 \cdot 11} = \dfrac{4 \cdot 11 \cdot 21}{3 \cdot 21 \cdot 11}$$
$$= \dfrac{4}{3}$$

(b) Because the quotient of a positive number and a negative number is negative, the quotient is negative. Remember, -10 may be written as $-\dfrac{10}{1}$.

$$\dfrac{2}{15} \div (-10) = \dfrac{2}{15} \div \left(-\dfrac{10}{1}\right)$$

Write the equivalent multiplication problem: $= \dfrac{2}{15} \cdot \left(-\dfrac{1}{10}\right)$

Multiply the fractions. Use the correct sign for the product: $= -\dfrac{2 \cdot 1}{15 \cdot 10}$

Write the products in factored form: $= -\dfrac{2 \cdot 1}{3 \cdot 5 \cdot 2 \cdot 5}$

Divide out common factors: $= -\dfrac{\cancel{2} \cdot 1}{3 \cdot 5 \cdot \cancel{2} \cdot 5}$

Multiply the remaining factors: $= -\dfrac{1}{75}$

(c) Because the quotient of a negative number and a positive number is negative, the quotient is negative. Remember, $-3 = -\dfrac{3}{1}$.

$$-3 \div \dfrac{1}{6} = -\dfrac{3}{1} \div \dfrac{1}{6}$$

Write the equivalent multiplication problem: $= -\dfrac{3}{1} \cdot \dfrac{6}{1}$

$$= -\dfrac{3 \cdot 6}{1 \cdot 1}$$

Multiply the factors: $= -\dfrac{18}{1}$

$\dfrac{a}{1} = a$: $= -18$

Quick ✓

In Problems 19–24, find each quotient, and write the answer in lowest terms.

19. $-\dfrac{16}{21} \div \dfrac{4}{7}$ 20. $-\dfrac{9}{25} \div \left(-\dfrac{24}{25}\right)$ 21. $\dfrac{2}{3} \div \left(-\dfrac{1}{15}\right)$

22. $-\dfrac{27}{6} \div (-9)$ 23. $15 \div \dfrac{25}{8}$ 24. $-\dfrac{50}{36} \div \dfrac{25}{18}$

1.IR4 Exercises   MyLab Statistics

Underlined exercises have complete video solutions in MyLab.

Problems **1–24** are the Quick ✓s that follow the EXAMPLES.

Building Skills

In Problems 25–34, find the product, and write the answer in lowest terms. See Objective 1.

25. $\dfrac{1}{3} \cdot \dfrac{1}{3}$ 26. $\dfrac{1}{4} \cdot \dfrac{1}{4}$ 27. $\dfrac{11}{5} \cdot \dfrac{25}{22}$ 28. $\dfrac{12}{25} \cdot \dfrac{15}{14}$

29. $\dfrac{18}{56} \cdot \dfrac{35}{27}$ 30. $\dfrac{9}{24} \cdot \dfrac{30}{21}$ 31. $\dfrac{7}{8} \cdot \dfrac{3}{14} \cdot \dfrac{12}{9}$ 32. $\dfrac{1}{2} \cdot \dfrac{3}{7} \cdot \dfrac{4}{9}$

33. $3 \cdot \dfrac{4}{9}$ 34. $\dfrac{5}{4} \cdot 2$

In Problems 35–42, find the product, and write the answer in lowest terms. See Objective 1.

35. $\left(-\dfrac{1}{5}\right)\left(\dfrac{1}{3}\right)$ 36. $\left(\dfrac{1}{4}\right)\left(-\dfrac{1}{2}\right)$

37. $\dfrac{3}{4} \cdot \left(-\dfrac{16}{27}\right)$ 38. $-\dfrac{15}{6} \cdot \dfrac{3}{8}$

39. $-27 \cdot \left(-\dfrac{2}{3}\right)$ 40. $-\dfrac{2}{3} \cdot (-36)$

41. $\left(-\dfrac{2}{3}\right)\left(-\dfrac{9}{4}\right)\left(-\dfrac{10}{3}\right)$ 42. $\left(-\dfrac{6}{5}\right)\left(-\dfrac{25}{3}\right)\left(-\dfrac{9}{10}\right)$

In Problems 43–50, find the reciprocal of each number. See Objective 2.

43. $\dfrac{9}{17}$ 44. $\dfrac{3}{8}$ 45. $-\dfrac{8}{5}$ 46. $-\dfrac{10}{7}$

47. -3 48. -8 49. $\dfrac{1}{7}$ 50. $\dfrac{1}{12}$

In Problems 51–58, find each quotient, and write the answer in lowest terms. See Objective 2.

51. $\dfrac{6}{17} \div \dfrac{15}{34}$ 52. $\dfrac{4}{13} \div \dfrac{12}{39}$ 53. $\dfrac{32}{7} \div \dfrac{8}{35}$

54. $\dfrac{25}{7} \div \dfrac{15}{14}$ 55. $\dfrac{3}{5} \div 45$ 56. $\dfrac{2}{7} \div 28$

57. $15 \div \dfrac{35}{6}$ 58. $8 \div \dfrac{20}{36}$

In Problems 59–66, find each quotient, and write the answer in lowest terms. See Objective 2.

59. $-\dfrac{4}{9} \div \left(-\dfrac{7}{12}\right)$ 60. $-\dfrac{7}{8} \div \left(-\dfrac{20}{14}\right)$

61. $\dfrac{3}{8} \div \left(-\dfrac{24}{32}\right)$ 62. $\dfrac{9}{5} \div \left(-\dfrac{54}{10}\right)$

63. $-\dfrac{4}{7} \div 28$ 64. $-\dfrac{3}{4} \div 36$

65. $-12 \div \left(-\dfrac{6}{5}\right)$ 66. $-9 \div \left(-\dfrac{3}{8}\right)$

Mixed Practice

In Problems 67–76, perform the indicated operation, and write the answer in lowest terms.

67. $4 \div 12$ 68. $9 \div 36$

69. $-\dfrac{4}{5} \cdot \dfrac{65}{40}$ 70. $\dfrac{4}{7} \cdot \left(-\dfrac{35}{50}\right)$

71. $\dfrac{2}{5} \div \left(-\dfrac{4}{35}\right)$ 72. $-\dfrac{3}{8} \div \dfrac{9}{24}$

73. $\dfrac{0}{36} \div \dfrac{7}{15}$ 74. $\dfrac{0}{36} \div \dfrac{1}{4}$

75. $-\dfrac{16}{5} \cdot \left(-\dfrac{35}{24}\right)$ 76. $-\dfrac{14}{3} \cdot \left(-\dfrac{27}{32}\right)$

Applying the Concepts

77. Find $\dfrac{2}{3}$ of 24. 78. Find $\dfrac{3}{4}$ of 36.

79. Find $\dfrac{1}{5}$ of $\dfrac{15}{4}$. 80. Find $\dfrac{1}{3}$ of $\dfrac{21}{8}$.

△ 81. Find the area of a square which is $\dfrac{9}{13}$ mile on a side. *Hint:* The formula for area of a square is $A = s \cdot s$.

△ 82. Find the area of a square that is $\dfrac{15}{8}$ inches on a side.

83. **Federal Withholding** Samuel has $\frac{3}{20}$ of his gross income withheld for federal income tax. If Samuel has gross income of $1500 one week, what is the amount of federal withholding from Samuel's paycheck?

84. **Retirement Savings** Yolanda always saves $\frac{3}{50}$ of her weekly pay for retirement. If Yolanda earned $2000 last week, what will she put away in her retirement savings?

85. **Mortgage Payments** In a recent survey conducted by Pew Research, three-fifths of 1200 adult Americans believe it is not acceptable for a homeowner to stop making mortgage payments. How many said it is not acceptable for a homeowner to stop making mortgage payments?

86. **Bachelor's Degree** A recent survey found that one-third of 1575 Americans age 25 to 29 have at least a bachelor's degree. How many of the 1575 Americans surveyed have at least a bachelor's degree?

87. **Road Trip** Rami is driving from Los Angeles to San Francisco to visit her boyfriend. If the driving distance is 380 miles and Rami has completed $\frac{3}{4}$ of the trip, how far has Rami driven?

88. **To the Moon** Apollo 13 was the third manned mission to the moon, which is 238,900 miles from Earth. When the space craft was near the moon, there was an explosion in one of the oxygen tanks, and the mission had to be abandoned. If the crew had completed $\frac{9}{10}$ of their journey to the moon before the accident, how far did the astronauts have to travel back to Earth?

Explaining the Concepts

89. Draw a visual representation of $\frac{1}{2} \cdot \frac{1}{4}$ as in Figure 16. Explain the meaning of multiplying fractions, using your picture to describe the process.

1.IR5 Adding and Subtracting Fractions

Objectives

1. Add or Subtract Fractions with Like Denominators
2. Find the Least Common Denominator and Equivalent Fractions
3. Add or Subtract Fractions with Unlike Denominators

Figure 17

1 Add or Subtract Fractions with Like Denominators

Like fractions are fractions with the same denominator. For example, $\frac{3}{4}$ and $\frac{5}{4}$ are like fractions.

In Figure 17, two of the eight equal regions are shaded. Each of the regions represents the fraction $\frac{1}{8}$. Together, the two shaded regions make up $\frac{1}{4}$ of the circle. This implies that

$$\frac{1}{8} + \frac{1}{8} = \frac{1+1}{8}$$
$$= \frac{2}{8}$$
$$= \frac{2 \cdot 1}{2 \cdot 2 \cdot 2}$$
$$= \frac{1}{4}$$

The idea is that if you have 1 part out of 8 and then add another 1 part out of 8, you end up with 2 parts out of 8 (which is the same as 1 part out of 4). Also, because $a - b = a + (-b)$, the same process works for subtraction.

In Words
To add (or subtract) like fractions, add (or subtract) the numerators and keep the denominator.

Adding or Subtracting Like Fractions

$$\frac{a}{c} + \frac{b}{c} = \frac{a+b}{c}, \text{ where } c \neq 0 \qquad \frac{a}{c} - \frac{b}{c} = \frac{a-b}{c}, \text{ where } c \neq 0$$

EXAMPLE 1 Adding Like Fractions

Find the sum and write the answer in lowest terms: $\dfrac{1}{8} + \dfrac{3}{8}$

Solution

Write the numerators as a sum over the common denominator: $\dfrac{1}{8} + \dfrac{3}{8} = \dfrac{1+3}{8}$

Add the numerators: $= \dfrac{4}{8}$

Write the numerator and denominator as a product of prime factors, and divide out common factors: $= \dfrac{2 \cdot 2 \cdot 1}{2 \cdot 2 \cdot 2}$

$= \dfrac{1}{2}$

EXAMPLE 2 Subtracting Like Fractions

Find the difference and write the answer in lowest terms: $\dfrac{7}{8} - \dfrac{5}{8}$

Solution

Write the numerators as a difference over the common denominator: $\dfrac{7}{8} - \dfrac{5}{8} = \dfrac{7-5}{8}$

Subtract the numerators: $= \dfrac{2}{8}$

Write the numerator and denominator as a product of prime factors, and divide out common factors: $= \dfrac{2 \cdot 1}{2 \cdot 2 \cdot 2}$

$= \dfrac{1}{4}$

Quick ✓

In Problems 1–6, perform the indicated operation, and write the answer in lowest terms.

1. $\dfrac{2}{9} + \dfrac{3}{9}$
2. $\dfrac{7}{16} - \dfrac{4}{16}$
3. $\dfrac{11}{12} - \dfrac{3}{12}$
4. $\dfrac{1}{4} + \dfrac{5}{4}$
5. $\dfrac{25}{33} + \dfrac{5}{33}$
6. $\dfrac{15}{36} - \dfrac{6}{36}$

When adding and subtracting fractions containing negatives, it is helpful to remember the following:

$$\dfrac{-a}{b} = \dfrac{a}{-b} = -\dfrac{a}{b}$$

When adding or subtracting fractions, place the negative in the numerator, as in $\dfrac{-a}{b}$.

EXAMPLE 3 Adding Like Fractions

Find the sum and write the answer in lowest terms: $-\dfrac{18}{35} + \dfrac{3}{35}$

Solution

Place the negative sign into the numerator: $-\dfrac{18}{35} + \dfrac{3}{35} = \dfrac{-18}{35} + \dfrac{3}{35}$

Write the numerators as a sum over the common denominator: $= \dfrac{-18 + 3}{35}$

Add the numerators: $= \dfrac{-15}{35}$

Factor and divide out the common factors: $= -\dfrac{3 \cdot 5}{5 \cdot 7}$

$= -\dfrac{3}{7}$

EXAMPLE 4 Subtracting Like Fractions

Find the difference and write the answer in lowest terms: $-\dfrac{5}{14} - \dfrac{2}{14}$

Solution

$-\dfrac{5}{14} = \dfrac{-5}{14}$; $\quad -\dfrac{5}{14} - \dfrac{2}{14} = \dfrac{-5}{14} - \dfrac{2}{14}$

Write the numerators as a difference over the common denominator:
$= \dfrac{-5 - 2}{14}$

$= \dfrac{-7}{14}$

Factor and divide out the common factors: $= -\dfrac{7 \cdot 1}{7 \cdot 2}$

$= -\dfrac{1}{2}$

Quick ✓

In Problems 7–14, perform the indicated operation, and write the answer in lowest terms.

7. $\dfrac{2}{15} + \left(-\dfrac{11}{15}\right)$ 8. $\dfrac{14}{6} - \left(-\dfrac{1}{6}\right)$ 9. $-\dfrac{7}{12} - \dfrac{8}{12}$

10. $-\dfrac{9}{20} + \dfrac{3}{20}$ 11. $\dfrac{4}{9} + \left(-\dfrac{1}{9}\right)$ 12. $-\dfrac{5}{4} - \left(-\dfrac{3}{4}\right)$

13. $-\dfrac{7}{15} - \left(-\dfrac{4}{15}\right)$ 14. $-\dfrac{5}{12} + \left(-\dfrac{5}{12}\right)$

EXAMPLE 5 Wasting Time Online

In a survey conducted by Salary.com of 800 American office workers, 192 reported Google as their biggest online time waster, 184 reported Facebook as their biggest time waster, and 8 reported ESPN as their biggest online time waster. What fraction of those surveyed report Google, Facebook, or ESPN as their biggest online time waster? Express your answer in lowest terms.

Solution

First, determine the fraction of workers who report each site as their biggest online time waster.

Google: 192 out of 800 or $\dfrac{192}{800}$

Facebook: 184 out of 800 or $\dfrac{184}{800}$

ESPN: 8 out of 800 or $\dfrac{8}{800}$

Now, add each of these fractions to determine the fraction of those surveyed who report Google, Facebook, or ESPN as their biggest online time waster. Then, reduce the fraction, if necessary.

$$\dfrac{192}{800} + \dfrac{184}{800} + \dfrac{8}{800} = \dfrac{192 + 184 + 8}{800}$$

$$= \dfrac{384}{800}$$

$$= \dfrac{32 \cdot 12}{32 \cdot 25}$$

$$= \dfrac{12}{25}$$

So, 12 out of 25 workers report Google, Facebook, or ESPN as their biggest online time waster.

EXAMPLE 6 **Calculating the Missing Quantity**

Erika is having her friends over for a party, and the group decides to order a pizza. They divide the pizza into 16 equal pieces. The first friend eats 4 slices, two girls have 1 slice each, and Erika has 5 slices. What fraction of the pizza is left over?

Solution
First, calculate the fraction of the pizza eaten.

$$\frac{4}{16} + \frac{1}{16} + \frac{1}{16} + \frac{5}{16} = \frac{4+1+1+5}{16}$$

$$= \frac{11}{16}$$

Work Smart
Remember that
$$\frac{a}{a} = 1, a \neq 0$$

Second, calculate how much was left. One whole pizza is $\frac{16}{16}$, so subtract the amount eaten from one whole to determine the amount left.

$$\frac{16}{16} - \frac{11}{16} = \frac{16-11}{16}$$

$$= \frac{5}{16}$$

Therefore, when the girls were finished eating, there was $\frac{5}{16}$ of the pizza left over.

Quick ✓

15. In a survey of 600 American teens (13 to 18 years of age) conducted by Common Sense, 60 reported reading as their favorite media activity, 180 reported listening to music as their favorite media activity, and 36 reported watching online videos as their favorite media activity. What fraction of those surveyed reported reading, listening to music, or watching online videos as their favorite media activity? Express your result as a fraction in lowest terms.

16. After competing in the Olympics, Michael's next adventure was to swim the English Channel. The distance across the Channel is approximately 24 miles, and Michael was hoping to finish the swim in 8 hours. He completed $\frac{7}{16}$ of the trip before he needed some food. Next, he swam another $\frac{5}{16}$ of the distance and then took a short break from swimming. What fraction of the trip was left for Michael to swim? If he was able to complete the trip in 8 hours, how much time elapsed when he took his second break?

❷ Find the Least Common Denominator and Equivalent Fractions

We now know how to add and subtract fractions with like denominators. How do we add and subtract **unlike fractions**—that is, fractions with different denominators? We rewrite each fraction with the *least common denominator*.

Recall that the least common multiple (LCM) of two or more numbers is the smallest number that is a multiple of each of the numbers. The LCM is the same number that we use for the *least common denominator (LCD)*.

Definition
The **least common denominator (LCD)** is the least common multiple of the denominators of a group of fractions.

For instance, to determine the least common denominator of $\frac{4}{9}$ and $\frac{5}{12}$, find the least common multiple (LCM) of 9 and 12. The LCM is the least common denominator (LCD).

To find the least common multiple of 9 and 12, we could list the multiples of 9 and 12 and then find the smallest multiple that they share.

Multiples of 9:   9, 18, 27, **36**, 45, 54, ...

Multiples of 12:   12, 24, **36**, 48, 60, ...

The LCM of 9 and 12 is 36. Therefore, the LCD of $\frac{4}{9}$ and $\frac{5}{12}$ is 36.

However, in the next example we present a second method for determining the LCD.

EXAMPLE 7 **How to Find the Least Common Denominator (LCD)**

Determine the LCD of $\frac{4}{9}$ and $\frac{5}{12}$.

Step-by-Step Solution

Step 1: Write each denominator as a product of prime factors.

$$9 = 3 \cdot 3$$
$$12 = 3 \cdot 2 \cdot 2$$

Step 2: Write down the shared factors and all remaining factors.

$$\text{LCD} = 3 \cdot 3 \cdot 2 \cdot 2$$

Step 3: Multiply the factors in Step 2.

$$= 36$$

The LCD of $\frac{4}{9}$ and $\frac{5}{12}$ is 36.

Finding the LCD of Two or More Fractions

Step 1: Write each denominator as the product of prime factors. Align the common factors of each denominator vertically.

Step 2: Write down the factor(s) that the denominators share, if any. Then write down each of the remaining factors.

Step 3: Multiply the factors listed in Step 2. The product is the LCD.

EXAMPLE 8 **Finding the Least Common Denominator (LCD)**

Determine the LCD of $\frac{3}{8}, \frac{11}{12}$, and $\frac{7}{18}$.

Solution

Write each denominator as a product of prime factors:

$$8 = 2 \cdot 2 \cdot 2$$
$$12 = 2 \cdot 2 \cdot 3$$
$$18 = 2 \cdot 3 \cdot 3$$

Write down the shared factors and all remaining factors: $\text{LCD} = 2 \cdot 2 \cdot 2 \cdot 3 \cdot 3$

Multiply the factors: $= 72$

The LCD of $\frac{3}{8}, \frac{11}{12}$, and $\frac{7}{18}$ is 72.

Quick ✓

In Problems 17–20, determine the least common denominator (LCD).

17. $\dfrac{3}{8}$ and $\dfrac{1}{5}$ **18.** $\dfrac{5}{12}$ and $\dfrac{1}{4}$ **19.** $\dfrac{8}{9}$ and $\dfrac{5}{6}$ **20.** $\dfrac{8}{15}$ and $\dfrac{7}{10}$

We now discuss how to rewrite fractions with the least common denominator (LCD). Remember, $\dfrac{a}{a} = 1$ as long as $a \neq 0$. Use this idea to rewrite fractions with the LCD.

EXAMPLE 9   Finding Equivalent Fractions with the LCD

Rewrite $\dfrac{4}{9}$ and $\dfrac{5}{12}$ as equivalent fractions with the LCD.

Solution

The LCD of $\dfrac{4}{9}$ and $\dfrac{5}{12}$ is 36. See Example 7.

Now, rewrite each fraction as an equivalent fraction with the LCD, 36.

Work Smart

To determine what to multiply each fraction by to get the LCD, compare the factored form of the denominator to the factored form of the LCD. For example, $9 = 3 \cdot 3$ and LCD $= 3 \cdot 3 \cdot 2 \cdot 2$, so $2 \cdot 2$ is "missing", therefore, multiply by 4.

$$\dfrac{4}{9} \cdot \dfrac{4}{4} = \dfrac{16}{36}$$

$$\dfrac{5}{12} \cdot \dfrac{3}{3} = \dfrac{15}{36}$$

So $\dfrac{4}{9} = \dfrac{16}{36}$ and $\dfrac{5}{12} = \dfrac{15}{36}$.

EXAMPLE 10   Finding Equivalent Fractions

Rewrite $\dfrac{3}{8}, \dfrac{11}{12},$ and $\dfrac{7}{18}$ as equivalent fractions with the LCD.

Solution

The LCD of $\dfrac{3}{8}, \dfrac{11}{12},$ and $\dfrac{7}{18}$ is 72. See Example 8.

Now, rewrite each fraction as an equivalent fraction with the LCD, 72.

$$\dfrac{11}{12} \cdot \dfrac{6}{6} = \dfrac{66}{72}$$

$$\dfrac{3}{8} \cdot \dfrac{9}{9} = \dfrac{27}{72}$$

$$\dfrac{7}{18} \cdot \dfrac{4}{4} = \dfrac{28}{72}$$

So $\dfrac{3}{8} = \dfrac{27}{72}, \dfrac{11}{12} = \dfrac{66}{72},$ and $\dfrac{7}{18} = \dfrac{28}{72}$.

Quick ✓

In Problems 21–24, rewrite each fraction as an equivalent fraction with the LCD.

21. $\dfrac{3}{8}$ and $\dfrac{1}{5}$ **22.** $\dfrac{5}{12}$ and $\dfrac{1}{4}$ **23.** $\dfrac{8}{9}$ and $\dfrac{5}{6}$ **24.** $\dfrac{8}{15}$ and $\dfrac{7}{10}$

▶ ❸ Add or Subtract Fractions with Unlike Denominators

How do we add or subtract rational numbers with different denominators? First, find the least common denominator of the two rational numbers. Then, add or subtract the fractions written with the LCD.

EXAMPLE 11 How to Add Fractions with Unlike Denominators

Find the sum: $\dfrac{5}{6} + \dfrac{3}{8}$

Step-by-Step Solution

Step 1: Find the least common denominator of the fractions.

Write each denominator as the product of prime factors, arranging like factors vertically:

$$6 = 2 \cdot 3$$
$$8 = 2 \cdot 2 \cdot 2$$
$$\text{LCD} = 2 \cdot 2 \cdot 2 \cdot 3$$

Multiply: $= 24$

Step 2: Write each rational number with the denominator found in Step 1.

Use $1 = \dfrac{4}{4}$ to change the denominator 6 to 24:

$$\dfrac{5}{6} = \dfrac{5}{6} \cdot \dfrac{4}{4} = \dfrac{20}{24}$$

Use $1 = \dfrac{3}{3}$ to change the denominator 8 to 24:

$$\dfrac{3}{8} = \dfrac{3}{8} \cdot \dfrac{3}{3} = \dfrac{9}{24}$$

Step 3: Add the numerators and write the result over the common denominator.

$$\dfrac{5}{6} + \dfrac{3}{8} = \dfrac{20}{24} + \dfrac{9}{24}$$

$\dfrac{a}{c} + \dfrac{b}{c} = \dfrac{a+b}{c}$:

$$= \dfrac{20 + 9}{24}$$

$$= \dfrac{29}{24}$$

Step 4: Write in lowest terms.

The rational number is in lowest terms, so $\dfrac{5}{6} + \dfrac{3}{8} = \dfrac{29}{24}$.

EXAMPLE 12 How to Subtract Fractions with Unlike Denominators

Find the difference: $\dfrac{9}{14} - \dfrac{1}{6}$

Step-by-Step Solution

Step 1: Find the least common denominator of the fractions.

Write each denominator as the product of prime factors, aligning like factors vertically:

$$14 = 2 \cdot 7$$
$$6 = 2 \cdot 3$$
$$\text{LCD} = 2 \cdot 7 \cdot 3$$
$$= 42$$

Step 2: Write each rational number with the denominator found in Step 1.

Use $1 = \dfrac{3}{3}$ to change the denominator 14 to 42:

$$\dfrac{9}{14} = \dfrac{9}{14} \cdot \dfrac{3}{3} = \dfrac{27}{42}$$

Use $1 = \dfrac{7}{7}$ to change the denominator 6 to 42:

$$\dfrac{1}{6} = \dfrac{1}{6} \cdot \dfrac{7}{7} = \dfrac{7}{42}$$

Step 3: Subtract the numerators and write the result over the common denominator.

$$\dfrac{9}{14} - \dfrac{1}{6} = \dfrac{27}{42} - \dfrac{7}{42}$$

$$= \dfrac{27 - 7}{42}$$

$$= \dfrac{20}{42}$$

(continued)

Step 4: Write in lowest terms.   Factor 20 and 42 and divide out like factors: $= \dfrac{2 \cdot 10}{2 \cdot 21}$

$= \dfrac{10}{21}$

> **Adding or Subtracting Fractions with Unlike Denominators**
>
> **Step 1:** Find the LCD of the rational numbers.
> **Step 2:** Write each rational number with the LCD.
> **Step 3:** Add or subtract the numerators and write the result over the common denominator.
> **Step 4:** Write the result in lowest terms.

Quick ✓

In Problems 25–28, find each sum or difference, and write in lowest terms.

25. $\dfrac{3}{14} + \dfrac{10}{21}$ **26.** $\dfrac{5}{12} - \dfrac{5}{18}$ **27.** $\dfrac{7}{12} + \dfrac{1}{6} - \dfrac{5}{24}$ **28.** $\dfrac{2}{15} + \dfrac{1}{12} - \dfrac{7}{60}$

EXAMPLE 13 **Subtracting a Fraction from a Whole Number**

Find the difference: $4 - \dfrac{2}{3}$

Solution
The key is to remember that $4 = \dfrac{4}{1}$.

$$4 - \dfrac{2}{3} = \dfrac{4}{1} - \dfrac{2}{3}$$

Rewrite each fraction with LCD = 3: $= \dfrac{4}{1} \cdot \dfrac{3}{3} - \dfrac{2}{3}$

$$= \dfrac{12}{3} - \dfrac{2}{3}$$

$$= \dfrac{10}{3}$$

Quick ✓

In Problems 29 and 30, find each sum or difference, and write in lowest terms.

29. $-2 + \dfrac{7}{16}$ **30.** $1 - \dfrac{5}{12}$

EXAMPLE 14 **Subtracting Fractions with Unlike Denominators**

Find the difference: $-\dfrac{9}{14} - \dfrac{1}{6}$

Solution

Write each denominator as the product of prime factors, aligning like factors vertically:

$14 = 2 \cdot 7$
$6 = 2 \cdot 3$

Multiply: LCD $= 2 \cdot 7 \cdot 3$

$= 42$

Write $-\dfrac{9}{14}$ as $\dfrac{-9}{14}$ since $-\dfrac{a}{b} = \dfrac{-a}{b}$: $\quad -\dfrac{9}{14} - \dfrac{1}{6} = \dfrac{-9}{14} - \dfrac{1}{6}$

Multiply $\dfrac{-9}{14}$ by $\dfrac{3}{3}$ and $\dfrac{1}{6}$ by $\dfrac{7}{7}$: $\quad = \dfrac{-9}{14} \cdot \dfrac{3}{3} - \dfrac{1}{6} \cdot \dfrac{7}{7}$

$$= \dfrac{-27}{42} - \dfrac{7}{42}$$

$$= \dfrac{-27 - 7}{42}$$

$$= \dfrac{-34}{42}$$

Factor -34 and 42, and divide out like factors: $\quad = -\dfrac{2 \cdot 17}{2 \cdot 21}$

$$= -\dfrac{17}{21}$$

Quick ✓

In Problems 31–35, perform the indicated operation, and write your answer in lowest terms.

31. $-\dfrac{3}{4} - \dfrac{1}{5}$ **32.** $-\dfrac{8}{9} + \dfrac{13}{15}$ **33.** $-\dfrac{7}{10} + \dfrac{8}{15}$

34. $\dfrac{7}{15} - \left(-\dfrac{4}{3}\right)$ **35.** $-\dfrac{2}{5} - \left(\dfrac{7}{15} - \dfrac{27}{35}\right)$

EXAMPLE 15   Dropping and Passing a Class

Juanita's algebra class began with 30 students. By the end of the semester, $\dfrac{1}{5}$ of the students had dropped the class and $\dfrac{2}{3}$ of the students who had begun the class had passed.

(a) What fraction of the class dropped the class or passed the class?

(b) What fraction of the students who began the class finished the course but did not pass?

(c) Of the 30 students, how many finished but did not pass?

Solution

(a) To find the fraction of the class that either dropped or passed, add $\dfrac{1}{5}$ and $\dfrac{2}{3}$.

$$\dfrac{1}{5} + \dfrac{2}{3} \stackrel{\text{LCD}=15}{=} \dfrac{1}{5} \cdot \dfrac{3}{3} + \dfrac{2}{3} \cdot \dfrac{5}{5}$$

$$= \dfrac{3}{15} + \dfrac{10}{15}$$

$$= \dfrac{13}{15}$$

(b) To find the fraction of the class that finished the course but did not pass, subtract $\dfrac{13}{15}$ from one whole.

$$1 - \dfrac{13}{15} = \dfrac{15}{15} - \dfrac{13}{15}$$

$$= \dfrac{2}{15}$$

Therefore, $\dfrac{2}{15}$ of the class finished the course but did not pass.

(c) To find how many of the starting 30 students finished but did not pass, find $\frac{2}{15}$ of 30. Remember, "of" means multiplication.

$$\frac{2}{15} \cdot \frac{30}{1} = \frac{2 \cdot 30}{15 \cdot 1}$$

$$= \frac{2 \cdot 3 \cdot 2 \cdot 5}{3 \cdot 5 \cdot 1}$$

$$= \frac{4}{1}$$

$$= 4$$

Out of a class of 30 students, 4 students finished the course but did not pass.

Quick ✓

36. In a study conducted on 300 participants for a new cold medicine, $\frac{1}{12}$ of the volunteers reported headaches as a side effect and $\frac{1}{15}$ reported drowsiness as a side effect. If none of the participants who reported headaches also reported drowsiness how many participants reported either headaches or drowsiness as a side effect?

1.IR5 Exercises   MyLab Statistics

Underlined exercises have complete video solutions in MyLab.

Problems 1–36 are the Quick ✓s that follow the EXAMPLES.

Building Skills

In Problems 37–46, perform the indicated operation, and write the answer in lowest terms. See Objective 1.

37. $\frac{5}{40} + \frac{8}{40}$

38. $\frac{9}{50} + \frac{11}{50}$

39. $\frac{11}{15} - \frac{6}{15}$

40. $\frac{18}{21} - \frac{12}{21}$

41. $\frac{57}{105} - \frac{32}{105}$

42. $\frac{35}{90} + \frac{75}{90}$

43. $\frac{9}{42} + \frac{15}{42} + \frac{6}{42}$

44. $\frac{19}{6} + \frac{13}{6} + \frac{2}{6}$

45. $\frac{20}{28} + \frac{21}{28} - \frac{17}{28}$

46. $\frac{12}{25} + \frac{9}{25} - \frac{1}{25}$

In Problems 47–58, perform the indicated operation, and write the answer in lowest terms. See Objective 2.

47. $\frac{11}{20} + \left(-\frac{1}{20}\right)$

48. $\frac{18}{13} + \left(-\frac{5}{13}\right)$

49. $-\frac{21}{48} + \frac{3}{48}$

50. $-\frac{35}{44} + \frac{7}{44}$

51. $-\frac{6}{15} + \left(-\frac{4}{15}\right)$

52. $-\frac{9}{25} + \left(-\frac{11}{25}\right)$

53. $\frac{75}{30} - \left(-\frac{15}{30}\right)$

54. $\frac{36}{42} - \left(-\frac{18}{42}\right)$

55. $-\frac{5}{9} - \left(-\frac{2}{9}\right)$

56. $-\frac{12}{18} - \left(-\frac{36}{18}\right)$

57. $-\frac{20}{15} + \frac{13}{15} - \left(-\frac{4}{15}\right)$

58. $-\frac{15}{12} - \frac{18}{12} + \frac{5}{12}$

In Problems 59–70, determine the least common denominator (LCD). See Objective 2.

59. $\frac{2}{3}$ and $\frac{3}{4}$

60. $\frac{5}{8}$ and $\frac{2}{3}$

61. $\frac{1}{6}$ and $\frac{2}{3}$

62. $\frac{1}{2}$ and $\frac{1}{4}$

63. $\frac{5}{6}$ and $\frac{-3}{4}$

64. $\frac{5}{9}$ and $\frac{-1}{6}$

65. $\frac{11}{24}, \frac{4}{27}$ and $\frac{-5}{18}$

66. $\frac{8}{15}, \frac{1}{20}$ and $\frac{-7}{18}$

67. $\frac{8}{9}, \frac{1}{15}$ and $\frac{-3}{20}$

68. $\frac{7}{8}, \frac{5}{12}$ and $\frac{-1}{6}$

69. $\frac{9}{96}$ and $\frac{5}{72}$

70. $\frac{5}{78}$ and $\frac{1}{52}$

In Problems 71–82, rewrite the fractions as equivalent fractions with the LCD. See Objective 2.

71. $\frac{5}{9}$ and $\frac{3}{4}$; 36

72. $\frac{7}{8}$ and $\frac{4}{5}$; 40

73. $\dfrac{11}{12}$ and $\dfrac{-5}{8}$; 24

74. $\dfrac{13}{18}$ and $\dfrac{-1}{4}$; 36

75. $\dfrac{5}{17}$ and $-\dfrac{5}{51}$; 51

76. $\dfrac{20}{23}$ and $-\dfrac{10}{69}$; 69

77. $\dfrac{4}{9}$ and $\dfrac{5}{8}$; 72

78. $\dfrac{8}{9}$ and $\dfrac{1}{15}$; 45

79. $-\dfrac{21}{60}$ and $-\dfrac{55}{90}$; 180

80. $-\dfrac{37}{40}$ and $-\dfrac{17}{60}$; 120

81. $\dfrac{4}{21}$ and $-\dfrac{3}{28}$; 84

82. $\dfrac{17}{36}$ and $-\dfrac{10}{27}$; 108

In Problems 83–118, perform the indicated operation, and write your answer in lowest terms. See Objective 3.

83. $\dfrac{7}{12} + \dfrac{3}{8}$

84. $\dfrac{3}{8} + \dfrac{1}{4}$

85. $\dfrac{7}{12} - \dfrac{2}{9}$

86. $\dfrac{7}{5} + \left(-\dfrac{23}{20}\right)$

87. $\dfrac{8}{15} - \dfrac{7}{10}$

88. $\dfrac{1}{12} - \dfrac{5}{28}$

89. $\dfrac{1}{4} + \dfrac{1}{8} + \dfrac{3}{16}$

90. $\dfrac{1}{5} + \dfrac{2}{15} + \dfrac{7}{20}$

91. $-\dfrac{5}{8} + \dfrac{7}{10}$

92. $-\dfrac{3}{5} + \dfrac{2}{3}$

93. $\dfrac{15}{28} - \dfrac{20}{21}$

94. $\dfrac{5}{12} - \dfrac{6}{8}$

95. $-\dfrac{28}{45} + \left(-\dfrac{19}{15}\right)$

96. $-\dfrac{11}{24} + \left(-\dfrac{5}{8}\right)$

97. $6 - \dfrac{7}{2}$

98. $3 - \dfrac{5}{3}$

99. $\dfrac{1}{2} + \left(-\dfrac{5}{4}\right) + \dfrac{3}{2}$

100. $\dfrac{8}{3} + \left(-\dfrac{8}{5}\right) - \dfrac{2}{9}$

101. $\dfrac{3}{4} + \left(-\dfrac{3}{4}\right)$

102. $-\dfrac{5}{8} + \dfrac{5}{8}$

103. $-\dfrac{5}{3} + 2$

104. $-\dfrac{7}{8} + 4$

105. $\dfrac{3}{8} + \dfrac{1}{4} - \dfrac{1}{16}$

106. $\dfrac{1}{3} + \dfrac{1}{2} - \dfrac{1}{12}$

107. $-\dfrac{5}{8} \div 15$

108. $-\dfrac{4}{3} \div 8$

109. $\dfrac{9}{16} \cdot \left(-\dfrac{7}{12} - \dfrac{3}{4}\right)$

110. $-\dfrac{15}{4} \cdot \left(-\dfrac{7}{45} - \dfrac{11}{15}\right)$

111. $-\dfrac{3}{7} - \left(-\dfrac{4}{5}\right)$

112. $-\dfrac{6}{25} - \left(-\dfrac{12}{10}\right)$

113. $\dfrac{8}{18} \cdot \left(-\dfrac{3}{2}\right) + \dfrac{16}{24}$

114. $-\dfrac{2}{3} \cdot \left(-\dfrac{27}{32}\right) - \dfrac{27}{48}$

115. $\dfrac{9}{80} + \dfrac{9}{40} - \dfrac{3}{20}$

116. $\dfrac{11}{120} + \dfrac{7}{40} - \dfrac{7}{60}$

117. $-\dfrac{7}{10} + \dfrac{18}{20}$

118. $-\dfrac{5}{7} + \dfrac{3}{14}$

Applying the Concepts

119. **Cougar Packaging** John owns $\dfrac{1}{5}$ of Cougar Packaging, Inc., and Armando owns $\dfrac{1}{3}$ of Cougar Packaging.
    (a) How much of Cougar Packaging do John and Armando own together?
    (b) How much of Cougar Packaging is owned by others?
    (c) Who owns more of Cougar Packaging, John or Armando? By how much?

120. **Painting a Room** Maria and Samuel decide to paint their kitchen. After the first day, Maria completed $\dfrac{3}{10}$ of the job, while Samuel completed $\dfrac{1}{4}$ of the job.
    (a) How much of the job did Maria and Samuel complete together?
    (b) How much of the job remains?
    (c) Who painted more of the kitchen on the first day? By how much?

121. **College Engagement** In a recent Community College Survey of Student Engagement, the following question was asked of 1500 full-time community college students: "About how many hours do you spend in a typical 7-day week preparing for class (Studying, reading, writing, doing homework)?" A total of 22 students said they spend 0 hours preparing for class, while 593 students said they spend 1 to 5 hours preparing for class.
    (a) What fraction of those surveyed spend 0 to 5 hours each week preparing for class? Express the result in lowest terms.
    (b) Determine the fraction of students who spend more than 5 hours each week preparing for class by subtracting the result in part (a) from 1.

122. **College Engagement** In a recent Community College Survey of Student Engagement, the following question was asked of 1500 full-time community college students: "During the current academic year, how often have you arrived at class without completing the readings or assignments?" A total of 141 students said they often arrived at class without preparing or doing homework, while 59 students said they very often arrived at class without preparing or doing homework.
    (a) What fraction of those surveyed often or very often arrive at class without preparing or doing homework? Express the result in lowest terms.
    (b) Determine the fraction of students who never or sometimes arrive at class without preparing or doing homework by subtracting the result in part (a) from 1.

123. **Side Effects** In a study for a new medicine suppose $\frac{1}{8}$ of the participants experienced headaches as a side effect, $\frac{1}{12}$ experienced nausea as a side effect, and $\frac{1}{24}$ experienced both headaches and nausea as a side effect.
   (a) Determine the fraction of study participants who experienced either a headache or nausea as a side effect by evaluating $\frac{1}{8} + \frac{1}{12} - \frac{1}{24}$. Express the result as a fraction in lowest terms.
   (b) If there were 1200 participants in the study, determine the number who experienced either headache or nausea as a side effect.

124. **Side Effects** In a study for a new medicine suppose $\frac{1}{15}$ of the participants experienced insomnia as a side effect, $\frac{1}{18}$ experienced dizziness as a side effect, and $\frac{2}{45}$ experienced both insomnia and dizziness as a side effect.
   (a) Determine the fraction of study participants who experienced either insomnia or dizziness as a side effect by evaluating $\frac{1}{15} + \frac{1}{18} - \frac{2}{45}$. Express the result as a fraction in lowest terms.
   (b) If there were 1800 participants in the study, determine the number who experienced either insomnia or dizziness as a side effect.

125. **Making Trail Mix** Gino bought $\frac{1}{2}$ pound of M&Ms, $\frac{3}{4}$ pounds of nuts, and $\frac{5}{8}$ pounds of granola to make a trail mix. How many pounds of trail mix did Gino prepare?

126. **Making Curtains** Grayson has $\frac{43}{4}$ yards of material to make curtains for her bedroom. If her curtains require $\frac{14}{3}$ yards of material, how much material does she have left for her next project?

127. **Pizza** Joseph left a whole pizza in the work room to answer a phone call. When he returned, he discovered that someone had eaten $\frac{1}{3}$ of his pizza. He sat down for lunch only to be called away to fix the copy machine. And while he was gone, another person came in and took $\frac{1}{4}$ of what was left! How much of the original pizza remains for Joseph?

128. **Grades** In Kathy's algebra class, $\frac{2}{3}$ of the students get a passing grade, and $\frac{3}{4}$ of those passing will go on to receive a degree in math or science. If there were 30 students in the class at the beginning of the term, how many will go on to get a degree in math or science?

Explaining the Concepts

129. A student puts her solution on the board and writes the following:
$$\frac{2}{3} + \frac{2}{3} = \frac{2+2}{3+3}$$
$$= \frac{4}{6}$$
Pretend you are the teacher and want to tell the class what is right and what is wrong with this presentation. What property do you think the student is misapplying?

130. How could you draw a visual representation of the problem $\frac{1}{4} + \frac{1}{2}$? Even though we have not studied this problem yet, looking at Figure 17 should help you explain how to do the problem.

131. Draw a visual representation of $\frac{1}{5} + \frac{2}{5}$.

1.IR6 Operations on Decimals

Objectives

1. Add or Subtract Decimals
2. Multiply Decimals
3. Divide a Decimal by a Whole Number
4. Divide a Decimal by a Decimal
5. Divide Decimals by Powers of 10
6. Convert a Fraction to a Decimal

1 Add or Subtract Decimals

Adding Decimals with the Same Sign

Adding two or more positive decimals is similar to adding whole numbers. Line up the numbers in a column with the decimal points aligned. Then add the digits with like place values and place the decimal point in the answer directly below the decimal point in the problem.

Section 1.IR6 Operations on Decimals   IR-43

> **EXAMPLE 1**   **Adding Positive Decimals**
>
> Find the sum: $14.276 + 12.12$
>
> **Solution**
> Align the decimal points vertically and add.
>
> $$\begin{array}{r} 14.276 \\ \text{Insert 0 so the digits line up:}\quad +12.120 \\ \hline 26.396 \end{array}$$

Work Smart

You can add zeros to the right of a decimal number without changing the value of the decimal.

The next example requires that we "carry"

> **EXAMPLE 2**   **Adding Positive Decimals**
>
> Find the sum: $254.076 + 31.98 + 8.6$
>
> **Solution**
>
> $$\begin{array}{r} \overset{1\;1\;1}{254.076} \\ \text{Insert one 0 so the digits line up:}\quad 31.980 \\ \text{Insert two 0s so the digits line up:}\quad +\;\;8.600 \\ \hline 294.656 \end{array}$$

> **EXAMPLE 3**   **Adding Positive Decimals**
>
> Find the sum: $42 + 6.53$
>
> **Solution**
>
> $$\begin{array}{r} 42.00 \\ \text{Insert the decimal point and two 0s:}\quad +\;6.53 \\ \hline 48.53 \end{array}$$

Work Smart

Recall that the sum is the answer to an addition problem.

Quick ✓

In Problems 1–4, find each sum.

1. $12.06 + 15.43$
2. $8.23 + 17.42 + 21.99$
3. $92 + 8.67$
4. $0.23 + 0.18 + 0.07$

To add two negative decimals, add the absolute values of the decimals and attach a negative sign on the sum.

> **EXAMPLE 4**   **Adding Negative Decimals**
>
> Find the sum: $-2.3 + (-4.6)$
>
> **Solution**
> To add -2.3 and -4.6, add the absolute values of the decimals and then attach the negative sign.
>
> We have $|-2.3| = 2.3$ and $|-4.6| = 4.6$, and $2.3 + 4.6 = 6.9$. Because both decimals are negative in the original problem, $-2.3 + (-4.6) = -6.9$.

> **EXAMPLE 5**   **Adding Negative Decimals**
>
> Find the sum: $-81.4 + (-3.25)$

Solution
Add the absolute values of the decimals and then use the common sign. We have $|-81.4| = 81.4$ and $|-3.25| = 3.25$. Now find the sum of 81.4 and 3.25.

$$\begin{array}{r} 81.40 \\ +3.25 \\ \hline 84.65 \end{array}$$

Both decimals in the original problem are negative, so $-81.4 + (-3.25) = -84.65$.

Quick ✓
In Problems 5–7, find each sum.

5. $-2.045 + (-193.26)$ **6.** $-46 + (-29.37)$

7. $-17.9 + (-32.1)$

Subtract Decimals with the Same Sign

To subtract decimals, use the same method we use with adding decimals: Align the decimal points vertically. Then subtract the digits with like place values, and place the decimal point in the answer directly below the decimal point in the problem.

EXAMPLE 6 Subtracting Decimals

Subtract: $13.96 - 8.22$

Solution

$$\begin{array}{r} 13.96 \\ -8.22 \\ \hline 5.74 \end{array}$$

Work Smart
Remember that the answer to a subtraction problem is called the difference.

Check: Recall that you can check your answer to a subtraction problem using addition.

$$\begin{array}{rr} \text{Difference:} & 5.74 \\ \text{Subtrahend:} & +8.22 \\ \hline \text{Minuend:} & 13.96 \end{array}$$

The next example illustrates "borrowing."

EXAMPLE 7 Subtracting Decimals

Find the difference: $76.93 - 14.517$

Solution
Add a 0 to 76.93 so there is a digit in each decimal place.

Work Smart
You may add zeros to the right of the last digit (as long as it is right of the decimal point).

Insert one 0 so the digits line up:

$$\begin{array}{r} \overset{2\,1}{76.930} \\ -14.517 \\ \hline 62.413 \end{array}$$

← Because we cannot subtract 7 from 0, we borrow.

Quick ✓
In Problems 8–10, compute the difference.

8. $54.83 - 41.62$ **9.** $58.6709 - 12.35$ **10.** $0.9 - 0.864$

Section 1.IR6 Operations on Decimals   IR-45

Adding Decimals with Different Signs

To add decimals that have different signs, subtract the smaller absolute value from the larger absolute value, and use the sign of the number with the larger absolute value.

EXAMPLE 8   **Adding Decimals with Different Signs**
Add: $6.14 + (-8.95)$

Solution
We have $|6.14| = 6.14$ and $|-8.95| = 8.95$. The smaller absolute value is 6.14, so we compute $8.95 - 6.14$.

$$\begin{array}{r} 8.95 \\ -6.14 \\ \hline 2.81 \end{array}$$

The larger absolute value is 8.95, which was a negative number in the original problem, so the sum is negative. Therefore, $6.14 + (-8.95) = -2.81$.

EXAMPLE 9   **Adding Decimals with Different Signs**
Find the sum: $-21.2 + 68.053$

Solution
We have $|-21.2| = 21.2$ and $|68.053| = 68.053$. The smaller absolute value is 21.2, so we compute $68.053 - 21.2$ as

$$\begin{array}{r} 7\,1 \\ 68.053 \\ -21.200 \quad \text{Insert two 0s} \\ \hline 46.853 \end{array}$$

The larger absolute value is 68.053, which was a positive number in the original problem, so the sum is positive. Therefore, $-21.2 + 68.053 = 46.853$.

Quick ✓
In Problems 11–14, find each sum.

11. $-12.54 + 21.811$ **12.** $-19.8 + 2.003$

13. $87 + (-100.56)$ **14.** $-65.223 + 567.89$

Subtract Decimals with Different Signs

To subtract decimals with different signs, rewrite the subtraction problem as an addition problem and use the addition rules. Recall that $a - b = a + (-b)$.

EXAMPLE 10   **Subtracting Decimals**
Find the difference: $-59.7 - 12.01$

Solution
First change the subtraction problem to an equivalent addition problem.
$$-59.7 - 12.01 = -59.7 + (-12.01)$$
Add the absolute value of the two numbers.
$$59.7 + 12.01 = 71.71$$
The sign of the sum is negative because both decimals in the original problem are negative. Therefore, $-59.7 - 12.01 = -71.71$.

IR-46 CHAPTER 1 Integrated Review: Getting Ready for Organizing and Summarizing Data

EXAMPLE 11 **Subtracting Decimals**

Find the difference: $-56.02 - (-8)$

Solution
First change the subtraction problem to an equivalent addition problem.
$$-56.02 - (-8) = -56.02 + 8$$
Now find the sum $-56.02 + 8$. The absolute value of -56.02 is 56.02, and the absolute value of 8 is 8, so compute $56.02 - 8$.

$$\begin{array}{r} \overset{4\;1}{56.02} \\ -8.00 \\ \hline 48.02 \end{array}$$

The larger absolute value is 56.02, which was a negative number in the original problem, so the sum is negative. Therefore, $-56.02 - (-8) = -48.02$.

EXAMPLE 12 **Subtracting Decimals**

Subtract 9.38 from 1.456.

Solution
The statement "subtract 9.38 from 1.456" translates to $1.456 - 9.38$. We write $1.456 - 9.38$ as addition: $1.456 + (-9.38)$. Now find the absolute value of each number. The absolute value of 1.456 is 1.456, and the absolute value of -9.38 is 9.38, so compute $9.38 - 1.456$.

$$\begin{array}{r} \overset{8\;17\;1}{9.380} \\ -1.456 \\ \hline 7.924 \end{array}$$

The larger absolute value is 9.38, which was a negative number in the original problem, so $1.456 - 9.38 = -7.924$.

Quick ✓

15. The subtraction problem $-3.2 - 11.4$ is equivalent to $-3.2 +$ _____.

In Problems 16–18, find each difference.

16. $23.7 - 32.33$ 17. $-15.2 - (-81.06)$

18. $1 - 0.816$

❷ Multiply Decimals

The rules for multiplying decimals come from the rules for multiplying fractions. For example, to find the product $(0.7)(0.03)$, write

$$\underbrace{0.7}_{\text{1 decimal place}} \cdot \underbrace{0.03}_{\text{2 decimal places}} = \frac{7}{10} \cdot \frac{3}{100} = \frac{21}{1000} = \underbrace{0.021}_{\text{3 decimal places}}$$

Notice that there are three digits to the right of the decimal point, which is the sum of the number of decimal places in the factors, $1 + 2 = 3$.

Here's another example: Find the product $(0.02)(0.004)$.

$$\underbrace{0.02}_{\text{2 decimal places}} \cdot \underbrace{0.004}_{\text{3 decimal places}} = \frac{2}{100} \cdot \frac{4}{1000} = \frac{8}{100{,}000} = \underbrace{0.00008}_{\text{5 decimal places}}$$

Again, the number of decimal places in the product is equal to the sum of the number of decimal places in the factors: 2 + 3 = 5.

When we multiply decimals, we do *not* need to line up decimal points (as we do for addition or subtraction).

EXAMPLE 13 **Multiplying Decimals**

Multiply:

(a) 34.65×0.7 (b) -6.52×0.35

Solution

(a) First, multiply 3465 by 7.

$$\begin{array}{r} {\scriptstyle 3\;4\;3} \\ 3465 \\ \times\;\;\;7 \\ \hline 24255 \end{array}$$

Work Smart

Recall the rules of signs when multiplying two numbers.
- the product of two positive real numbers is positive.
- the product of two negative real numbers is positive.
- the product of a positive and negative real number is negative.

The two factors, 34.65 and 0.7, have a total of three digits to the right of the decimal point. Therefore, move the decimal point in 24255 three places to the left.

$$\begin{array}{r} 34.65 \\ \times\;\;0.7 \\ \hline 24.255 \end{array}$$ 2 decimal places
1 decimal place

Thus $34.65 \times 0.7 = 24.255$.

(b)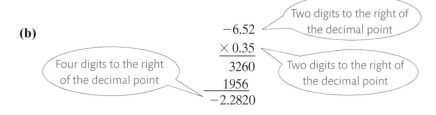

-2.2820

Work Smart

The number of digits to the right of the decimal point in the product is the *sum* of the numbers of digits to the right of each decimal point in the factors.

Multiplying Decimals

Step 1: Multiply the factors as though they were integers.

Step 2: Place the decimal point so that the number of digits to the right of the decimal point in the product equals the *sum* of the number of digits to the right of each decimal point in the factors.

Quick ✓

In Problems 19 and 20, find each product.

19. 4.73×5.2 **20.** -2.14×0.32

Sometimes it is necessary to add zeros as placeholders in a product.

EXAMPLE 14 **Writing Zeros as Placeholders**

Multiply: 0.052×0.09

Work Smart

If needed, put a zero to the left of the decimal point so that the decimal point will not get "lost."

Solution

First multiply the decimals as though they were integers.

$$\begin{array}{r} {\scriptstyle 1} \\ 52 \\ \times\;9 \\ \hline 468 \end{array}$$

The product of 52 and 9 is 468. We need to move the decimal point *five* places to the left. To do this, write two zeros on the left side of the answer.

$$0.052 \quad \text{Three digits to the right of the decimal point}$$
$$\times\, 0.09 \quad \text{Two digits to the right of the decimal point}$$

Write two zeros on left side of 468: 0.00468 Move decimal point $3 + 2 = 5$ places left

So, $0.052 \times 0.09 = 0.00468$.

Quick ✓

In Problems 21–23, find each product.

21. 0.03×0.071 **22.** 0.012×0.16 **23.** 0.176×0.003

EXAMPLE 15 **How Many Grams of Fat?**

The average café latte contains 4.5 grams of fat, and you drink 5 café lattes each week. How many grams of fat do you consume from the 5 lattes each week?

Solution
Let t represent the total number of fat grams. To find t, we multiply:

total number fat grams equals number grams of fat number of days
t = 4.5 × 5
= 22.5

There are 22.5 fat grams in 5 café lattes.

Quick ✓

24. Laura is attending college part-time and working part-time. Laura worked 30 hours last week. Calculate her pay before taxes for last week if her hourly wage is $19.76.

Multiply Decimals by Powers of 10

Patterns occur when a number is multiplied by 10, 100, 1000, and so on. Several numbers multiplied by 10 are shown in Table 3

Table 3

$3.24 \times 10 = 32.4$	$6.5 \times 10 = 65$
$0.98 \times 10 = 9.8$	$74.9 \times 10 = 749$
$824.8 \times 10 = 8248$	$0.021 \times 10 = 0.21$

Do you recognize the pattern? When we multiply a number by 10, the decimal point moves *one* place to the right.

There is a similar pattern for multiplying by 100. What happens when you multiply 21.398 by 100? Because $21.398 \times 100 = 2139.8$, the decimal place moves *two* places to the right.

The pattern continues for multiplying by 1000, 10,000, and so on:

$21.398 \times 1000 = 21{,}398$ Decimal point moves three places right.
$21.398 \times 10{,}000 = 213{,}980$ Decimal point moves four places right.

What is the rule? Move the decimal point to the *right* the same number of places that there are zeros in the power of 10.

Section 1.IR6 Operations on Decimals **IR-49**

Multiplying a Number by 10, 100, 1000, ...
To multiply a number by 10, 100, 1000, ..., count the number of zeros in the power of 10 and move the decimal point in the factor that many places to the right.

EXAMPLE 16 **Multiplying by a Power of 10**
Multiply:

(a) 97.23×100 (b) $1.7 \times 10{,}000$

Solution

(a) To multiply 97.23 by 100, move the decimal point two places to the right. Thus

$$97.23 \times 100 = 97.23$$
$$= 9723$$

Work Smart
You may need to add zeros to the right of the decimal point when multiplying by a power of 10.

(b) To find the product $1.7 \times 10{,}000$, move the decimal point four places to the right. Be sure to add zeros to the right of the decimal point, as needed.

$$1.7 \times 10{,}000 = 1.7000$$
$$= 17{,}000$$

Quick ✓
In Problems 25–28, find each product.

25. 72.87×10 **26.** 0.453×1000

27. 2076.3×100 **28.** $342.9 \times 10{,}000$

In Problems 29 and 30, convert each dollar amount to cents by multiplying by 100 (because 100¢ equals $1.)

29. $7.21   **30.** $105.88

There is also a pattern for multiplying by *decimal* powers of 10, such as 0.1, 0.01, 0.001, and so on. See Table 4.

Table 4

Product	Rule
$3.24 \times 0.1 = 0.324$	Move decimal point one place left.
$824.8 \times 0.01 = 8.248$	Move decimal point two places left.
$74.9 \times 0.001 = 0.0749$	Move decimal point three places left.

Multiplying a Number by 0.1, 0.01, 0.001, ...
To multiply a number by 0.1, 0.01, 0.001, ..., move the decimal point to the *left* the same number of places as there are *decimal places* in the power of 10.

Notice that when we multiply by 10, 100, 1000, and so on, the product is larger than the other factor (because we move the decimal point to the right). When we multiply by 0.1, 0.01, 0.001, and so on, the product is smaller than the other factor because we are multiplying by a number between 0 and 1.

EXAMPLE 17 **Multiplying by a Decimal Power of 10**
Multiply:

(a) 213.84×0.01 (b) 1.93×0.0001

Solution

(a) To multiply 213.84 by 0.01, move the decimal point two places to the left. Thus

$$213.84 \times 0.01 = 213.84$$
$$= 2.1384$$

(b) To find the product 1.93×0.0001, move the decimal point four places to the left. Be sure to add zeros to the left of the decimal point.

$$1.93 \times 0.0001 = 0001\,93$$
$$= 0.000193$$

Quick ✓
In Problems 31–34, find each product.

31. 185.98×0.1 **32.** $678{,}095 \times 0.001$

33. 0.456×0.01 **34.** 0.3×0.001

③ Divide a Decimal by a Whole Number

The rules for dividing decimals are similar to the rules for dividing whole numbers. The only difference is that when we divide a decimal by a whole number, we have a decimal point in the quotient.

We begin with a review of some language used in division. In the division problem $2\overline{)7.94}^{\,3.97}$, the number 2 is the *divisor*, 7.94 is the *dividend*, and 3.97 is the *quotient*. Notice that when we divide by a whole number, we line up the decimal points in the quotient and the dividend. We can also write the division problem $2\overline{)7.94}^{\,3.97}$ as $\frac{7.94}{2} = 3.97$.

Recall that $\frac{\text{dividend}}{\text{divisor}} = \text{quotient}$, so divisor $\times$ quotient $=$ dividend. This allows us to check the answer of a division problem. For example, because $\frac{0.245}{7} = 0.035$, we know that $7 \times 0.035 = 0.245$.

Quick ✓
35. In $7.5 \div 2.5 = 3$, the number 7.5 is called the _____, the number 2.5 is called the _____, and the number 3 is called the _____.

36. To check a division problem, multiply: _____ · quotient = _____.

EXAMPLE 18   **Dividing a Decimal by a Whole Number**

Divide: $148.5 \div 9$. Check your answer.

Solution
Write $148.5 \div 9$ as $9\overline{)148.5}$. First, determine the number of times the divisor, 9, divides into the digit in the dividend with the highest place value, which is 1. Because 9 does not divide into 1, we include the next digit and ask, "How many times does 9 divide into 14?" The answer is 1 time, so we place a 1 above the 4 in the dividend.

$$9\overline{)148.5}^{\,1}$$

Now multiply the divisor, 9, by 1, place the product below 14, and subtract. Bring down the next digit.

$$\begin{array}{r} 1 \\ 9{\overline{\smash{\big)}\,148.5}} \\ -9\downarrow \\ \hline 58 \end{array} \quad 1 \times 9 = 9$$

Repeat this process until there are no more digits in the original dividend, and the newest dividend is less than the divisor. Be sure to align the decimal in the quotient above the decimal in the dividend.

$$\begin{array}{r} 16.5 \\ 9{\overline{\smash{\big)}\,148.5}} \\ -9\downarrow \\ \hline 58 \\ -54\downarrow \\ \hline 45 \\ -45 \\ \hline 0 \end{array}$$

Write the decimal point.
Step 1: 9 divides into 14, 1 time.
Step 2: $1 \times 9 = 9$
Step 3: 9 divides into 58, 6 times.
Step 4: $6 \times 9 = 54$
Step 5: 9 divides into 45, 5 times.
Step 6: $5 \times 9 = 45$

Check: Verify that the product of the quotient and the divisor is the dividend. Because $16.5 \times 9 = 148.5$, our answer is correct. The quotient is 16.5.

EXAMPLE 19 **Dividing a Decimal by a Whole Number**

Find the quotient: $\dfrac{3.25}{25}$

Solution

$$\begin{array}{r} 0.13 \\ 25{\overline{\smash{\big)}\,3.25}} \\ -25\downarrow \\ \hline 75 \\ -75 \\ \hline 0 \end{array}$$

Step 1: 25 divides into 32, 1 time
Step 2: 25 divides into 75, 3 times

The remainder is 0, so we are finished.

Check: Verify that the product of the quotient and the divisor is the dividend. Because $0.13 \times 25 = 3.25$, our answer is correct. The quotient is 0.13.

EXAMPLE 20 **Dividing a Decimal by a Whole Number**

Divide: $-8.432 \div 124$

Solution
Recall that a negative number divided by a positive number gives a negative quotient. Begin by dividing as we do with whole numbers. Because 124 does not divide into 8 or into 84, place 0s above each digit in the quotient. Don't forget the decimal point!

$$\begin{array}{r} 0.068 \\ 124{\overline{\smash{\big)}\,8.432}} \\ -744\downarrow \\ \hline 992 \\ -992 \\ \hline 0 \end{array}$$

Step 1: 124 divides into 843, 6 times
$6 \times 124 = 744$
Step 2: 124 divides into 992, 8 times

The quotient is negative, so $-8.432 \div 124 = -0.068$. We leave the check to you.

EXAMPLE 21   Finding an Hourly Wage

Juan earned $621.20 for working 40 hours last week. What was Juan's hourly wage?

Solution
To find Juan's hourly wage, divide his weekly earnings by the number of hours he worked. That is, find 621.20 ÷ 40.

$$\require{enclose} \begin{array}{r} 15.53 \\ 40\enclose{longdiv}{621.20} \\ \underline{-40} \\ 221 \\ \underline{-200} \\ 212 \\ \underline{-200} \\ 120 \\ \underline{-120} \\ 0 \end{array}$$

To check, verify that $15.53 \times 40 = \$621.20$. Juan earned $15.53 per hour last week.

Quick ✓

In Problems 37–39, find the quotient.

37. $18.25 \div 73$ **38.** $\dfrac{1360.8}{56}$ **39.** $\dfrac{592.9}{-7}$

40. Find Amanda's hourly wage if she earned $437.50 for working 25 hours.

EXAMPLE 22   Dividing a Decimal by a Whole Number and Rounding the Quotient

Divide: $\dfrac{252.65}{32}$. Round your answer to the nearest hundredth.

Solution
Divide until the quotient is expressed to the thousandth.

$$\require{enclose} \begin{array}{r} 7.895 \\ 32\enclose{longdiv}{252.650} \\ \underline{-224} \\ 286 \\ \underline{-256} \\ 305 \\ \underline{-288} \\ 170 \\ \underline{-160} \\ 10 \end{array}$$

Work Smart
Remember that the symbol ≈ means "approximately equal to."

Because we are rounding to the nearest hundredth, carry out division to the *thousandths* place. Round 7.895 to 7.90. So, $\dfrac{252.65}{32} \approx 7.90$.

Quick ✓

In Problems 41 and 42 divide. Round your answer to the nearest hundredth.

41. $369.24 \div 26$ **42.** $\dfrac{4879.6}{60}$

▶ ④ Divide a Decimal by a Decimal

If the divisor is not a whole number, multiply the dividend and the divisor by a power of 10 so that the divisor is a whole number. Then divide as described above.

EXAMPLE 23 **Dividing a Decimal by a Decimal**

Divide: $\dfrac{7.74}{3.6}$

Work Smart

Remember $\dfrac{a}{a} = 1$ if $a \neq 0$. So, multiplying by $\dfrac{10}{10}$ is like multiplying by 1.

Solution
Because the divisor is 3 and 6 *tenths*, multiply both the dividend and the divisor by *10* to make the divisor a whole number.

$$\frac{7.74}{3.6} \cdot \frac{10}{10} = \frac{77.4}{36}$$

Now we divide.

$$\begin{array}{r} 2.1 \\ 36{\overline{\smash{\big)}\,77.4}} \\ \underline{-72} \\ 54 \\ \underline{-36} \\ 18 \end{array}$$

The remainder is not equal to zero, so we are not finished. We can add zeros to the right of the last digit of the decimal in the dividend. This does not change its value.

$$\begin{array}{r} 2.15 \\ 36{\overline{\smash{\big)}\,77.40}} \\ \underline{-72} \\ 54 \\ \underline{-36} \\ 180 \\ \underline{-180} \\ 0 \end{array}$$

The remainder is 0, so we are finished. The quotient is 2.15. We leave the check to you. ●

We summarize the process for dividing decimals.

> **Dividing Decimals**
> **Step 1:** Multiply both the dividend and the divisor by a power of 10 that will make the divisor a whole number.
> **Step 2:** Divide as though dividing by a whole number.

Quick ✓

In Problems 43 and 44, find the quotient.

43. $180.67 \div 62.3$

44. $\dfrac{25.48}{0.52}$

⑤ Divide Decimals by Powers of 10

Remember, there is a pattern when we multiply by powers of 10. For each power of 10, move the decimal point one place to the right. For example,

$$3.8745 \times 10 = 38.745 \quad 3.8745 \times 100 = 387.45 \quad 3.8745 \times 1000 = 3874.5$$

What about dividing by a power of 10?

$$\frac{38.745}{10} = 3.8745 \quad \frac{38.745}{100} = 0.38745 \quad \frac{38.745}{1000} = 0.038745$$

For each power of 10, the decimal point moves one place to the *left*.

Dividing by a Power of 10

Count the number of zeros in the divisor, and move the decimal point in the dividend that many places to the left to form the quotient.

EXAMPLE 24 **Dividing a Decimal by a Power of 10**

Divide.

(a) $\dfrac{86.3}{10}$ (b) $\dfrac{45.132}{1000}$

Solution

(a) The divisor is 10, which has one zero. Move the decimal point in the dividend one place to the left. Therefore, $\dfrac{86.3}{10} = 8.63$.

(b) The divisor is 1000, which has three zeros. Move the decimal point in the dividend three places to the left. Therefore, $\dfrac{45.132}{1000} = 0.045132$. Notice we inserted zeros in front of the quotient as we moved the decimal left.

Quick ✓

In Problems 45 and 46, divide.

45. $\dfrac{2750.43}{100}$ 46. $\dfrac{91.06}{1000}$

There is also a pattern when we divide by decimal powers of 10, such as 0.1, 0.01, 0.001, and so on. See Table 5.

Work Smart

$\dfrac{98.1}{0.1} = \dfrac{98.1}{\frac{1}{10}}$

$= 98.1 \cdot \dfrac{10}{1}$

$= 981$

Table 5

Quotient	Rule
$\dfrac{98.1}{0.1} = 981$	Move decimal point one place right.
$\dfrac{7.25}{0.01} = 725$	Move decimal point two places right.
$\dfrac{0.0364}{0.001} = 36.4$	Move decimal point three places right.

Dividing a Number by 0.1, 0.01, 0.001, . . .

To divide a number by 0.1, 0.01, 0.001, . . . , move the decimal point to the *right* the same number of places as there are *decimal places* in the power of 10.

Notice that when we divide by 10, 100, 1000, and so on, the quotient is smaller than the dividend. When we divide by 0.1, 0.01, 0.001, and so on, the quotient is larger than the dividend because we are multiplying by the reciprocal of the denominator; that is, we are multiplying by 10, 100, 1000, and so on.

Quick ✓

In Problems 47 and 48, divide.

47. $\dfrac{14.235}{0.01}$

48. $\dfrac{0.983}{0.0001}$

▶ ⑥ Convert a Fraction to a Decimal

To convert a fraction to a decimal, divide the numerator of the fraction by the denominator of the fraction until the remainder is 0 or the remainder repeats.

EXAMPLE 25 **Converting a Fraction to a Decimal**

Convert each number to a decimal.

(a) $\dfrac{3}{4}$

(b) $-\dfrac{7}{8}$

Solution

Work Smart

The fraction bar means division. Order matters:

$\dfrac{3}{4} = 4\overline{)3}$ but $\dfrac{4}{3} = 3\overline{)4}$.

Be sure to place the dividend and divisor correctly.

(a)
```
    0.75
4)3.00
  -28
   20
  -20
    0
```

So $\dfrac{3}{4} = 0.75$.

(b)
```
    0.875
8)7.000
  -64
   60
  -56
   40
  -40
    0
```

So $-\dfrac{7}{8} = -0.875$.

EXAMPLE 26 **Converting a Fraction to a Decimal**

Convert $\dfrac{2}{3}$ to a decimal.

Solution

```
    0.666
3)2.000
  1 8
   20
   18
    20    The pattern (20, then 18) continues
    18
     2
```

Notice that the remainder, 2, repeats. So $\dfrac{2}{3} = 0.666\ldots$.

The decimals in Example 25 are called **terminating decimals** because the remainder is 0. The quotient in Example 26, 0.666..., is called a **repeating decimal** because the 6 continues repeating indefinitely. We use three dots, (...), called an **ellipsis,** to indicate that the pattern repeats. The decimal 0.666... can also be written as $0.\overline{6}$. That is, place a bar above the repeating portion of the quotient.

Quick ✓
In Problems 49–52, write the fraction as a decimal.

49. $\dfrac{3}{5}$

50. $\dfrac{159}{424}$

51. $\dfrac{5}{6}$

52. $\dfrac{5}{9}$

EXAMPLE 27 **Comparing a Fraction to a Decimal**

Insert the symbol $>$, $<$, or $=$ to make a true sentence.

(a) $\dfrac{3}{8}$ _____ 0.4

(b) 0.78 _____ $\dfrac{7}{9}$

Solution

(a) Convert $\dfrac{3}{8}$ to a decimal by evaluating $8\overline{)3}$, and then compare.

$$\begin{array}{r} 0.375 \\ 8\overline{)3.000} \\ \underline{-24} \\ 60 \\ \underline{-56} \\ 40 \\ \underline{-40} \\ 0 \end{array}$$

Because $\dfrac{3}{8} = 0.375 < 0.4$, we find that $\dfrac{3}{8} < 0.4$.

(b) Convert $\dfrac{7}{9}$ to a decimal and compare.

$$\begin{array}{r} 0.777 \\ 9\overline{)7.000} \\ \underline{-63} \\ 70 \\ \underline{-63} \\ 70 \\ \underline{-63} \\ 7 \end{array}$$

The pattern (70, then 63) continues.

We see that $\dfrac{7}{9}$ is equal to the repeating decimal 0.777... and 0.78 > 0.777....

Therefore, $0.78 > \dfrac{7}{9}$.

Quick ✓
In Problems 53 and 54, insert the symbol $>$, $<$, or $=$ to make a true statement.

53. $\dfrac{8}{11}$ _____ 0.6

54. $\dfrac{3}{50}$ _____ 0.1

1.IR6 Exercises   MyLab Statistics

Underlined exercises have complete video solutions in MyLab.

Problems 1–54 are the Quick ✓s that follow the EXAMPLES.

Skill Building

In Problems 55–64, find the sum or difference. See Objective 1.

55. $0.45 + 0.21$ **56.** $0.64 + 0.13$

57. $0.245 + 0.396$ **58.** $0.014 + 0.748$

59. $0.058 + 0.284 + 0.533$ **60.** $0.384 + 0.211 + 0.349$

61. $0.96 - 0.43$ **62.** $0.75 - 0.23$

63. $0.482 - 0.038$ **64.** $0.584 - 0.285$

In Problems 65–78, find the product. See Objective 2.

65. 0.4×0.3 **66.** 0.7×0.2

67. 0.45×0.9 **68.** 0.86×0.3

69. 0.057×0.82 **70.** 0.278×0.64

71. 0.45×100 **72.** 0.93×100

73. 0.823×1000 **74.** 0.492×1000

75. 451×0.1 **76.** 349×0.1

77. 0.357×0.01 **78.** 0.873×0.01

In Problems 79–82, find the quotient. See Objective 3.

79. $\dfrac{205.45}{35}$ **80.** $\dfrac{168.3}{45}$

81. $\dfrac{10741.5}{1050}$ **82.** $\dfrac{7129.8}{1020}$

In Problems 83–86, find the quotient. Round your answer to the indicated value. See Objective 3.

83. $\dfrac{90.83}{28}$; thousandth **84.** $\dfrac{78.28}{23}$; thousandth

85. $\dfrac{157.2}{19}$; hundredth **86.** $\dfrac{254.1}{13}$; hundredth

In Problems 87 and 88, find the quotient. See Objective 4.

87. $\dfrac{16.252}{3.4}$ **88.** $\dfrac{16.644}{5.7}$

In Problems 89–96, find the quotient. See Objective 5.

89. $\dfrac{32}{100}$ **90.** $\dfrac{3}{100}$

91. $\dfrac{803}{1000}$ **92.** $\dfrac{304}{1000}$

93. $\dfrac{5}{0.01}$ **94.** $\dfrac{14}{0.01}$

95. $\dfrac{4.93}{0.001}$ **96.** $\dfrac{10.931}{0.001}$

In Problems 97–102, write each fraction as a decimal. See Objective 6.

97. $\dfrac{182}{520}$ **98.** $\dfrac{156}{650}$

99. $\dfrac{672}{1050}$ **100.** $\dfrac{594}{1100}$

101. $\dfrac{60}{144}$ **102.** $\dfrac{72}{176}$

In Problems 103–112, insert the symbol $>$, $<$, or $=$ to make a true statement. See Objective 4.

103. 0.9 ____ $\dfrac{9}{11}$ **104.** 1.25 ____ $\dfrac{5}{4}$

105. 0.96 ____ 0.959 **106.** 0.724 ____ 0.73

107. 3.14 ____ $\dfrac{22}{7}$ **108.** $\dfrac{15}{6}$ ____ 2.5

109. 0.1 ____ 0.099 **110.** 0.02 ____ 0.19

111. $-\dfrac{2}{3}$ ____ -0.67 **112.** $-\dfrac{1}{3}$ ____ -0.3

Applying the Concepts

113. Budgeting Carey's monthly take-home pay is $2315.86. She pays the following bills each month:

Rent	$1250
Utilities	$75.16
Car payment	$256.83
Insurance	$107.95

How much does Carey have left each month for food, entertainment, and other expenses?

114. Shopping Spree Ralph is going clothes shopping at a high-end department store and purchases the following items (including taxes): polo shirt: $59.99; shorts: $69.50; and jeans on sale for $49.99. He took two hundred-dollar bills with him to the store. Does he have enough money to pay for his purchases? What is the difference between what he has and what he wants to spend?

115. Use the equation $F = 1.8C + 32$ to find the Fahrenheit temperature F that corresponds to 12°C.

116. Use the equation $F = 1.8C + 32$ to find the Fahrenheit temperature F that corresponds to 6°C.

117. Cab Fare A taxi charges $9.25 plus $0.75 per mile traveled. Use the formula $C = 9.25 + 0.75m$, where m is the number of miles traveled and C is the cab fare, to find the fare for traveling 16 miles from the airport to your hotel.

118. Cab Fare A taxi charges $11.00 plus $0.95 per mile traveled. Use the formula $C = 11 + 0.95m$, where m is the number of miles traveled and C is the cab fare, to find the fare for traveling 12 miles from your house to the airport.

119. Simple Interest Find the amount of interest on a principal of $P = \$1000$ at an interest rate of $r = 4\% = 0.04$ for $t = 2$ years using the formula $I = Prt$.

120. Simple Interest Find the amount of interest on a principal of $P = \$2500$ at an interest rate of $r = 3\% = 0.03$ for $t = 4$ years using the formula $I = Prt$.

121. Wages Graham worked the following hours during one week: 8.6 hours, 8.3 hours, 7 hours, 7.5 hours, and 7.75 hours. How many hours did Graham work during the week? If Graham is paid $36.00 an hour, what is Graham's pay for the week?

122. Wages Megan worked the following hours during one week: 9.5 hours, 6.75 hours, 5.75 hours, 8.4 hours, and 7.75 hours. How many hours did Megan work during the week? If Megan is paid $43.00 an hour, what is Megan's pay for the week?

123. Earnings Find Carly's hourly wage if she worked 15 hours last week and earned $222.30.

124. Earnings Find Pablo's hourly wage if he worked 32 hours last week and earned $624.64.

125. Better Buy While grocery shopping, Ginger has the choice of buying canned tomatoes in a 16-oz can for $1.29 or in a 10-oz can for $0.89. Which size is the better buy? *Hint:* Calculate the cost per ounce. The smaller cost per ounce is the better buy.

126. Better Buy Gino can buy pizza in two different sizes: the 12-inch pizza contains 113.1 sq. inches of pizza and sells for $8.50, and the 18-inch pizza contains 254.5 sq. inches of pizza and sells for $19.75. Which pizza size is the better buy?

1.IR7 Fundamentals of Percent Notation

Objectives

1. Define Percent
2. Convert Percents to Decimals and Decimals to Percents
3. Convert Percents to Fractions and Fractions to Percents

1 Define Percent

We use percentages everyday. "Today only! 40% off!" or, "Your final exam counts as 30% of your course grade." **Percent** means "divided by 100" or "per hundred." We use the symbol % to denote percent, so 79% means 79 out of 100 or $\frac{79}{100}$.

Suppose your statistics class will drop your lowest exam score if you have excellent attendance. You took 4 exams and had the following scores: 65 out of 80 or $\frac{65}{80}$, 46 out of 50 or $\frac{46}{50}$, 57 out of 73 or $\frac{57}{73}$, and 56 out of 64 or $\frac{56}{64}$. How do you know which is the low score to be dropped? Percents allow us to compare such quantities by writing all scores using an equivalent score out of 100.

In Words
Typically we use the word *percent* in numerical expressions and the word *percentage* to indicate the portion obtained from calculating with a percent.

EXAMPLE 1 **Writing a Percent**

Write a percent that represents the shaded region in Figure 18.

Figure 18

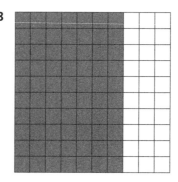

Solution

There are 100 squares, and 70 of the 100 are shaded. The fraction that is shaded is $\frac{70}{100}$. Because percent means "per hundred," the shaded region is 70% of the total.

EXAMPLE 2 **Writing a Percent**

Write each of the following ratios as percents.

(a) Rainforests represent about 2 acres out of every 100 acres of Earth's surface.

(b) Rainforests are home to 50 plants and animals out of every 100 plants and animals on Earth.

(c) At the current rate of tropical forest destruction, 10 species out of every 100 species from the tropical rainforest will become extinct each decade. *Source:* The Nature Conservancy

Solution

(a) The rainforests cover about $\frac{2}{100} = 2\%$ of Earth's surface.

(b) Rainforests are home to $\frac{50}{100} = 50\%$ of Earth's plant and animal species.

(c) At the current rate of rainforest destruction, $\frac{10}{100} = 10\%$ of the tropical rainforest species will become extinct each decade.

Quick ✓

In Problems 1 and 2, write the ratio as a percent.

1. 38 out of 100 students complete their college education within 4 years. *Source:* National Center for Education Statistics
2. Sockeye salmon make up about 25 out of 100 salmon caught on the West Coast. *Source:* U.S. Fish and Wildlife Service

❷ Convert Percents to Decimals and Decimals to Percents

Percents to Decimals

Percent means "out of 100" or "divided by 100." To convert a percent, use the fact that 1 whole = 100%.

EXAMPLE 3 **Writing a Percent as a Decimal**

Write 70% as a decimal.

Solution

$$70\% = 70\% \div 1$$
$$= \frac{70\%}{1}$$
$$1 = 100\%: \quad = \frac{70\%}{100\%}$$
$$= \frac{70}{100}$$
$$= 0.70$$
$$= 0.7$$

To convert from a percent to a decimal, we could "drop" the % symbol and shift the decimal point two places left. Since no decimal point appears in the number 70, the decimal point is after the zero. Start with 70, and then move two places left.

$$70\% = 70.$$
$$= 0.70$$
$$= 0.7$$

> **Steps for Converting a Percent to a Decimal**
>
> **Step 1:** Remove the % sign.
> **Step 2:** Shift the decimal point in the number two places left.
> **Step 3:** Write the decimal number with the decimal point in its new location.

EXAMPLE 4 **Writing a Percent as a Decimal**

Write each percent as a decimal.

(a) 89%   (b) 3%   (c) 100%   (d) 7.65%   (e) 0.47%

Solution

(a) $89\% = 89. = 0.89$
(b) $3\% = 03. = 0.03$
(c) $100\% = 100. = 1.00 = 1$
(d) $7.65\% = 07.65 = 0.0765$
(e) $0.47\% = 000.4 = 0.004$

Work Smart

Notice, in Example (4c), $100\% = 1$

Did you notice in parts (b) and (d) that when we do not have enough digits to shift the decimal point, we added zeros to the left of the number?

> **Quick ✓**
>
> *In Problems 3–6, write the percent as a decimal.*
>
> **3.** 90%   **4.** 2%   **5.** 4.5%   **6.** 0.25%

Decimals to Percents

In Examples 3 and 4, to change a percent to a decimal we divided by 100, or shifted the decimal point two places to the left. Reverse the process to write a decimal as a percent. To convert a decimal to a percent, multiply by 100, or shift the decimal point two places to the right, and add the percent sign.

EXAMPLE 5 **Writing a Decimal as a Percent**

Write 0.75 as a percent.

Solution

$$0.75 = 0.75 \cdot 1$$
$$= 0.75 \cdot 100\%$$
$$= 75\%$$

In Example 5, notice that we shift the decimal point two places to the right and add the percent sign.

$$0.75 = 075.\%$$
$$= 75\%$$

Section 1.IR7 Fundamentals of Percent Notation   **IR-61**

> **Converting a Decimal to a Percent**
>
> **Step 1:** Shift the decimal point two places to the right.
> **Step 2:** Add the percent sign.

EXAMPLE 6 **Writing a Decimal as a Percent**

Work Smart

If you forget which way to move the decimal point, you can either recall some basic facts, such as 75% = 0.75, or write the problem you are doing (decimal to % or % to decimal) and draw an arrow in the direction of the % sign. The direction the arrow is pointing tells you which direction to move the decimal point. Remember that you always move the decimal point two places.

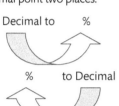

Decimal to   %

%   to Decimal

Write each decimal as a percent.

(a) 0.48   (b) 0.6   (c) 0.015   (d) 2

Solution

(a) $0.48 = 048.\% = 48\%$

(b) $0.60 = 060.\% = 60\%$

(c) $0.015 = 001.5\% = 1.5\%$

(d) $2.00 = 200.\% = 200\%$

If there are not enough digits to shift the decimal point 2 places, add zeros to the right of the number.

Quick ✓

In Problems 7–10, write the decimal as a percent.

7. 0.72   **8.** 1.8   **9.** 0.012   **10.** 0.35

▶ ❸ Convert Percents to Fractions and Fractions to Percents

Percent to Fraction

To write a percent as a fraction, recall that the % sign means "divided by 100."

EXAMPLE 7 **Writing a Percent as a Fraction**

Write 75% as a fraction.

Solution

Remember, % means "per 100" or "divided by 100."

$$75\% = \frac{75}{100}$$

Factor: $= \dfrac{25 \cdot 3}{25 \cdot 4}$

Divide out common factors: $= \dfrac{3}{4}$

So $75\% = \dfrac{3}{4}$.

> **Converting a Percent to a Fraction**
>
> **Step 1:** Remove the % sign and write the number divided by 100.
> **Step 2:** Write the fraction in lowest terms.

EXAMPLE 8 **Writing a Percent as a Fraction**

Write 3.5% as a fraction.

Solution

Remove the % sign and write the number divided by 100: $3.5\% = \dfrac{3.5}{100}$

Multiply by a factor of 1, written as $\dfrac{10}{10}$: $= \dfrac{3.5}{100} \cdot \dfrac{10}{10}$

$\dfrac{a}{b} \cdot \dfrac{c}{d} = \dfrac{a \cdot c}{b \cdot d}$: $= \dfrac{35}{1000}$

Factor: $= \dfrac{5 \cdot 7}{5 \cdot 200}$

Divide out common factors: $= \dfrac{7}{200}$

EXAMPLE 9 **Writing a Percent as a Fraction**

Write $33\dfrac{1}{3}\%$ as a fraction.

Solution

Remove the % sign and write the number over 100: $33\dfrac{1}{3}\% = \dfrac{33\frac{1}{3}}{100}$

Convert $33\dfrac{1}{3}$ to an improper fraction. $\dfrac{a}{b} = a \div b$: $= \dfrac{100}{3} \div \dfrac{100}{1}$

$\dfrac{a}{b} \div \dfrac{c}{d} = \dfrac{a}{b} \cdot \dfrac{d}{c}$: $= \dfrac{100}{3} \cdot \dfrac{1}{100}$

$\dfrac{a}{b} \cdot \dfrac{d}{c} = \dfrac{a \cdot d}{b \cdot c}$: $= \dfrac{100 \cdot 1}{3 \cdot 100}$

Divide out common factors: $= \dfrac{1}{3}$

Work Smart

To convert $33\dfrac{1}{3}$ to a fraction, multiply 33 by 3, then add 1. Write this result over 3. In general,

$$a\dfrac{b}{c} = \dfrac{a \cdot c + b}{c}$$

Quick ✓

In Problems 11–14, write the percent as a fraction in lowest terms.

11. 60%   **12.** 12.5%   **13.** 125%   **14.** $5\dfrac{3}{4}\%$

Fraction to Decimal to Percent

To write a fraction as a percent, write the fraction as a decimal and then convert the decimal to a percent.

EXAMPLE 10 **Writing a Fraction as a Percent**

Write $\dfrac{17}{20}$ as a percent.

Solution

Divide to convert the fraction to a decimal: $\dfrac{17}{20} \rightarrow 20\overline{)17.00}$

$$\begin{array}{r} 0.85 \\ 20\overline{)17.00} \\ -\,160 \\ \hline 100 \\ -\,100 \\ \hline 0 \end{array}$$

$8 \cdot 20 = 160$

$5 \cdot 20 = 100$

Section 1.IR7 Fundamentals of Percent Notation   IR-63

$$\text{Write the fraction as an equivalent decimal:} \quad \frac{17}{20} = 0.85$$

$$\text{Move the decimal point two places to the right, and add the \% sign:} \quad = 85\%$$

The numbers may not divide out as nicely as the ones in Example 10.

EXAMPLE 11 **Writing a Fraction as a Percent**

Write $\frac{8}{9}$ as a percent, rounded to the nearest tenth of a percent.

Work Smart

$\frac{8}{9}$ means $9\overline{)8}$

```
    0.8888
9)8.0000
   -72
     80
    -72
     80
    -72
     80
    -72
      8
```

Solution

$$\text{Divide to convert the fraction to a decimal:} \quad \frac{8}{9} \approx 0.8888$$

$$\text{Move the decimal point two places to the right:} \quad = 88.88\%$$

$$\text{Round to the nearest tenth:} \quad = 88.9\%$$

Converting a Fraction to a Percent

Divide the denominator into the numerator to get the equivalent decimal. Next, shift the decimal point two places to the right to convert the decimal to a percent.

Quick ✓

In Problems 15–18, write the fraction as a decimal and a percent.

15. $\frac{1}{20}$ **16.** $\frac{11}{4}$ **17.** $\frac{3}{8}$ **18.** $\frac{8}{5}$

In Problems 19 and 20, write the fraction as a decimal and a percent. Round to the nearest tenth of a percent.

19. $\frac{5}{12}$ **20.** $\frac{83}{120}$

1.IR7 Exercises MyLab Statistics

Underlined exercises have complete video solutions in MyLab.

*Problems **1–20** are the Quick ✓s that follow the EXAMPLES.*

Building Skills

In Problems 21–24, write the percent represented by the shaded region. See Objective 1.

21.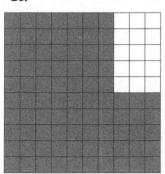

22.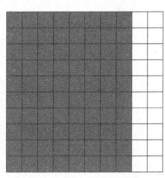

23.

24.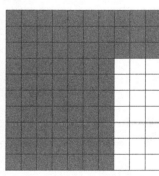

In Problems 25–28, write the ratio as a percent. See Objective 1.

25. Overall, 46 minutes out of 100 minutes of television violence occurs in cartoons. *Source: American Academy of Family Physicians*

26. Ice covers about 10.4 square miles out of every 100 square miles of Earth's land surface. *Source: Yahoo Answers*

27. In a survey of 100 Joliet College full-time students, 84 were employed at least part-time.

28. In 2009 among Americans aged 15 and older, 30 out of 100 had not been married. *Source: U.S. Census*

In Problems 29–42, write the percent as a decimal. See Objective 2.

29. 45%   30. 52%   31. 60%
32. 80%   33. 400%   34. 600%
35. 1%   36. 5%   37. 25.5%
38. 34.8%   39. 0.2%   40. 0.9%
41. 7.25%   42. 6.25%

In Problems 43–52, write the decimal as a percent. See Objective 2.

43. 0.96   44. 0.28   45. 0.3   46. 0.4
47. 0.035   48. 0.085   49. 1.6   50. 2.3
51. 0.0003 52. 0.005

In Problems 53–64, write the percent as a fraction in lowest terms. See Objective 3.

53. 40%   54. 20%   55. 225%
56. 275%   57. $66\frac{2}{3}$%   58. $83\frac{1}{3}$%
59. 62.5%   60. 87.5%   61. 1.8%
62. 4.5%   63. $18\frac{3}{4}$%   64. $43\frac{3}{4}$%

In Problems 65–78, write the fraction as a decimal and a percent. See Objective 3.

65. $\frac{3}{20}$ 66. $\frac{19}{20}$ 67. $\frac{7}{25}$ 68. $\frac{4}{5}$
69. $\frac{19}{50}$ 70. $\frac{38}{50}$ 71. $\frac{23}{200}$ 72. $\frac{158}{300}$
73. $\frac{138}{500}$ 74. $\frac{245}{800}$

In Problems 75–84, write the fraction as a decimal and a percent, rounding to the indicated place. See Objective 3.

75. $\frac{7}{12}$; nearest hundredth of a percent

76. $\frac{13}{17}$; nearest hundredth of a percent

77. $\frac{4}{7}$; nearest tenth of a percent

78. $\frac{6}{7}$; nearest tenth of a percent

79. $\frac{530}{675}$; nearest hundredth of a percent

80. $\frac{58}{120}$; nearest hundredth of a percent

81. $\frac{875}{1050}$; nearest tenth of a percent

82. $\frac{425}{1030}$; nearest tenth of a percent

83. $\frac{2}{375}$; nearest hundredth of a percent

84. $\frac{1}{150}$; nearest hundredth of a percent

Mixed Practice

In Problems 85 and 86, complete the chart.

85.

	Fraction	Decimal	Percent
(a)	$\frac{3}{50}$		
(b)		0.25	
(c)			85%
(d)		0.4	
(e)	$\frac{19}{200}$		
(f)			12.5%

86.

	Fraction	Decimal	Percent
(a)			60%
(b)	$\frac{1}{2}$		
(c)		1.1	
(d)	$\frac{21}{20}$		
(e)		0.002	
(f)			54%

Applying the Concepts

87. **Worried about Retirement Finances** In a recent survey of 2500 adult Americans, 950 said they are "not too" or "not at all" confident that they will have enough income and assets for their retirement. What percent of adult Americans are "not too" or "not at all" confident that they will have enough income and assets for their retirement? Note: In 2009 this percentage was 25%.

88. **Video Games and Violence** In a recent survey of 2200 adult Americans conducted by the Harris Poll, 1276 felt there is a link between playing

violent video games and teenagers showing violent behavior. What percent of adult Americans believe there is a link between playing violent video games and teenagers showing violent behavior?

89. **Fast Food** In a recent survey of 2496 adult Americans conducted by the Harris Poll, 1595 indicated that they dined at a fast food chain in the past month. What percent of adult Americans have dined at a fast food chain in the past month? Round your answer to the nearest tenth of a percent.

90. **Donating to Charity** The percent of people of who said they no longer donate to charities doubled from 2009 to 2010. If 3 out of 50 people did not donate to a charity in 2009, what percent of people did not donate to a charity in 2010? *Source:* Online Harris Interactive Poll

1.IR8 Language Used in Modeling

Objectives

1. Use Models That Involve Addition and Subtraction
2. Use Models That Involve Multiplication
3. Use Models That Involve Division

1 Use Models That Involve Addition and Subtraction

It is important to be able to apply your math skills to real-world problems. The process of taking a verbal description and developing a mathematical problem from the words in the description is called **mathematical modeling**.

Key words can help you decide which operation to use. Some of the words that we use for addition and subtraction are listed in Table 6.

Table 6 Words and the Math Symbols They Represent

Words for *Addition*	Examples (English)	Examples (math notation)
sum	the **sum** of 6 and 9	6 + 9
plus	14 **plus** 25	14 + 25
greater than	5 **greater than** 10	10 + 5
more than	21 **more than** 50	50 + 21
increased by	100 **increased by** 25	100 + 25
total	the **total** of 3, 7, and 10	3 + 7 + 10
Words for *Subtraction*	**Examples (English)**	**Examples (math notation)**
difference	the **difference** of 9 and 2	9 − 2
minus	15 **minus** 6	15 − 6
subtracted from	35 **subtracted from** 100	100 − 35
less	20 **less** 12	20 − 12
less than	30 **less than** 50	50 − 30
decreased by	16 **decreased by** 10	16 − 10
fewer than	7 **fewer than** 9	9 − 7

EXAMPLE 1 **Pizza Preference**

Harris Interactive asked, "What is your favorite topping on pizza?" Among those asked, 57 responded sausage, 64 responded pepperoni, 25 responded cheese only, and 12 responded bacon. What is the total number of people surveyed by Harris Interactive?

Solution
The word "total" implies finding the sum of all responses. The total is

$$57 + 64 + 25 + 12 = 158$$

A total of 158 people were surveyed.

EXAMPLE 2 Quality of Neighborhood

The Pew Research Center asked parents to rate their neighborhood as Fair/Poor, Good, or Excellent in terms of a place to raise their kids. A total of 1030 parents were surveyed with 185 responding Fair/Poor and 206 responding Good.

(a) What is the difference between the number who responded Good and the number who responded Fair/Poor?

(b) How many responded Excellent?

Solution

(a) The word "difference" implies subtraction. Therefore, subtract the number responding Fair/Poor from the number responding Good.

$$206 - 185 = 21$$

So, 21 more parents feel their neighborhood is good for raising their kids than feel their neighborhood is Fair/Poor for raising their kids.

(b) To find the number who responded Excellent, subtract the sum of the parents who responded Fair/Poor and Good from the total.

$$1030 - 185 - 206 = 639$$

So, 639 parents feel their neighborhood is Excellent for raising their kids.

Quick ✓

1. Writing a mathematical problem from a verbal description is called _____ _____.

In Problems 2–8, write the English phrase using math notation.

2. The difference of 12 and 8
3. 6 more than 15
4. The sum of 80 and 90
5. 25 decreased by 10
6. 18 less than 30
7. 36 subtracted from 100
8. 85 increased by 45
9. Individuals were asked to rate their health as Poor, Fair, Good, or Excellent. Of the individuals in the survey, 78 responded Poor, 159 responded Fair, 503 responded Good.

   (a) What is the difference in the number of individuals who responded Good and Fair?

   (b) The number who responded Excellent was 151 more than the number who responded Fair. How many individuals rated their health Excellent?

   (c) What was the total number of individuals surveyed?

❷ Use Models That Involve Multiplication

Table 7 lists some words used for multiplication.

Table 7

Word for Multiplication	Examples (English)	Examples (math notation)
multiplied	5 multiplied by 7	5×7 or $(5)(7)$ or $5 \cdot 7$
times	0.04 times 0.8	0.04×0.8 or $(0.04)(0.8)$ or $0.04 \cdot 0.8$
of	60% of 100	0.6×100 or $(0.6)(100)$ or $0.6 \cdot 100$
twice or doubled	Twice 70	2×70 or $(2)(70)$ or $2 \cdot 70$
triple	Triple 50	3×50 or $(3)(50)$ or $3 \cdot 50$
product	The product of 0.3 and 0.1	0.3×0.1 or $(0.3)(0.1)$ or $0.3 \cdot 0.1$

Section 1.IR8 Language Used in Modeling IR-67

EXAMPLE 3 **Survey Responses**

In a recent poll conducted by Pew Research, 60% of 1050 parents with teenagers aged 13 to 17 have checked their teen's social media profile. How many of those polled have checked their teen's social media profile?

Solution
The use of word "of" suggests multiplication. Because 60% = 0.6, we find 0.6 × 1050 = 630. Therefore, 630 of the 1050 parents have checked their teens social profile.

Quick ✓

In Problems 10–12, write the English phrase using math notation.

10. Twice 18   **11.** The product of 0.45 and 18   **12.** 27 times 100

13. In a recent poll conducted by Pew Research, 16% of 1050 parents with teenagers aged 13 to 17 used parental controls to restrict their teen's cellphone use. How many of those polled use parental controls to restrict their teen's cellphone use?

❸ Use Models That Involve Division

Table 8 lists some words used in division.

Table 8

Word for Division	Examples (English)	Examples (math notation)
divided	100 divided by 5	100 ÷ 5 or 100/5 or $\frac{100}{5}$
quotient	The quotient of 28 and 4	28 ÷ 4 or 28/4 or $\frac{28}{4}$
per	$100 per 5 days	$100 ÷ 5 days or $100/5 days or $\frac{\$100}{5 \text{ days}}$
out of	530 out of 1020	530 ÷ 1020 or 530/1020 or $\frac{530}{1020}$

Definition
The arithmetic mean is calculated by adding all the values and dividing this sum by the number of values.

EXAMPLE 4 **Finding the Arithmetic Mean Income**

In Wasilla, Alaska, five families have the following annual household incomes: $76,000, $40,000, $12,000, $55,000, and $57,000. Find the arithmetic mean household income for these five families.

Solution
First find the sum.

$$\$76{,}000 + \$40{,}000 + \$12{,}000 + \$55{,}000 + \$57{,}000 = \$240{,}000$$

Then divide the sum by the number of families, 5.

$$\frac{\$240{,}000}{5} = \$48{,}000$$

As a check, we note that if each of the five families earned $48,000, the total would be 5 · $48,000 = $240,000. For these five families, the arithmetic mean household income is $48,000 per family.

Quick ✓

In Problems 14–16, write the English phrase using math notation.

14. 130 divided by 18 **15.** The quotient of 500 and 12

16. 100 miles per 3 hours

17. In a recent poll of the residents of Springfield, 830 out of 2400 individuals wanted to increase property taxes to fund a new library. Express this result as a fraction and a decimal rounded to the nearest thousandth.

18. The following data represent the asking price of used 2015 Ford Fusions:

$$\$25{,}888, \$24{,}352, \$23{,}777, \$22{,}988$$

Find the arithmetic mean asking price.

1.IR8 Exercises — MyLab Statistics

Underlined exercises have complete video solutions in MyLab.

Problems 1–18 are the Quick ✓s that follow the EXAMPLES.

Applying the Concepts

19. Counting Calories Adrianna consumed 364 calories for breakfast, 583 calories for lunch, and 832 calories for dinner. If she also had a snack of 153 calories, what was her total calorie intake for the day? If the recommended calorie intake for women of her age is 1800 calories, was Adrianna over or under the recommended number of calories? By how much?

20. Banking Yaritza is a teller at First National Bank. During her shift the following cash transactions occurred: deposit of $36, deposit of $835, withdrawal of $380, deposit of $74, and withdrawal of $173. If her cash drawer began the shift with $1200, how much cash should be in the drawer at the end of her shift?

In Problems 21–24, use the following bar graph. It represents the educational attainment in 1990 and 2015 of adults 25 years of age or older who are residents of the United States. The data are in thousands, so 39,344 represents 39,344,000 adults aged 25 or older who did not graduate from high school in 1990.

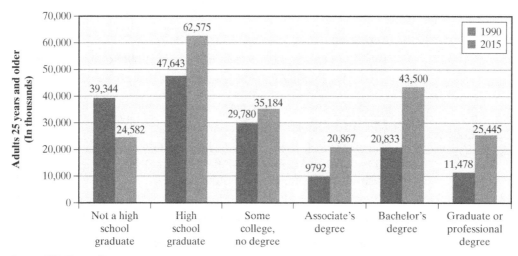

Source: U.S. Census Bureau

21. How many more adults were high school graduates in 2015 than in 1990?

22. How many more adults did not graduate from high school in 1990 than in 2015?

23. How many adults had Associate's, Bachelor's, and Graduate degrees in 2015?

24. Find the difference between the number of people who had Associate's degrees in 1990 and the number of people who had Graduate degrees in 1990.

25. **Employment** A sample of community college students was asked to report their employment status as full-time, part-time, or not employed. Of the individuals surveyed, 48 reported full-time employment and 153 reported part-time employment.
    (a) The number reporting that they were not employed was 25 more than those reporting part-time employment. How many of those surveyed reported they were not employed?
    (b) What was the total number of students surveyed?

26. **Party Affiliation** A sample of residents of a town were asked to report whether they were Democrats, Republicans, or Independents. Of the individuals surveyed, 230 reported they were Democrats and 212 reported they were Republicans.
    (a) The number who reported they were independents was 17 fewer than those who reported they were Republicans. How many reported they were independents?
    (b) What was the total number of residents surveyed?

27. **Text and Drive** In a survey of drivers in the United States between the ages of 18 and 64, 31% of 1200 drivers reported they had read or sent text messages or email on the cell phone while driving in the past 30 days. How many drivers in the survey had read or sent text messages or email on the cell phone while driving in the past 30 days?

28. **Seat Belt Use** In a survey of drivers in Arkansas, 72% of 800 drivers reported they always use a seat belt when driving. How many drivers in the survey always use their seat belt when driving?

29. **Comfort Food** In a poll conducted by The Harris Poll, 338 of 2252 Americans reported pizza as their favorite comfort food. Express this result as a fraction and as a decimal rounded to the nearest thousandth.

30. **Spanking** In a poll conducted by The Harris Poll, 434 of 2286 adult Americans stated that parents spanking their children is never appropriate. Express this result as a fraction and as a decimal rounded to the nearest thousandth.

31. **Donations** Marnie made the following donations to the Red Cross: $30.25, $51.75, and $50. What was the arithmetic mean amount of Marnie's donations?

32. **Rainfall** At a certain location in India, the following rainfall totals were recorded for a five-year period: 13.472 meters, 9.07 meters, 11.415 meters, 12.647 meters, and 8.734 meters. Find the arithmetic mean annual rainfall over this period of time.

33. **Football Players** The Cleveland Browns have six wide receivers on their roster. Their listed weights are 175 lb, 190 lb, 215 lb, 225 lb, 220 lb, and 207 lb. To the nearest pound, find the arithmetic mean weight of a Cleveland Browns wide receivers. (*Source:* Cleveland Browns)

34. **Football Players** The Chicago Bears have nine linebackers on their preseason roster. Their listed weights are 242 lb, 247 lb, 244 lb, 235, lb, 226 lb, 234 lb, 236 lb, 244 lb, and 238 lb. To the nearest tenth of a pound, find the arithmetic mean weight of a Chicago Bears linebacker. (*Source:* Chicago Bears)

CHAPTER 2

Integrated Review: Getting Ready for Numerically Summarizing Data

Outline

2.IR1 Exponents and the Order of Operations
2.IR2 Square Roots
2.IR3 Simplifying Algebraic Expressions; Summation Notation
2.IR4 Solving Linear Equations
2.IR5 Using Linear Equations to Solve Problems

2.IR1 Exponents and the Order of Operations

Objectives

① Evaluate Exponential Expressions
② Apply the Rules for Order of Operations

① Evaluate Exponential Expressions

If we wanted to multiply 3 to itself eight times, we would write $3 \cdot 3 \cdot 3 \cdot 3 \cdot 3 \cdot 3 \cdot 3 \cdot 3$. That's a lot of writing! To reduce the amount of writing needed to show repeated multiplication, use **exponential notation**, where $3 \cdot 3 \cdot 3 \cdot 3 \cdot 3 \cdot 3 \cdot 3 \cdot 3$ is written 3^8. In 3^8, 3 is called the **base** and 8 is called the **exponent**.

Exponential Notation

If n is a natural number and a is a real number, then

$$a^n = \underbrace{a \cdot a \cdot a \cdot \ldots \cdot a}_{n \text{ factors}}$$

where a is the **base** and n is the **exponent** or **power**. The exponent tells the number of times the base is used as a factor.

An expression written in the form a^n is said to be in **exponential form**. The expression 6^2 is read "six squared," 8^3 is read "eight cubed," and the expression 11^4 is read "eleven to the fourth power." In general, we read a^n as "a to the nth power."

EXAMPLE 1 Writing a Numerical Expression in Exponential Form

Write each expression in exponential form.

(a) $5 \cdot 5 \cdot 5$ (b) $(-4)(-4)(-4)(-4)(-4)(-4)$

Solution

(a) The expression $5 \cdot 5 \cdot 5$ contains three factors of 5, so $5 \cdot 5 \cdot 5 = 5^3$.

(b) The expression $(-4)(-4)(-4)(-4)(-4)(-4)$ contains six factors of -4, so $(-4)(-4)(-4)(-4)(-4)(-4) = (-4)^6$.

Quick ✓

1. In the expression 3^6, 3 is the _____ and 6 is the _____ or _____.

In Problems 2 and 3, write each expression in exponential form.

2. $11 \cdot 11 \cdot 11 \cdot 11 \cdot 11$

3. $(-7)(-7)(-7)(-7)$

To evaluate an exponential expression, write the expression in **expanded form.** For example, 2^8 in expanded form is $2 \cdot 2 \cdot 2 \cdot 2 \cdot 2 \cdot 2 \cdot 2 \cdot 2$.

EXAMPLE 2 **Evaluating an Exponential Expression**

Evaluate each exponential expression:

(a) 6^4

(b) $\left(\dfrac{5}{3}\right)^5$

Solution

(a) $6^4 = 6 \cdot 6 \cdot 6 \cdot 6$
$= 1296$

(b) $\left(\dfrac{5}{3}\right)^5 = \left(\dfrac{5}{3}\right)\left(\dfrac{5}{3}\right)\left(\dfrac{5}{3}\right)\left(\dfrac{5}{3}\right)\left(\dfrac{5}{3}\right)$
$= \dfrac{5 \cdot 5 \cdot 5 \cdot 5 \cdot 5}{3 \cdot 3 \cdot 3 \cdot 3 \cdot 3}$
$= \dfrac{3125}{243}$ ●

EXAMPLE 3 **Evaluating an Exponential Expression–Odd Exponent**

Evaluate each exponential expression:

(a) $(-5)^3$

(b) -5^3

Solution

(a) $(-5)^3 = (-5)(-5)(-5)$
$= -125$

(b) $-5^3 = -(5 \cdot 5 \cdot 5)$
$= -125$ ●

EXAMPLE 4 **Evaluating an Exponential Expression–Even Exponent**

Evaluate each exponential expression:

(a) $(-5)^4$

(b) -5^4

Solution

(a) $(-5)^4 = (-5)(-5)(-5)(-5)$
$= 625$

(b) $-5^4 = -(5 \cdot 5 \cdot 5 \cdot 5)$
$= -625$ ●

Work Smart

There is a difference between $(-5)^4$ and -5^4.

The parentheses in $(-5)^4$ tell us to use four factors of -5. However, in the expression -5^4, use 5 as a factor four times and then multiply the result by -1. We could also read -5^4 as "Take the opposite of the quantity 5^4."

Quick ✓

In Problems 4–9, evaluate each exponential expression.

4. 2^4

5. $(-7)^2$

6. $\left(-\dfrac{1}{6}\right)^3$

7. $(0.9)^2$

8. -2^4

9. $(-2)^4$

▶ ❷ Apply the Rules for Order of Operations

To evaluate $3 \cdot 5 + 4$, do you multiply first and then add to get $15 + 4 = 19$ *or* do you add first and then multiply to get $3 \cdot 9 = 27$?

Because $3 \cdot 5$ is equivalent to $5 + 5 + 5$, we have

$$3 \cdot 5 + 4 = 5 + 5 + 5 + 4$$
$$= 19$$

In Words
Multiply first, and then add.

Based on this, **whenever addition and multiplication appear in the same expression, always multiply first and then add.**

Because any division problem can be written as a multiplication problem, divide before adding as well. Also, because any subtraction problem can be written as an addition problem, always multiply and divide before adding and subtracting.

EXAMPLE 5 Evaluating an Expression Containing Multiplication, Division, and Addition

Evaluate each expression:

(a) $11 + 2 \cdot (-6)$

(b) $7 + 12 \div 3 \cdot 5$

Solution

(a)   Multiply first: $11 + 2 \cdot (-6) = 11 + (-12)$
         Add: $= -1$

(b)   Multiply/divide left to right: $7 + 12 \div 3 \cdot 5 = 7 + 4 \cdot 5$
         Multiply: $= 7 + 20$
                                                     $= 27$

Quick ✓

In Problems 10–13, evaluate each expression.

10. $1 + 7 \cdot 2$

11. $-3 \div \left(-\dfrac{1}{2}\right) + 18$

12. $9 \cdot 4 \div 2 + 5$

13. $\dfrac{15}{2} \div (-5) \cdot 8 - 7$

▶ Parentheses

If we want to add two numbers first and then multiply, use parentheses and write $(3 + 5) \cdot 4$. In other words, always evaluate **the expression in parentheses first.**

EXAMPLE 6 Evaluating an Expression Containing Parentheses

Evaluate each expression:

(a) $(5 + 3) \cdot 2$

(b) $\left(\dfrac{3}{2} - \dfrac{5}{2}\right)\left(\dfrac{7}{3} + \dfrac{2}{3}\right)$

Solution

(a) $(5 + 3) \cdot 2 = 8 \cdot 2$
               $= 16$

(b) $\left(\dfrac{3}{2} - \dfrac{5}{2}\right)\left(\dfrac{7}{3} + \dfrac{2}{3}\right) = \left(-\dfrac{2}{2}\right)\left(\dfrac{9}{3}\right)$
$= (-1)(3)$
$= -3$

Quick ✓

In Problems 14–16, evaluate each expression.

14. $8(2 + 3)$ 　　**15.** $(2 - 9)(5 + 4)$ 　　**16.** $\left(\dfrac{6}{7} + \dfrac{8}{7}\right)\left(\dfrac{11}{8} + \dfrac{5}{8}\right)$

Work Smart

The division bar acts like a set of parentheses.

The Division Bar

If an expression contains a division bar, treat the terms above and below the division bar as if they were in parentheses. For example,

$$\frac{3+5}{9+7} = \frac{(3+5)}{(9+7)} = \frac{8}{16} = \frac{8 \cdot 1}{8 \cdot 2} = \frac{1}{2}$$

EXAMPLE 7 **Finding the Value of an Expression That Contains a Division Bar**

Evaluate each expression:

(a) $\dfrac{7 \cdot 3}{3 + 9 \cdot 2}$ 　　(b) $\dfrac{1 + 7 \div \dfrac{1}{5}}{-6 \cdot 2 + 8}$

Solution

(a) Multiply: $\dfrac{7 \cdot 3}{3 + 9 \cdot 2} = \dfrac{21}{3 + 18}$

　　Add: $= \dfrac{21}{21}$

　　　　$= 1$

(b) Write division as multiplication: $\dfrac{1 + 7 \div \dfrac{1}{5}}{-6 \cdot 2 + 8} = \dfrac{1 + 7 \cdot 5}{-6 \cdot 2 + 8}$

　　Multiply: $= \dfrac{1 + 35}{-12 + 8}$

　　Add: $= \dfrac{36}{-4}$

　　　　$= \dfrac{9 \cdot 4}{-1 \cdot 4}$

　　　　$= -9$

Quick ✓

In Problems 17–19, evaluate each expression.

17. $\dfrac{2 + 5 \cdot 6}{-3 \cdot 8 - 4}$ 　　**18.** $\dfrac{(12 + 14) \cdot 2}{13 \cdot 2 + 13 \cdot 5}$ 　　**19.** $\dfrac{4 + 3 \div \dfrac{1}{7}}{2 \cdot 9 - 3}$

▶ Multiple Grouping Symbols

Grouping symbols include parentheses ( ), brackets [ ], braces { }, and absolute value symbols, | |. Operations within the grouping symbols are performed first. **When multiple grouping symbols occur, evaluate the expression in the innermost grouping symbols first and work outward.**

EXAMPLE 8 Finding the Value of an Expression Containing Grouping Symbols

Evaluate each expression:

(a) $2[3(6 + 3) - 7]$

(b) $\left[4 + \left(\dfrac{2}{3} \cdot (-9)\right)\right] \cdot 3$

Solution

(a) Perform the operation in parentheses first: $2[3(6 + 3) - 7] = 2[3 \cdot 9 - 7]$

Perform the operations in brackets, multiply first: $= 2[27 - 7]$

$= 2[20]$

$= 40$

(b) Perform the operation in parentheses first: $\left[4 + \left(\dfrac{2}{3} \cdot (-9)\right)\right] \cdot 3 = [4 + (-6)] \cdot 3$

Perform the operation in brackets: $= -2 \cdot 3$

$= -6$

Quick ✓

In Problems 20 and 21, evaluate each expression.

20. $4[2(3 + 7) - 15]$ **21.** $2\{4[26 - (9 + 7)] - 15\} - 10$

When do we evaluate exponents in the order of operations? In $2 \cdot 4^3$, do we multiply first and then evaluate the exponent to obtain $2 \cdot 4^3 = 8^3 = 512$, or do we evaluate the exponent first and then multiply to obtain $2 \cdot 4^3 = 2 \cdot 64 = 128$? Because $2 \cdot 4^3 = 2 \cdot 4 \cdot 4 \cdot 4 = 128$, **evaluate exponents before multiplication.** Don't forget to evaluate the expression in the grouping symbol first.

EXAMPLE 9 Finding the Value of an Expression Containing Exponents

Evaluate each of the following:

(a) $2 + 7(-4)^2$

(b) $\dfrac{2 \cdot 3^2 + 4}{3(2 - 6)}$

Solution

Evaluate the exponent
↓

(a) $2 + 7(-4)^2 = 2 + 7 \cdot 16$

Multiply: $= 2 + 112$

Add: $= 114$

Parentheses first
↓

(b) $\dfrac{2 \cdot 3^2 + 4}{3(2 - 6)} = \dfrac{2 \cdot 3^2 + 4}{3(-4)}$

Evaluate the exponent: $= \dfrac{2 \cdot 9 + 4}{3(-4)}$

Find products: $= \dfrac{18 + 4}{-12}$

Add terms in numerator: $= \dfrac{22}{-12}$

Write in lowest terms: $= \dfrac{2 \cdot 11}{2 \cdot -6}$

$= -\dfrac{11}{6}$

Quick ✓

In Problems 22–24, evaluate each of the following:

22. $\dfrac{7-5^2}{2}$
23. $3(7-3)^2$
24. $\dfrac{(-3)^2 + 7(1-3)}{3 \cdot 2^3 + 6}$

Order of Operations

Step 1: Perform all operations within *grouping symbols* first. When an expression has more than one set of grouping symbols, begin within the innermost grouping symbols and work outward.

Step 2: Evaluate expressions containing exponents.

Step 3: Perform *multiplication and division*, working from *left to right*.

Step 4: Perform *addition and subtraction*, working from *left to right*.

EXAMPLE 10   How to Evaluate an Expression Using Order of Operations

Evaluate: $18 + 7(2^3 - 26) + 5^2$

Step-by-Step Solution

Steps 1 and 2: Evaluate the expression in the parentheses first. In the parentheses, evaluate the expression containing the exponent first.

$$18 + 7(2^3 - 26) + 5^2 = 18 + 7(8 - 26) + 25$$

Evaluate $8 - 26$ in the parentheses: $= 18 + 7(-18) + 25$

Step 3: Perform multiplication and division, working from left to right.

Multiply $7(-18)$: $= 18 - 126 + 25$

Step 4: Perform addition and subtraction, working from left to right.

$$= -108 + 25$$
$$= -83$$

EXAMPLE 11   Evaluating a Numerical Expression Using Order of Operations

Evaluate: $\left(\dfrac{2^3 - 6}{10 - 2 \cdot 3}\right)^2$

Solution

Evaluate the exponential expression inside the parentheses

$$\left(\dfrac{2^3 - 6}{10 - 2 \cdot 3}\right)^2 = \left(\dfrac{8 - 6}{10 - 2 \cdot 3}\right)^2$$

Multiply inside the parentheses: $= \left(\dfrac{8 - 6}{10 - 6}\right)^2$

Add/subtract inside the parentheses: $= \left(\dfrac{2}{4}\right)^2$

Simplify: $= \left(\dfrac{1}{2}\right)^2$

Evaluate the exponential expression: $= \dfrac{1}{4}$

Quick ✓

In Problems 25–28, evaluate each expression.

25. $\dfrac{(4-10)^2}{2^3-5}$

26. $-3[(-4)^2 - 5(8-6)]^2$

27. $\dfrac{(2.9+7.1)^2}{5^2-15}$

28. $\left(\dfrac{4^2 - 4(-3)(1)}{7\cdot 2}\right)^2$

2.IR1 Exercises MyLab Statistics

Underlined exercises have complete video solutions in MyLab.

Problems **1–28** are the Quick ✓s that follow the EXAMPLES.

Building Skills

In Problems 29–32, write in exponential form. See Objective 1.

29. $5 \cdot 5$

30. $4 \cdot 4 \cdot 4 \cdot 4 \cdot 4$

31. $\left(-\dfrac{3}{5}\right)\left(-\dfrac{3}{5}\right)\left(-\dfrac{3}{5}\right)$

32. $(-8)(-8)(-8)$

In Problems 33–54, evaluate each exponential expression. See Objective 1.

33. 8^2

34. 4^2

35. $(-8)^2$

36. $(-4)^2$

37. 10^3

38. 2^5

39. -10^3

40. -2^5

41. $(-10)^3$

42. $(-2)^5$

43. $(1.5)^2$

44. $(0.04)^2$

45. -8^2

46. -4^2

47. -1^{20}

48. $(-1)^{19}$

49. 0^4

50. 1^6

51. $\left(-\dfrac{1}{2}\right)^6$

52. $\left(-\dfrac{3}{2}\right)^5$

53. $\left(-\dfrac{1}{3}\right)^3$

54. $\left(-\dfrac{3}{4}\right)^2$

In Problems 55–86, evaluate each expression. See Objective 2.

55. $2 + 3 \cdot 4$

56. $12 + 8 \cdot 3$

57. $-5 \cdot 3 + 12$

58. $-3 \cdot 12 + 9$

59. $100 \div 2 \cdot 50$

60. $50 \div 5 \cdot 4$

61. $156 - 3 \cdot 2 + 10$

62. $86 - 4 \cdot 3 + 6$

63. $(2 + 3) \cdot 4$

64. $(7 - 5) \cdot \dfrac{5}{2}$

65. $8 \div 4 \cdot 2$

66. $4 \div 7 \cdot 21$

67. $\dfrac{4 + 2}{2 + 8}$

68. $\dfrac{5 + 3}{3 + 15}$

69. $\dfrac{14 - 6}{6 - 14}$

70. $\dfrac{15 - 7}{7 - 15}$

71. $13 - [3 + (-8)4]$

72. $12 - [7 + (-6)3]$

73. $(-8.75 - 1.25) \div (-2)$

74. $(-11.8 - 15.2) \div (-2)$

75. $4 - 2^3$

76. $10 - 4^2$

77. $(13 - 9)^2 + (10 - 9)^2 + (4 - 9)^2$

78. $(20 - 13)^2 + (12 - 13)^2 + (7 - 13)^2$

79. $-2^3 + 3^2 \div (2^2 - 1)$

80. $-5^2 + 3^2 \div (3^2 + 9)$

81. $\left(\dfrac{4^2 - 3}{12 - 2 \cdot 5}\right)^2$

82. $\left(\dfrac{7 - 5^2}{8 + 4 \cdot 2}\right)^2$

83. $-2[5(9 - 3) - 3 \cdot 6]$

84. $3[6(5 - 2) - 2 \cdot 5]$

85. $\left(\dfrac{4}{3} + \dfrac{5}{6}\right)\left(\dfrac{2}{5} - \dfrac{9}{10}\right)$

86. $\left(\dfrac{3}{4} + \dfrac{1}{2}\right)\left(\dfrac{2}{3} - \dfrac{1}{2}\right)$

Mixed Practice

In Problems 87–110, evaluate each expression.

87. $4^2 - 3 \cdot 4 + 7$

88. $(-2)^2 + 4(-2) + 11$

89. $4 + 2(6 - 2)$

90. $3 + 6(9 - 5)$

91. $\dfrac{2.4 + 1.6 + 0.3 + 2.5}{4}$

92. $\dfrac{10.3 + 8.7 + 10.1}{3}$

93. $\left(\dfrac{2 - (-4)^3}{5^2 - 7 \cdot 2}\right)^2$

94. $\left(\dfrac{9 \cdot 2 - (-2)^3}{4^2 + 3(-1)^5}\right)^2$

95. $1 - (0.21 + 0.13 + 0.47)$

96. $1 - (0.15 + 0.36 + 0.09)$

97. $|6(5 - 3^2)|$

98. $-6(2 + |2 \cdot 3 - 4^2|)$

99. $\dfrac{4 - (-6)}{1 - (-1)}$

100. $\dfrac{-2 - 12}{3 - (-4)}$

101. $1 - 0.95^2$

102. $1 - 0.99^2$

103. $\dfrac{21 - 3^2}{1 + 3}$

104. $\dfrac{5 + 3^2}{2 + 5}$

105. $\dfrac{3}{4} \cdot \left[\dfrac{5}{4} \div \left(\dfrac{3}{8} - \dfrac{1}{8} \right) - 3 \right]$

106. $\left[\dfrac{9}{10} \div \left(\dfrac{2}{5} + \dfrac{1}{5} \right) + \dfrac{7}{2} \right] \cdot \dfrac{1}{10}$

107. $\dfrac{(12 - 9)^2 + (15 - 9)^2 + (4 - 9)^2 + (5 - 9)^2}{4}$

108. $\dfrac{(18 - 12)^2 + (12 - 12)^2 + (6 - 12)^2}{3}$

109. $\dfrac{5^2 - 3^3}{|4 - 4^2|}$

110. $\dfrac{3 \cdot 2^3 - 2^2 \cdot 12}{3 + 3^2}$

Applying the Concepts

In Problems 111–114, express each number as the product of prime factors. Write the answer in exponential form.

111. 72

112. 675

113. 48

114. 200

In Problems 115–120, insert grouping symbols so that the expression has the desired value.

115. $4 \cdot 3 + 6 \cdot 2$ results in 36

116. $4 \cdot 7 - 4^2$ results in -36

117. $4 + 3 \cdot 4 + 2$ results in 42

118. $6 - 4 + 3 - 1$ results in 0

119. $6 - 4 + 3 - 1$ results in 4

120. $4 + 3 \cdot 2 - 1 \cdot 6$ results in 42

121. Cost of a TV The total amount paid for a flat-screen television that costs $479, plus sales tax of 7.5%, is found by evaluating the expression $479 + 0.075(479)$. Evaluate this expression rounded to the nearest cent.

122. Manufacturing Cost Evaluate the expression $3000 + 6(100) - \dfrac{100^2}{1000}$ to find the weekly production cost of manufacturing 100 calculators.

123. Surface Area The surface area of a right circular cylinder whose radius is 6 inches and height is 10 inches is given approximately by $2 \cdot 3.1416 \cdot 6^2 + 2 \cdot 3.1416 \cdot 6 \cdot 10$. Evaluate this expression. Round the answer to two decimal places.

124. Volume of a Cone The volume of a cone whose radius is 3 centimeters and whose height is 12 centimeters is given approximately by $\dfrac{1}{3} \cdot 3.1416 \cdot 3^2 \cdot 12$. Evaluate this expression. Round the answer to two decimal places.

125. Investing If $1000 is invested at 3% annual interest and remains untouched for 2 years, the amount of money that is in the account after 2 years is given by the expression $1000(1 + 0.03)^2$. Evaluate this expression, rounded to the nearest cent.

126. Investing If $5000 is invested at 4.5% annual interest and remains untouched for 5 years, the amount of money that is in the account after 5 years is given by the expression $5000(1 + 0.045)^5$. Evaluate this expression, rounded to the nearest cent.

Extending the Concepts

The Angle Addition Postulate from geometry states that the measure of an angle is equal to the sum of the measures of its parts. Refer to the figure. Use the Angle Addition Postulate to answer Problems 127 and 128.

127. If the measure of $\angle XYQ = 46.5°$ and the measure of $\angle QYZ = 69.25°$, find the measure of $\angle XYZ$.

128. If the measure of $\angle QYZ = 18°$ and the measure of $\angle XYZ = 57°$, find the measure of $\angle XYQ$.

Explaining the Concepts

129. Explain the difference between -3^2 and $(-3)^2$. Identify the distinguishing characteristics of the two problems, and explain how to evaluate each expression.

2.IR2 Square Roots

Objectives

1. Evaluate Square Roots
2. Determine Whether a Square Root Is Rational, Irrational, or Not a Real Number

Exponents indicate repeated multiplication. For example, 4^2 means $4 \cdot 4$, so $4^2 = 16; (-6)^2 = (-6)(-6) = 36$. In this section, we will "undo" the process of raising a number to the second power and ask questions such as, "What number, or numbers, when squared, give me 16?"

1 Evaluate Square Roots

You know that 5^2 and $(-5)^2$ both equal 25. Thus, if you were asked, "What number, when squared, equals 25?", you would respond, "Either -5 or 5."

In Words
Taking the square root of a number "undoes" raising a number to the second power.

Definition

For any real numbers a and b, b is a **square root** of a if $b^2 = a$.

Put another way, the square roots of a are the numbers that, when squared, give a. For example, the square roots of 25 are -5 and 5 because $(-5)^2 = 25$ and $5^2 = 25$.

EXAMPLE 1 **Finding the Square Roots of Numbers**

Find the square roots: **(a)** 36 **(b)** 0

Solution

(a) Because $6^2 = 36$ and $(-6)^2 = 36$, the square roots of 36 are -6 and 6.

(b) 0 is the only square root of 0 because only $0^2 = 0$.

EXAMPLE 2 **Finding the Square Roots of Numbers**

Work Smart
When finding square roots of fractions, look at the numerator and denominator separately.

Find the square roots: **(a)** $\dfrac{16}{49}$ **(b)** 0.09

Solution

(a) $-\dfrac{4}{7}$ and $\dfrac{4}{7}$ are the square roots of $\dfrac{16}{49}$ because $\left(-\dfrac{4}{7}\right)^2 = \dfrac{16}{49}$ and $\left(\dfrac{4}{7}\right)^2 = \dfrac{16}{49}$.

(b) -0.3 and 0.3 are the square roots of 0.09 because $(-0.3)^2 = 0.09$ and $(0.3)^2 = 0.09$.

Quick ✓

1. For any real numbers a and b, b is a square root of a if _____ = _____.

In Problems 2–4, find the square roots of each real number.

2. 16   **3.** $\dfrac{9}{100}$ **4.** 0.04

Notice in Examples 1 and 2 that every positive number has a positive and a negative square root. The symbol $\sqrt{}$, called a **radical**, is used to denote the **principal square root**, or nonnegative (zero or positive) square root. For example, if we want the positive square root of 25, write $\sqrt{25} = 5$. Read $\sqrt{25} = 5$ as "The positive square root of 25 is 5." But what if we want the negative square root of a real number? In that case, use the expression $-\sqrt{25} = -5$.

Work Smart

Symbols and notation in math enable us to express our thoughts using shorthand. The symbol $\sqrt{100}$ means "the positive number whose square is 100." If you want the negative number whose square is 100, write $-\sqrt{100}$.

> **Properties of Square Roots**
> - Every positive real number has two square roots, one positive and one negative.
> - Use the symbol $\sqrt{}$, called a radical, to denote the nonnegative square root of a real number. The nonnegative square root is called the principal square root.
> - The square root of 0 is 0. In symbols, $\sqrt{0} = 0$.
> - The number under the radical is called the **radicand.** For example, the radicand in $\sqrt{25}$ is 25.

EXAMPLE 3 **Evaluating Square Roots**

Evaluate: (a) $\sqrt{64}$ (b) $-\sqrt{81}$

Solution

(a) The notation $\sqrt{64}$ means "the positive number whose square is 64." Because $8^2 = 64$,
$$\sqrt{64} = 8$$

(b) The notation $-\sqrt{81}$ means "the opposite of the positive number whose square is 81." Thus
$$-\sqrt{81} = -1 \cdot \sqrt{81} = -1 \cdot 9 = -9$$

EXAMPLE 4 **Evaluating Square Roots**

Evaluate: (a) $\sqrt{\dfrac{1}{4}}$ (b) $-\sqrt{0.01}$

Solution

(a) $\sqrt{\dfrac{1}{4}} = \dfrac{1}{2}$ because $\left(\dfrac{1}{2}\right)^2 = \dfrac{1}{4}$.

(b) $-\sqrt{0.01} = -0.1$ because $0.1^2 = 0.01$.

Quick ✓

5. The nonnegative square root of a real number a, represented by $\sqrt{a}$, is called the _____ of a.

6. In the expression $\sqrt{27}$, the 27 is called the _____.

In Problems 7–10, evaluate each square root.

7. $\sqrt{100}$ 8. $-\sqrt{9}$ 9. $\sqrt{\dfrac{25}{49}}$ 10. $\sqrt{0.36}$

Figure 1

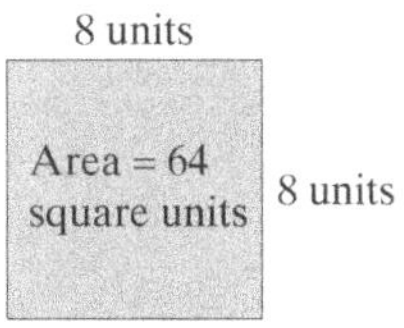

8 units

Area = 64 square units   8 units

A rational number is a **perfect square** if it is the square of a rational number. Examples 3 and 4 show that 64, 81, $\dfrac{1}{4}$, and 0.01 are all perfect squares because their square roots are rational numbers. We can think of perfect squares geometrically as in Figure 1, which shows a square whose area is 64 square units. The square root of the area, $\sqrt{64}$, gives the length of each side of the square, 8 units.

EXAMPLE 5 **Evaluating an Expression Containing Square Roots**

Evaluate: $-4\sqrt{25}$

Solution

The expression $-4\sqrt{25}$ means "-4 times the positive square root of 25." First, find the positive square root of 25, and then multiply this result by -4.

$$-4\sqrt{25} = -4 \cdot 5$$
$$= -20$$

EXAMPLE 6 Evaluating an Expression Containing Square Roots

Work Smart

In Examples 6(a) and (b), notice that $\sqrt{9} + \sqrt{16} \neq \sqrt{9+16}$. In general,

$$\sqrt{a} + \sqrt{b} \neq \sqrt{a+b}$$

The radical acts like a grouping symbol, so always simplify the radicand before taking the square root.

Evaluate:

(a) $\sqrt{9} + \sqrt{16}$ (b) $\sqrt{9+16}$ (c) $\sqrt{8^2 - 4 \cdot 7 \cdot 1}$

Solution

(a) $\sqrt{9} + \sqrt{16} = 3 + 4$
$= 7$

(b) $\sqrt{9+16} = \sqrt{25}$
$= 5$

(c) $\sqrt{8^2 - 4 \cdot 7 \cdot 1} = \sqrt{64 - 28}$
$= \sqrt{36}$
$= 6$

Quick ✓

In Problems 11–14, evaluate each expression.

11. $2\sqrt{36}$ 12. $\sqrt{25 + 144}$ 13. $\sqrt{25} + \sqrt{144}$ 14. $\sqrt{5^2 - 4 \cdot 3 \cdot (-2)}$

❷ Determine Whether a Square Root Is Rational, Irrational, or Not a Real Number

Because there is no rational number whose square is 5, $\sqrt{5}$ is not a rational number. In fact, $\sqrt{5}$ is an *irrational* number. Recall that an irrational number is a number that cannot be written as the quotient of two integers.

What if we wanted to evaluate $\sqrt{-16}$? Because any positive real number squared is positive, any negative real number squared is also positive, and 0 squared is 0, there is no real number whose square is -16. We conclude: **Negative real numbers do not have square roots that are real numbers!**

Work Smart

The perfect squares of positive integers are

$1^2 = 1$ $5^2 = 25$
$2^2 = 4$ $6^2 = 36$
$3^2 = 9$ $7^2 = 49$
$4^2 = 16$ $8^2 = 64$

and so on.

Work Smart

The square roots of negative real numbers are not real.

The following comments regarding square roots are important.

> **More Properties of Square Roots**
>
> - The square root of a perfect square number is a rational number.
> - The square root of a positive rational number that is not a perfect square is an irrational number. For example, $\sqrt{20}$ is an irrational number because 20 is not a perfect square.
> - The square root of a negative real number is not a real number. For example, $\sqrt{-2}$ is not a real number.

If needed, find decimal approximations for irrational square roots using a calculator.

EXAMPLE 7 Approximating Square Roots

Approximate $\sqrt{5}$ by writing it rounded to two decimal places.

Solution

Work Smart

The notation $\approx$ means "is approximately equal to."

Because $\sqrt{4} = 2$ and $\sqrt{9} = 3$, $\sqrt{5}$ will be between 2 and 3 because 5 is between 4 and 9. Use a calculator and find $\sqrt{5} \approx 2.23607$. Rounded to two decimal places, $\sqrt{5} \approx 2.24$.

Quick ✓

In Problems 15 and 16, express each square root as a decimal rounded to two decimal places.

15. $\sqrt{35}$ 16. $-\sqrt{6}$

EXAMPLE 8   Determining Whether a Square Root of an Integer Is Rational, Irrational, or Not a Real Number

Determine whether each square root is rational, irrational, or not a real number. Evaluate each real square root, if possible. Express any irrational square root as a decimal rounded to two decimal places.

(a) $\sqrt{51}$ (b) $\sqrt{169}$ (c) $\sqrt{-81}$

Work Smart

$\sqrt{-81}$ is not a real number, but $-\sqrt{81}$ is because $-\sqrt{81} = -9$. Note the placement of the negative sign!

Solution

(a) $\sqrt{51}$ is irrational because 51 is not a perfect square. There is no rational number whose square is 51. Using a calculator, we find $\sqrt{51} \approx 7.14$.

(b) $\sqrt{169}$ is a rational number because $13^2 = 169$. So, $\sqrt{169} = 13$.

(c) $\sqrt{-81}$ is not a real number because there is no real number whose square is -81.

Quick ✓

In Problems 17–20, determine whether each square root is rational, irrational, or not a real number. Evaluate each square root, if possible. For each square root that is irrational, round your answer to two decimal places.

17. $\sqrt{49}$ 18. $\sqrt{71}$ 19. $\sqrt{-25}$ 20. $-\sqrt{16}$

2.IR2 Exercises   MyLab Statistics

Underlined exercises have complete video solutions in MyLab.

Problems **1–20** are the Quick ✓s that follow the EXAMPLES.

Building Skills

In Problems 21–28, find the value(s) of each expression. See Objective 1.

21. the square roots of 1
22. the square roots of 4
23. the square roots of $\frac{1}{9}$
24. the square roots of $\frac{9}{4}$
25. the square roots of 169
26. the square roots of 49
27. the square roots of 0.25
28. the square roots of 0.16

In Problems 29–52, find the exact value of each square root without a calculator. See Objective 1.

29. $\sqrt{144}$
30. $\sqrt{4}$
31. $-\sqrt{9}$
32. $-\sqrt{1}$
33. $\sqrt{225}$
34. $\sqrt{169}$
35. $\sqrt{\dfrac{1}{121}}$
36. $\sqrt{\dfrac{1}{49}}$
37. $\underline{\sqrt{0.04}}$
38. $\sqrt{0.09}$
39. $-6\sqrt{9}$
40. $-4\sqrt{4}$
41. $\sqrt{\dfrac{16}{81}}$
42. $\sqrt{\dfrac{25}{9}}$
43. $\underline{5\sqrt{49}}$
44. $4\sqrt{16}$
45. $\sqrt{36 + 64}$
46. $\sqrt{81 + 144}$
47. $\sqrt{36} + \sqrt{64}$
48. $\sqrt{81} + \sqrt{144}$
49. $\sqrt{7^2 - (4)(15)(-2)}$
50. $\sqrt{10^2 - (4)(8)(3)}$
51. $\sqrt{\dfrac{(20-11)^2 + (13-11)^2 + (4-11)^2 + (8-11)^2 + (10-11)^2}{4}}$
52. $\sqrt{\dfrac{(3-9)^2 + (6-9)^2 + (12-9)^2 + (14-9)^2 + (10-9)^2}{5}}$

In Problems 53–58, use a calculator to find the approximate value of the square root, rounded to the indicated place. See Objective 2.

53. $\sqrt{8}$ to 3 decimal places
54. $\sqrt{12}$ to 3 decimal places
55. $\sqrt{30}$ to the nearest hundredth
56. $\sqrt{18}$ to the nearest hundredth
57. $\sqrt{57}$ to the nearest tenth
58. $\sqrt{42}$ to the nearest tenth

In Problems 59–68, tell if the square root is rational, irrational, or not a real number. If the square root is rational, find the exact value; if the square root is irrational, write the approximate value rounded to two decimal places. See Objective 2.

59. $\sqrt{-4}$

60. $\sqrt{-100}$

61. $\sqrt{400}$

62. $\sqrt{900}$

63. $\sqrt{\dfrac{1}{4}}$

64. $\sqrt{\dfrac{49}{64}}$

65. $\sqrt{54}$

66. $\sqrt{24}$

67. $\sqrt{50}$

68. $\sqrt{12}$

Mixed Practice

In Problems 69–84, evaluate each square root if possible. If the square root is irrational, write the approximate value rounded to two decimal places. If the square root is not a real number, write "not a real number."

69. $\sqrt{3}$

70. $\sqrt{2}$

71. $\sqrt{0.4}$

72. $\sqrt{\dfrac{4}{81}}$

73. $\sqrt{-2}$

74. $\sqrt{-10}$

75. $3\sqrt{4}$

76. $-10\sqrt{9}$

77. $3\sqrt{\dfrac{25}{9}} - \sqrt{169}$

78. $2\sqrt{\dfrac{9}{4}} - \sqrt{4}$

79. $\sqrt{7^2 - (4)(-2)(-3)}$

80. $\sqrt{13^2 - (4)(-6)(-6)}$

81. $\sqrt{(11-8)^2 + (11-5)^2}$

82. $\sqrt{(-3-5)^2 + (-5-(-1))^2}$

83. $\sqrt{\dfrac{(5-4)^2 + (7-4)^2 + (4-4)^2 + (3-4)^2 + (1-4)^2}{4}}$

84. $\sqrt{\dfrac{(13-11)^2 + (14-11)^2 + (12-11)^2 + (7-11)^2 + (9-11)^2}{5}}$

Applying the Concepts

The area, A, of a square whose side has length s is given by $A = s^2$. We can calculate the length, s, of the side of a square as the positive square root of the area, A, using $s = \sqrt{A}$. In Problems 85–88, find the length of the side of the square whose area is given.

△ 85. 625 square feet △ 86. 256 square meters

△ 87. 256 square kilometers △ 88. 400 square inches

The area, A, of a circle whose radius is r is given by the formula $A = \pi r^2$. We can calculate the radius when the area is given using $r = \sqrt{\dfrac{A}{\pi}}$. In Problems 89–92, find the radius of the circle with the following areas.

△ 89. 49π square meters △ 90. 25π square feet

△ 91. 196π square inches △ 92. 289π square centimeters

△ 93. **Great Pyramid at Giza** The volume V of a pyramid with a square base s and height h is $V = \dfrac{1}{3}s^2 h$. If we solve this formula for s, we obtain $s = \sqrt{\dfrac{3V}{h}}$.
The Great Pyramid at Giza, built around 2500 B.C., has a volume of approximately 7,700,000 cubic meters. Find the length, to the nearest meter, of the side of the Great Pyramid if you know that the height is approximately 146 meters.

△ 94. **Sun Pyramid** Use the formula given in Problem 93 to the find the length, to the nearest foot, of the side of the Sun Pyramid if the volume of the pyramid is approximately 114,000,000 cubic feet and the height is 210 feet. The Sun Pyramid was built in the ancient city of Teotihuacán, around 200 A.D.

Explaining the Concepts

95. Explain why $-\sqrt{9}$ is a real number but $\sqrt{-9}$ is not a real number.

96. You don't have a calculator handy but need an estimate for $\sqrt{20}$. Explain how you might find an approximate value for this number without a calculator.

2.IR3 Simplifying Algebraic Expressions; Summation Notation

Objectives

1. Evaluate Algebraic Expressions
2. Identify Like Terms and Unlike Terms
3. Use the Distributive Property
4. Simplify Algebraic Expressions by Combining Like Terms
5. Use Summation Notation

The word "algebra" is derived from the Arabic word *al-jabr*, which means "restoration." **Algebra** uses symbols to represent quantities and to express general relationships.

① Evaluate Algebraic Expressions

In arithmetic, we work with numbers. In algebra, letters such as $x, y, a, b,$ and c are used to represent numbers.

Definition

When a letter represents any number from a set of numbers, it is called a **variable.**

For most applications in statistics, we use the set of real numbers.

Definition

A **constant** is either a fixed number, such as 5, or a letter or symbol that represents a fixed number.

For example, in Einstein's Theory of Relativity, $E = mc^2$, the E and m are variables that represent total energy and mass, respectively, and c is a constant that represents the speed of light (299,792,458 meters per second).

Definition

An **algebraic expression** is any combination of variables, constants, grouping symbols, and mathematical operations such as addition, subtraction, multiplication, division, and exponents.

Some examples of algebraic expressions are

$$x - 5 \qquad \frac{1}{2}x \qquad 2y - 7 \qquad z^2 + 3 \qquad \text{and} \qquad \frac{b-1}{b+1}$$

Recall that a variable represents any number from a set of numbers. One of the procedures we perform on algebraic expressions is *evaluating an algebraic expression*.

Definition

To **evaluate an algebraic expression,** substitute a numerical value for each variable in the expression and simplify the result.

EXAMPLE 1 **Evaluating an Algebraic Expression**

Evaluate each expression for the given value of the variable.

(a) $2x + 5$ for $x = 8$ \qquad (b) $a^2 - 2a + 4$ for $a = -3$

Solution

(a) Substitute 8 for x in the expression $2x + 5$:

$$2(8) + 5 = 16 + 5$$
$$= 21$$

(b) Substitute -3 for a in $a^2 - 2a + 4$:

$$(-3)^2 - 2(-3) + 4 = 9 + 6 + 4$$
$$= 19$$

EXAMPLE 2 An Algebraic Expression for Revenue

The expression $4.50x + 2.50y$ represents the total amount of money, in dollars, received at a school play, where x represents the number of adult tickets sold and y represents the number of student tickets sold. Evaluate $4.50x + 2.50y$ for $x = 50$ and $y = 82$. Interpret the result.

Solution

Substitute 50 for x and 82 for y in the expression $4.50x + 2.50y$.

$$4.50(50) + 2.50(82) = 225 + 205$$
$$= 430$$

So $430 was collected by selling 50 adult tickets and 82 student tickets.

Quick ✓

1. When a letter represents any number from a set of numbers, it is called a _____.

2. To _____ an algebraic expression, substitute a numerical value for each variable into the expression and simplify the result.

In Problems 3 and 4, evaluate each expression for the given value of the variable.

3. $-3k + 5$ for $k = 4$
4. $-2y^2 - y + 8$ for $y = -2$
5. The Amadeus Coffee Shop creates a breakfast blend of two types of coffee. They mix x pounds of a mild coffee that sells for $7.00 per pound with y pounds of a robust coffee that sells for $10.00 per pound. An algebraic expression that represents the value of the breakfast blend, in dollars, is $7x + 10y$. Evaluate this expression for $x = 8$ pounds and $y = 16$ pounds

❷ Identify Like Terms and Unlike Terms

Algebraic expressions consist of *terms*.

Definition

A **term** is a constant or the product of a constant and one or more variables raised to a power.

In algebraic expressions, the terms are separated by addition signs.

EXAMPLE 3 Identifying the Terms in an Algebraic Expression

Identify the terms in the following algebraic expressions.

(a) $4a^3 + 5b^2 - 8c + 12$ (b) $\dfrac{x}{4} - 7y + 8z$

Solution

(a) Rewrite $4a^3 + 5b^2 - 8c + 12$ so it contains only addition signs.

$$4a^3 + 5b^2 + (-8c) + 12$$

The four terms are $4a^3, 5b^2, -8c$, and 12.

(b) The algebraic expression $\dfrac{x}{4} - 7y + 8z$ has three terms: $\dfrac{x}{4}, -7y$, and $8z$.

Quick ✓

6. *True or False* A constant by itself can be a term.

In Problems 7–9, identify the terms in each algebraic expression.

7. $5x^2 + 3xy$

8. $9ab - 3bc + 5ac - ac^2$

9. $\dfrac{2mn}{5} - \dfrac{3n}{7}$

Definition

The **coefficient** of a term is the numerical factor of the term.

For example, the coefficient of $7x$ is 7; the coefficient of $-2x^2y$ is -2. Terms that have no number as a factor, such as mn, have a coefficient of 1 since $mn = 1 \cdot mn$. The coefficient of $-y$ is -1 since $-y = -1 \cdot y$. If a term consists of just a constant, the coefficient is the number itself. For example, the coefficient of 14 is 14.

EXAMPLE 4 Determining the Coefficient of a Term

Determine the coefficient of each term:

(a) $\dfrac{1}{2}xy^2$ (b) $-\dfrac{t}{12}$ (c) ab^3 (d) 12

Solution

(a) The coefficient of $\dfrac{1}{2}xy^2$ is $\dfrac{1}{2}$.

(b) The coefficient of $-\dfrac{t}{12}$ is $-\dfrac{1}{12}$ because $-\dfrac{t}{12}$ can be written as $-\dfrac{1}{12} \cdot t$.

(c) The coefficient of ab^3 is 1 because ab^3 can be written as $1 \cdot ab^3$.

(d) The coefficient of 12 is 12 because the coefficient of a constant is the number itself.

Quick ✓

In Problems 10–14, determine the coefficient of each term.

10. $2z^2$ **11.** xy **12.** $-b$

13. 5   **14.** $-\dfrac{2}{3}z$

Sometimes we can simplify algebraic expressions by combining *like terms*.

Work Smart

Like terms can have different coefficients, but they cannot have different variables or different exponents on those variables.

Definition

Terms that have the same variable factor(s) with the same exponent(s) are called **like terms**.

For example, $3x^2$ and $-7x^2$ are like terms because both contain x^2, but $3x^2$ and $-7x^3$ are not like terms because the variable x is raised to different powers. Constant terms such as -9 and 6 are like terms.

EXAMPLE 5 Classifying Terms as Like or Unlike

Classify the following pairs of terms as *like* or *unlike*.

(a) $2p^3$ and $-5p^3$ (b) $7kr$ and $\frac{1}{4}k^2r$ (c) 5 and 8

Solution

(a) $2p^3$ and $-5p^3$ are *like* terms because both contain p^3.

(b) $7kr$ and $\frac{1}{4}k^2r$ are *unlike* terms because k is raised to different powers.

(c) 5 and 8 are *like* terms because both are constants.

Quick ✓

15. *True or False* Like terms can have different coefficients or different exponents on the same variable.

In Problems 16–20, tell whether the terms are like or unlike.

16. $-\frac{2}{3}p^2$ and $\frac{4}{5}p^2$ 17. $\frac{m}{6}$ and $4m$ 18. $3a^2b$ and $-2ab^2$

19. $8a$ and 11   20. -7 and 12

❸ Use the Distributive Property

The *Distributive Property* is used to simplify algebraic expressions that contain parentheses.

The Distributive Property

If a, b, and c are real numbers, then

$$a \cdot (b + c) = a \cdot b + a \cdot c$$
$$(a + b) \cdot c = a \cdot c + b \cdot c$$

That is, multiply each of the terms inside the parentheses by the factor on the outside.

Because $b - c = b + (-c)$, it is also true that $a(b - c) = a \cdot b - a \cdot c$.

EXAMPLE 6 Using the Distributive Property to Remove Parentheses

Use the Distributive Property to remove the parentheses.

(a) $3(x + 5)$ (b) $-\frac{1}{3}(6x - 12)$

Solution

(a) To use the Distributive Property, multiply each term in the parentheses by 3:

$$3(x + 5) = 3 \cdot x + 3 \cdot 5$$
$$= 3x + 15$$

Work Smart

The long name for the Distributive Property is the Distributive Property of Multiplication over Addition. This name helps to remind us that we do not distribute across multiplication. For example,

$6x(5xy) \neq 6x \cdot 5x \cdot 6x \cdot y$

(b) Multiply each term in the parentheses by $-\frac{1}{3}$:

$$-\frac{1}{3}(6x - 12) = -\frac{1}{3} \cdot 6x - \left(-\frac{1}{3}\right) \cdot 12$$
$$= -2x + 4$$

Quick ✓

21. $a \cdot (b + c) = a \cdot \underline{} + a \cdot \underline{}$

In Problems 22–25, use the Distributive Property to remove the parentheses.

22. $6(x + 2)$

23. $-5(x + 2)$

24. $-2(k - 7)$

25. $(8x + 12)\frac{3}{4}$

▶ ❹ Simplify Algebraic Expressions by Combining Like Terms

An algebraic expression that contains the sum or difference of like terms may be simplified using the Distributive Property "in reverse." When we use the Distributive Property to add coefficients of like terms, we say that we are **combining like terms.**

EXAMPLE 7 **Using the Distributive Property to Combine Like Terms**

Combine like terms:

(a) $2x + 7x$

(b) $x^2 - 5x^2$

Solution

(a) $2x + 7x = (2 + 7)x$
$= 9x$

(b) $x^2 - 5x^2 = (1 - 5)x^2$
$= -4x^2$

Look carefully at the results of Example 7. Notice that when we combine like terms, we add the coefficients of the like terms and keep the variables and exponents the same.

Quick ✓

26. $4x^2 + 9x^2 = (\underline{} + \underline{})x^2$

In Problems 27–29, combine like terms.

27. $3x - 8x$

28. $-5x^2 + x^2$

29. $-7x - x + 6 - 3$

Sometimes we must rearrange terms using the Commutative Property of Addition to combine like terms.

EXAMPLE 8 **Combining Like Terms Using the Commutative Property**

Combine like terms: $4x + 5y + 12x - 7y$

Solution

Use the Commutative Property to rearrange the terms.

$$4x + 5y + 12x - 7y = 4x + 12x + 5y - 7y$$

Use the Distributive Property "in reverse":
$= (4 + 12)x + (5 - 7)y$
$= 16x + (-2y)$

Write the answer in simplest form:
$= 16x - 2y$

Quick ✓

In Problems 30–33, combine like terms.

30. $3a + 2b - 5a + 7b - 4$

31. $5ac + 2b + 7ac$

32. $5ab^2 + 7a^2b + 3ab^2 - 8a^2b$

33. $\frac{4}{3}rs - \frac{3}{2}r^2 + \frac{2}{3}rs - 5$

Often, we need to remove parentheses by using the Distributive Property before we can combine like terms. Recall that the rules for order of operations on real numbers place multiplication before addition or subtraction. In this section, the direction **simplify** will mean to remove all parentheses and combine like terms.

EXAMPLE 9 Combining Like Terms Using the Distributive Property

Simplify the algebraic expression: $3 - 4(2x + 3) - (5x + 1)$

Solution

First, we use the Distributive Property to remove parentheses.

$$3 - 4(2x + 3) - (5x + 1) = 3 - 8x - 12 - 5x - 1$$

Rearrange terms using the Commutative Property of Addition: $\quad = -8x - 5x + 3 - 12 - 1$

Combine like terms: $\quad = -13x - 10$ ●

Work Smart

Remember that multiplication comes before subtraction. In the first step of Example 9, **do not** compute $3 - 4$ first to obtain $-1(2x + 3)$.

Quick ✓

34. Explain what it means to simplify an algebraic expression.

In Problems 35–38, simplify each expression.

35. $3x + 2(x - 1) - 7x + 1$

36. $m + 2n - 3(m + 2n) - (7 - 3n)$

37. $2(a - 4b) - (a + 4b) + b$

38. $\frac{1}{2}(6x + 4) - \frac{1}{3}(12 - 9x)$

> **Simplifying an Algebraic Expression**
>
> **Step 1:** Remove any parentheses using the Distributive Property.
> **Step 2:** Combine any like terms.

⑤ Use Summation Notation

In statistics, the symbol Σ (the Greek letter capital sigma) is used to represent repeated addition or to sum. For example, suppose we wanted to add n values, $a_1, a_2, \ldots, a_n$. Each of the values a_i represent a different quantity and the subscripts $1, 2, \ldots, n$ serve no mathematical operation. Subscripts are used to denote different quantities so that a_1 may be different from a_2, for example. The addition could be represented using **summation notation** as follows:

$$a_1 + a_2 + a_3 + \ldots + a_n = \sum_{i=1}^{n} a_i$$

In this notation, i is a positive integer starting with 1 and ending with n. So, $i = 1, 2, 3, \ldots, n$. Each a_i represents a different quantity. The expression $\sum_{i=1}^{n} a_i$ is read as "the sum of a_i from $i = 1$ to $i = n$."

EXAMPLE 10 Using Summation Notation

Evaluate each sum.

(a) $\sum_{i=1}^{5} 2i$ (b) $\sum_{i=1}^{4} \dfrac{(5i - 12.5)^2}{4}$

Solution

(a) The notation $\sum_{i=1}^{5} 2i$ means to evaluate $2i$ at $i = 1, 2, 3, 4,$ and 5. Then add up these values.

$$\sum_{i=1}^{5} 2i = 2(1) + 2(2) + 2(3) + 2(4) + 2(5)$$
$$= 2 + 4 + 6 + 8 + 10$$
$$= 30$$

(b) $\sum_{i=1}^{4} \dfrac{(5i - 12.5)^2}{4} = \dfrac{(5(1) - 12.5)^2}{4} + \dfrac{(5(2) - 12.5)^2}{4} + \dfrac{(5(3) - 12.5)^2}{4}$
$$+ \dfrac{(5(4) - 12.5)^2}{4}$$
$$= \dfrac{(5 - 12.5)^2}{4} + \dfrac{(10 - 12.5)^2}{4} + \dfrac{(15 - 12.5)^2}{4} + \dfrac{(20 - 12.5)^2}{4}$$
$$= \dfrac{56.25}{4} + \dfrac{6.25}{4} + \dfrac{6.25}{4} + \dfrac{56.25}{4}$$
$$= \dfrac{56.25 + 6.25 + 6.25 + 56.25}{4}$$
$$= \dfrac{125}{4}$$
$$= 31.25$$

Quick ✓

In Problems 39–42, evaluate each sum.

39. $\sum_{i=1}^{8} (2i)$ **40.** $\sum_{i=1}^{5} (4i + 3)$ **41.** $\sum_{i=1}^{6} [(2i - 5)(4i + 1)]$ **42.** $\sum_{i=1}^{4} \dfrac{(4i - 10)^2}{4}$

2.IR3 Exercises   MyLab Statistics

Underlined exercises have complete video solutions in MyLab.

Problems 1–42 are the Quick ✓ s that follow the EXAMPLES.

Building Skills

In Problems 43–54, evaluate each expression using the given values of the variables. See Objective 1.

43. $2x + 5$ for $x = 4$

44. $3x + 7$ for $x = 2$

45. $x^2 + 3x - 1$ for $x = 3$

46. $n^2 - 4n + 3$ for $n = 2$

47. $4 - k^2$ for $k = -5$

48. $-2p^2 + 5p + 1$ for $p = -3$

49. $\dfrac{9x - 5y}{x + y}$ for $x = 3, y = 5$

50. $\dfrac{3y + 2z}{y - z}$ for $y = 4, z = -2$

51. $(x + 3y)^2$ for $x = 3, y = -4$

52. $(a - 2b)^2$ for $a = 1, b = 2$

53. $b^2 - 4ac$ for $a = 1, b = 4, c = 3$

54. $\dfrac{a^2 - 4}{a^2 + 5a - 14}$ for $a = -3$

In Problems 55–58, for each expression, identify the terms and then name the coefficient of each term. See Objective 2.

55. $2x^3 + 3x^2 - x + 6$ **56.** $3m^4 - m^3n^2 + 4n - 1$

57. $z^2 + \dfrac{2y}{3}$ **58.** $t^3 - \dfrac{t}{4}$

In Problems 59–66, determine whether the terms are like or unlike. See Objective 2.

59. $8x$ and 8 **60.** $11p$ and 11

61. 54 and -21 **62.** -13 and 38

63. $12b$ and $-b$ **64.** $6a^2$ and $-3a^2$

65. r^2s and rs^2 **66.** x^2y^3 and y^2x^3

In Problems 67–74, use the Distributive Property to remove the parentheses. See Objective 3.

67. $3(m + 2)$ **68.** $3(4s + 2)$

69. $(3n^2 + 2n - 1)6$ **70.** $(6a^4 - 4a^2 + 2)3$

71. $-(x - y)$ **72.** $-5(k - n)$

73. $\left(-\dfrac{1}{2}\right)(8x - 6y)$ **74.** $\left(-\dfrac{2}{5}\right)(20a - 15b)$

In Problems 75–102, simplify each expression by using the Distributive Property to remove parentheses and combining like terms. See Objective 4.

75. $5x - 2x$ **76.** $14k - 11k$

77. $4z - 6z + 8z$ **78.** $9m - 8m + 2m$

79. $2m + 3n + 8m + 7n$ **80.** $x + 2y + 5x + 7y$

81. $0.3x^7 + x^7 + 0.9x^7$ **82.** $1.7n^4 - n^2 + 2.1n^4$

83. $-3y^6 + 13y^6$ **84.** $-7p^5 + 2p^5$

85. $-(6w + 12y - 13z)$ **86.** $-(-6m + 9n - 8p)$

87. $5(k + 3) - 8k$ **88.** $3(7 - z) - z$

89. $7n - (3n + 8)$ **90.** $18m - (6 + 9m)$

91. $(7 - 2x) - (x + 4)$ **92.** $(3k + 1) - (4 - k)$

93. $(7n - 8) - (3n - 6)$

94. $(5y - 6) - (11y + 8)$

95. $-6(n - 3) + 2(n + 1)$

96. $-9(7r - 6) + 9(10r + 3)$

97. $\dfrac{2}{3}x + \dfrac{1}{6}x$ **98.** $\dfrac{3}{5}y + \dfrac{7}{10}y$

99. $\dfrac{1}{2}(8x + 5) - \dfrac{2}{3}(6x + 12)$

100. $\dfrac{1}{5}(60 - 15x) + \dfrac{3}{4}(12 - 4x)$

101. $2(0.5x + 9) - 3(1.5x + 8)$

102. $3(0.2x + 6) - 5(1.6x + 1)$

In Problems 103–108, evaluate each sum.

103. $\displaystyle\sum_{i=1}^{7}(i + 3)$ **104.** $\displaystyle\sum_{i=1}^{8}(2i + 1)$

105. $\dfrac{\displaystyle\sum_{i=1}^{6}(4i - 3)}{6}$ **106.** $\dfrac{\displaystyle\sum_{i=1}^{8}(5i - 2)}{8}$

107. $\dfrac{\displaystyle\sum_{i=1}^{5}(4i - 12)^2}{4}$ **108.** $\dfrac{\displaystyle\sum_{i=1}^{6}(3i - 10.5)^2}{5}$

Mixed Practice

In Problems 109–120, (a) evaluate the expression for the given value(s) of the variable(s) before combining like terms, (b) simplify the expression by combining like terms and then evaluate the expression for the given value(s) of the variable(s). Compare your results.

109. $5x + 3x;\ x = 4$ **110.** $8y + 2y;\ y = -3$

111. $-2a^2 + 5a^2;\ a = -3$ **112.** $4b^2 - 7b^2;\ b = 5$

113. $4z - 3(z + 2);\ z = 6$

114. $8p - 3(p - 4);\ p = 3$

115. $5y^2 + 6y - 2y^2 + 5y - 3;\ y = -2$

116. $3x^2 + 8x - x^2 - 6x;\ x = 5$

117. $\dfrac{1}{2}(4x - 2) - \dfrac{2}{3}(3x + 9);\ x = 3$

118. $\dfrac{1}{5}(5x - 10) - \dfrac{1}{6}(6x + 12);\ x = -2$

119. $3a + 4b - 7a + 3(a - 2b);\ a = 2,\ b = 5$

120. $-4x - y + 2(x - 3y);\ x = 3,\ y = -2$

Applying the Concepts

In Problems 121–132, evaluate each expression using the given values of the variables.

121. $\dfrac{1}{2}h(b + B);\ h = 4,\ b = 5,\ B = 17$

122. $\dfrac{1}{2}h(b + B);\ h = 9,\ b = 3,\ B = 12$

123. $\dfrac{a - b}{c - d};\ a = 6,\ b = 3,\ c = -4,\ d = -2$

124. $\dfrac{a - b}{c - d};\ a = -5,\ b = -2,\ c = 7,\ d = 1$

125. $b^2 - 4ac;\ a = 7,\ b = 8,\ c = 1$

126. $b^2 - 4ac;\ a = 2,\ b = 5,\ c = 3$

127. $\dfrac{x - \mu}{\sigma};\ x = 120,\ \mu = 100,\ \sigma = 15$

128. $z\sigma + \mu;\ z = 2,\ \sigma = 10,\ \mu = 50$

129. $4p^3(1 - p);\ p = 0.4$

130. $10p^3(1-p)^2$; $p = 0.2$

131. $\sum_{i=1}^{3}(x_i - \mu)^2$; $x_1 = 4$, $x_2 = 8$, $x_3 = 9$, $\mu = 7$

132. $\dfrac{\sum_{i=1}^{4}(x_i - \mu)^2}{4}$; $x_1 = 3$, $x_2 = 4$, $x_3 = 7$, $x_4 = 2$, $\mu = 4$

133. **Renting a Truck** The cost of renting a truck from Hamilton Truck Rental is $59.95 per day plus $0.15 per mile. The expression $59.95 + 0.15m$ represents the cost of renting a truck for one day and driving it m miles. Evaluate $59.95 + 0.15m$ for $m = 125$ miles.

134. **Renting a Car** The cost of renting a compact car for one day from CMH Auto is $29.95 plus $0.17 per mile. The expression $29.95 + 0.17m$ represents the total daily cost. Evaluate the expression $29.95 + 0.17m$ for $m = 245$ miles.

135. **Ticket Sales** The Center for Science and Industry sells adult tickets for $12 and children's tickets for $7. The expression $12a + 7c$ represents the total revenue from selling a adult tickets and c children's tickets. Evaluate the algebraic expression $12a + 7c$ for $a = 156$ and $c = 421$.

136. **Ticket Sales** A community college theatre group sold tickets to a recent production. Student tickets cost $5, and nonstudent tickets cost $8. The algebraic expression $5s + 8n$ represents the total revenue from selling s student tickets and n nonstudent tickets. Evaluate $5s + 8n$ for $s = 76$ and $n = 63$.

△ 137. **Rectangle** The width of a rectangle is w yards, and the length of the rectangle is $(3w - 4)$ yards. The perimeter of the rectangle is given by the algebraic expression $2w + 2(3w - 4)$.
   (a) Simplify the algebraic expression $2w + 2(3w - 4)$.
   (b) Determine the perimeter of a rectangle whose width w is 5 yards.

△ 138. **Rectangle** The length of a rectangle is l meters, and the width of the rectangle is $(l - 11)$ meters. The perimeter of the rectangle is given by the algebraic expression $2l + 2(l - 11)$.
   (a) Simplify the expression $2l + 2(l - 11)$.
   (b) Determine the perimeter of a rectangle whose length l is 15 meters.

139. **Finance** Novella invested some money in two investment funds. She placed s dollars in stocks that yield 5.5% annual interest and b dollars in bonds that yield 3.25% annual interest. Evaluate the expression $0.055s + 0.0325b$ for $s = \$2950$ and $b = \$2050$. Round your answer to the nearest cent.

140. **Finance** Jonathan received an inheritance from his grandparents. He invested x dollars in a certificate of deposit that pays 2.95% and y dollars in an off-shore oil drilling venture that is expected to pay 12.8%. Evaluate the algebraic expression $0.0295x + 0.128y$ for $x = \$2500$ and $y = \$1000$.

Extending the Concepts

141. Simplify the algebraic expression (cleverly!!) using the Distributive Property—in reverse! $2.75(-3x^2 + 7x - 3) - 1.75(-3x^2 + 7x - 3)$

142. Simplify the algebraic expression using the Distributive Property in reverse. $11.23(7.695x + 81.34) + 8.77(7.695x + 81.34)$

Explaining the Concepts

143. Explain why the sum $2x^2 + 4x^2$ is *not* equivalent to $6x^4$. What is the correct answer?

144. Use $x = 4$ and $y = 5$ to answer parts (a), (b), and (c).
   (a) Evaluate $x^2 + y^2$.
   (b) Evaluate $(x + y)^2$.
   (c) Are the results the same? Is $(x + y)^2$ equal to $x^2 + y^2$? Explain your response.

2.IR4 Solving Linear Equations

Objectives

① Determine Whether a Number Is a Solution to an Equation

② Solve Linear Equations

③ Determine Whether an Equation Is a Conditional Equation, an Identity, or a Contradiction

① Determine Whether a Number Is a Solution to an Equation

An **equation in one variable** is a statement made up of two expressions that are equal, where at least one of the expressions contains the variable. The expressions are called the **sides** of the equation. Examples of equations in one variable are

$$2y + 5 = 0 \qquad 4x + 5 = -2x + 10 \qquad \frac{3}{z + 2} = 9$$

In this section, we will concentrate on solving *linear equations in one variable*.

> **In Words**
> In the equation $2y + 5 = 0$, the expression $2y + 5$ is the *left* side of the equation and 0 is the *right* side. In this equation, only the left side contains the variable, y. In the equation $4x + 5 = -2x + 10$, the expression $4x + 5$ is the *left side* of the equation and $-2x + 10$ is the right side. In this equation, both sides have expressions that contain the variable, x.

Definition

A **linear equation in one variable** is an equation with one unknown that is written to the first power. Linear equations in one variable can be written in the form

$$ax + b = 0$$

where a and b are real numbers and $a \neq 0$.

The following are all examples of linear equations in one variable because they can be written in the form $ax + b = 0$ with a little algebraic manipulation.

$$4x - 3 = 12 \qquad \frac{2}{3}y + \frac{1}{5} = \frac{2}{15} \qquad -0.73p + 1.23 = 1.34p + 8.05$$

Because an equation is a statement, it can be either true or false, depending on the value of the variable. Any value of the variable that results in a true statement is called a **solution** of the equation. If a value of the variable results in a true statement, we say that the value **satisfies** the equation. To determine whether a number satisfies an equation, replace the variable with the number. If the left side of the equation equals the right side, then the statement is true and the number is a solution.

EXAMPLE 1 **Determining Whether a Number Is a Solution to a Linear Equation**

Determine whether the following numbers are solutions to the equation

$$3(x - 1) = -2x + 12$$

(a) $x = 5$ (b) $x = 3$

Solution

(a) Let $x = 5$ in the equation and simplify.

$$3(x - 1) = -2x + 12$$
$$3(5 - 1) \stackrel{?}{=} -2(5) + 12$$
Simplify: $\quad 3(4) \stackrel{?}{=} -10 + 12$
$$12 \neq 2$$

> **In Words**
> The symbol $\stackrel{?}{=}$ is used to indicate that we are unsure whether the left side of the equation equals the right side of the equation.

Because the left side of the equation does not equal the right side of the equation, we do not have a true statement. Therefore, $x = 5$ is not a solution.

(b)
$$3(x - 1) = -2x + 12$$
Let $x = 3$: $\quad 3(3 - 1) \stackrel{?}{=} -2(3) + 12$
Simplify: $\quad 3(2) \stackrel{?}{=} -6 + 12$
$$6 = 6 \quad \text{True}$$

Because the left and right sides of the equation are equal, we have a true statement. Therefore, $x = 3$ is a solution to the equation.

Quick ✓

1. The equation $3x + 5 = 2x - 3$ is a _____ equation in one variable. The expressions $3x + 5$ and $2x - 3$ are called _____ of the equation.

2. The _____ of a linear equation is the value or values of the variable that satisfy the equation.

In Problems 3–5, determine which of the given numbers are solutions to the equation.

3. $-5x + 3 = -2; x = -2, x = 1, x = 3$
4. $3x + 2 = 2x - 5; x = 0, x = 6, x = -7$
5. $-3(z + 2) = 4z + 1; z = -3, z = -1, z = 2$

Work Smart

The directions "solve," "simplify," and "evaluate" are different! We *solve* equations. We *simplify* algebraic expressions to form equivalent algebraic expressions. We *evaluate* algebraic expressions to find the value of the expression for a specific value of the variable.

❷ Solve Linear Equations

To **solve an equation** means to find ALL the solutions of the equation. The set of all solutions to the equation is called the **solution set** of the equation.

One method for solving equations uses a series of *equivalent equations* beginning with the original equation and ending with a solution.

Definition

Two or more equations with the same solutions are **equivalent equations.**

But how do we find equivalent equations? One way is to use the *Addition Property of Equality*.

Addition Property of Equality

The **Addition Property of Equality** states that for real numbers a, b, and c,

$$\text{if } a = b, \text{ then } a + c = b + c$$

In Words

The Addition Property says that whatever you add to one side of an equation, you must also add to the other side.

For example, if $x = 3$, then $x + 2 = 3 + 2$ (we added 2 to both sides of the equation). Because $a - b$ is equivalent to $a + (-b)$, the Addition Property can be used to add a real number to each side of an equation or to subtract a real number from each side of an equation.

We can also find an equivalent equation by using the *Multiplication Property of Equality*.

Multiplication Property of Equality

The **Multiplication Property of Equality** states that for real numbers a, b, and c where $c \neq 0$,

$$\text{if } a = b, \text{ then } ac = bc$$

In Words

The Multiplication Property says that whenever you multiply one side of an equation by a nonzero expression, you must also multiply the other side by the same nonzero expression.

For example, if $5x = 30$, then $\frac{1}{5} \cdot 5x = \frac{1}{5} \cdot 30$. Also, because $\frac{a}{b} = a \cdot \frac{1}{b}$, the Multiplicative Property of Equality says that we may divide both sides of an equation by the same nonzero number and the result will be an equivalent equation. Thus if $5x = 30$, then it is also true that $\frac{5x}{5} = \frac{30}{5}$. So the Multiplication Property can be used to multiply or divide each side of the equation by some nonzero number.

EXAMPLE 2 Using the Addition and Multiplication Properties to Solve a Linear Equation

Solve the linear equation: $\frac{1}{3}x - 2 = 4$

Solution

The goal in solving any linear equation is to get the variable by itself with a coefficient of 1—that is, to **isolate the variable.**

Work Smart

The number in front of the variable expression is the coefficient. For example, the coefficient in the expression $2x$ is 2.

$$\frac{1}{3}x - 2 = 4$$

Addition Property of Equality; add 2 to both sides: $\left(\frac{1}{3}x - 2\right) + 2 = 4 + 2$

Simplify: $\frac{1}{3}x = 6$

Multiplication Property of Equality; multiply both sides by 3: $3\left(\frac{1}{3}x\right) = 3 \cdot 6$

Simplify: $x = 18$

Check

Let $x = 18$ in the original equation: $\frac{1}{3}x - 2 = 4$

$\frac{1}{3}(18) - 2 \stackrel{?}{=} 4$

$6 - 2 \stackrel{?}{=} 4$

$4 = 4$ True

The solution set is $\{18\}$.

Quick ✓

6. *True or False* The Addition Property of Equality says that whatever you add to one side of an equation you must also add to the other side.
7. To _____ the variable means to get the variable by itself with a coefficient of 1.

In Problems 8–10, solve each equation and verify your solution.

8. $3x + 8 = 17$
9. $-4a - 7 = 1$
10. $5y + 1 = 2$

Often, we must combine like terms or use the Distributive Property to eliminate parentheses before we can use the Addition or Multiplication Properties. As usual, our goal is to isolate the variable.

EXAMPLE 3 **Solving a Linear Equation by Combining Like Terms**

Solve the linear equation: $3y - 2 + 5y = 2y + 5 + 4y + 3$

Solution

$$3y - 2 + 5y = 2y + 5 + 4y + 3$$

Combine like terms: $\quad 8y - 2 = 6y + 8$

Subtract 6y from both sides: $\quad 8y - 2 - 6y = 6y + 8 - 6y$

$$2y - 2 = 8$$

Add 2 to both sides: $\quad 2y - 2 + 2 = 8 + 2$

$$2y = 10$$

Divide both sides by 2: $\quad \dfrac{2y}{2} = \dfrac{10}{2}$

$$y = 5$$

Check $\quad 3y - 2 + 5y = 2y + 5 + 4y + 3$

Let $y = 5$ in the original equation: $\quad 3(5) - 2 + 5(5) \stackrel{?}{=} 2(5) + 5 + 4(5) + 3$

$15 - 2 + 25 \stackrel{?}{=} 10 + 5 + 20 + 3$

$38 = 38$ True

The solution set is $\{5\}$.

Quick ✓

In Problems 11–13, solve each linear equation and verify your solution.

11. $2x + 3 + 5x + 1 = 4x + 10$
12. $4b + 3 - b - 8 - 5b = 2b - 1 - b - 1$
13. $2w + 8 - 7w + 1 = 3w - 1 + 2w - 5$

EXAMPLE 4   Solving a Linear Equation Using the Distributive Property

Solve the linear equation: $4(x + 3) = x - 3(x - 2)$

Solution

Use the Distributive Property to remove parentheses:	$4(x + 3) = x - 3(x - 2)$
	$4x + 12 = x - 3x + 6$
Combine like terms:	$4x + 12 = -2x + 6$
Add $2x$ to both sides:	$4x + 12 + 2x = -2x + 6 + 2x$
	$6x + 12 = 6$
Subtract 12 from both sides:	$6x + 12 - 12 = 6 - 12$
	$6x = -6$
Divide both sides by 6:	$\dfrac{6x}{6} = \dfrac{-6}{6}$
	$x = -1$

Check
$$4(x + 3) = x - 3(x - 2)$$
Let $x = -1$ in the original equation:
$$4(-1 + 3) \stackrel{?}{=} -1 - 3(-1 - 2)$$
$$4(2) \stackrel{?}{=} -1 - 3(-3)$$
$$8 \stackrel{?}{=} -1 + 9$$
$$8 = 8 \quad \text{True}$$

Because $x = -1$ satisfies the equation, the solution of the equation is -1, or the solution set is $\{-1\}$.

Quick ✓

In Problems 14–17, solve each linear equation. Be sure to verify your solution.

14. $4(x - 1) = 12$
15. $-2(x - 4) - 6 = 3(x + 6) + 4$
16. $4(x + 3) - 8x = 3(x + 2) + x$
17. $5(x - 3) + 3(x + 3) = 2x - 3$

We now summarize the steps for solving a linear equation. Not every step is always necessary.

Solving a Linear Equation

Step 1: Remove any parentheses using the Distributive Property.
Step 2: Combine like terms on each side of the equation.
Step 3: Use the Addition Property of Equality to get the terms with the variable on one side of the equation and the constants on the other side.
Step 4: Use the Multiplication Property of Equality to get the coefficient of the variable to equal 1.
Step 5: Check the solution to verify that it satisfies the original equation.

Linear Equations with Fractions or Decimals

To rewrite a linear equation that contains fractions as an equivalent equation without fractions, multiply both sides of the equation by the Least Common Denominator (LCD) of all the fractions in the equation.

EXAMPLE 5 How to Solve a Linear Equation That Contains Fractions

Solve the linear equation: $\dfrac{y+1}{4} + \dfrac{y-2}{10} = \dfrac{y+7}{20}$

Step-by-Step Solution

Before we follow the summary steps, rewrite the equation without fractions by multiplying both sides by the LCD, which is 20.

$$20 \cdot \left(\dfrac{y+1}{4} + \dfrac{y-2}{10}\right) = 20 \cdot \left(\dfrac{y+7}{20}\right)$$

Now follow Steps 1–5 for solving a linear equation.

Step 1: Remove all parentheses using the Distributive Property.

$$20 \cdot \left(\dfrac{y+1}{4} + \dfrac{y-2}{10}\right) = 20 \cdot \left(\dfrac{y+7}{20}\right)$$

Use the Distributive Property: $20 \cdot \dfrac{y+1}{4} + 20 \cdot \dfrac{y-2}{10} = 20 \cdot \dfrac{y+7}{20}$

Divide out common factors: $5(y+1) + 2(y-2) = y+7$

Use the Distributive Property: $5y + 5 + 2y - 4 = y + 7$

Step 2: Combine like terms on each side of the equation.

$$7y + 1 = y + 7$$

Step 3: Use the Addition Property of Equality to get the terms with the variable on one side of the equation and the constants on the other side.

Subtract y from both sides: $7y + 1 - y = y + 7 - y$

$$6y + 1 = 7$$

Subtract 1 from both sides: $6y + 1 - 1 = 7 - 1$

$$6y = 6$$

Step 4: Use the Multiplication Property of Equality to get the coefficient on the variable to equal 1.

Divide both sides by 6: $\dfrac{6y}{6} = \dfrac{6}{6}$

$$y = 1$$

Step 5: Check: Verify the solution.

$$\dfrac{y+1}{4} + \dfrac{y-2}{10} = \dfrac{y+7}{20}$$

Let $y = 1$ in the original equation: $\dfrac{1+1}{4} + \dfrac{1-2}{10} \overset{?}{=} \dfrac{1+7}{20}$

$$\dfrac{2}{4} + \dfrac{-1}{10} \overset{?}{=} \dfrac{8}{20}$$

Rewrite each rational number with LCD = 20: $\dfrac{2}{4} \cdot \dfrac{5}{5} + \dfrac{-1}{10} \cdot \dfrac{2}{2} \overset{?}{=} \dfrac{8}{20}$

$$\dfrac{10}{20} + \dfrac{-2}{20} \overset{?}{=} \dfrac{8}{20}$$

$$\dfrac{8}{20} = \dfrac{8}{20} \quad \text{True}$$

The solution of the equation is 1, or the solution set is $\{1\}$.

Quick ✓

18. To solve a linear equation containing fractions, we can multiply each side of the equation by the _____ _____ _____ to clear the fractions.

In Problems 19–22, solve each linear equation and verify your solution.

19. $\dfrac{3y}{2} + \dfrac{y}{6} = \dfrac{10}{3}$

20. $\dfrac{3x}{4} - \dfrac{5}{12} = \dfrac{5x}{6}$

21. $\dfrac{x+2}{6} + 2 = \dfrac{5}{3}$

22. $\dfrac{4x+3}{9} - \dfrac{2x+1}{2} = \dfrac{1}{6}$

To rewrite a linear equation that contains decimals as an equivalent equation without decimals, multiply both sides of the equation by a power of 10 that removes the decimals. For example, multiplying 0.7 $\left(\text{or } \dfrac{7}{10}\right)$ by 10 "eliminates" the decimal, and multiplying 0.03 $\left(\text{or } \dfrac{3}{100}\right)$ by 100 "eliminates" the decimal.

EXAMPLE 6 **Solving a Linear Equation That Contains Decimals**

Solve the linear equation: $0.5x - 0.4 = 0.3x + 0.2$

Solution

Multiplying both sides of the equation by 10 will "eliminate" the decimal because each of the decimals is written to the tenths position.

$$10(0.5x - 0.4) = 10(0.3x + 0.2)$$

Use the Distributive Property: $10(0.5x) - 10(0.4) = 10(0.3x) + 10(0.2)$

$$5x - 4 = 3x + 2$$

Subtract 3x from both sides: $5x - 4 - 3x = 3x + 2 - 3x$

$$2x - 4 = 2$$

Add 4 to both sides: $2x - 4 + 4 = 2 + 4$

$$2x = 6$$

Divide both sides by 2: $\dfrac{2x}{2} = \dfrac{6}{2}$

$$x = 3$$

Check $0.5x - 0.4 = 0.3x + 0.2$

Let x = 3 in the original equation: $0.5(3) - 0.4 \stackrel{?}{=} 0.3(3) + 0.2$

$$1.5 - 0.4 \stackrel{?}{=} 0.9 + 0.2$$

$$1.1 = 1.1 \quad \text{True}$$

Because $x = 3$ satisfies the equation, the solution of the equation is 3, or the solution set is $\{3\}$.

Quick ✓

In Problems 23 and 24, solve each linear equation. Be sure to verify your solution.

23. $0.07x - 1.3 = 0.05x - 1.1$

24. $0.4(y + 3) = 0.5(y - 4)$

▶ ❸ Determine Whether an Equation Is a Conditional Equation, an Identity, or a Contradiction

Each linear equation we have studied so far has had one solution. Linear equations may have one solution, no solution, or infinitely many solutions. We can classify equations based on the number of their solutions.

The equations that we have solved so far are called *conditional equations*.

Definition
A **conditional equation** is an equation that is true for some values of the variable and false for other values of the variable.

For example, the equation $x + 7 = 10$ is a conditional equation because it is true when $x = 3$ and false for every other real number x.

Definition
An equation that is false for every value of the variable is a **contradiction.**

For example, the equation $3x + 8 = 3x + 6$ is a contradiction because it is false for any value of x. We recognize contradictions by creating equivalent equations. For example, subtracting $3x$ from both sides of $3x + 8 = 3x + 6$, gives $8 = 6$, which is clearly false. Contradictions have no solution, so the solution set is the empty set and is written as $\varnothing$ or $\{\ \}$.

Definition
An equation that is satisfied for every value of the variable for which both sides of the equation are defined is called an **identity.**

In Words
Conditional equations are true for some values of the variable and false for others. Contradictions are false for all values of the variable. Identities are true for all allowed values of the variable.

For example, $2x + 3 + x + 8 = 3x + 11$ is an identity because any real number x satisfies the equation. We recognize identities by creating equivalent equations. For example, if we combine like terms in the equation $2x + 3 + x + 8 = 3x + 11$, we get $3x + 11 = 3x + 11$, which is true no matter what value of x we choose. Therefore, the solution set of linear equations that are identities is the set of all real numbers, which is expressed as either $\{x \mid x \text{ is any real number}\}$ or $\mathbb{R}$.

EXAMPLE 7 **Classifying a Linear Equation**

Solve the linear equation $3(x + 3) - 6x = 5(x + 1) - 8x$. Then state whether it is an identity, a contradiction, or a conditional equation.

Solution

As with any linear equation, the goal is to isolate the variable.

$$3(x + 3) - 6x = 5(x + 1) - 8x$$

Use the Distributive Property: $\quad 3x + 9 - 6x = 5x + 5 - 8x$

Combine like terms: $\quad -3x + 9 = -3x + 5$

Add $3x$ to both sides: $\quad -3x + 9 + 3x = -3x + 5 + 3x$

$$9 = 5$$

Work Smart
In the solution to Example 7, we obtained the equation
$$-3x + 9 = -3x + 5$$
You may recognize at this point that the equation is a contradiction and express the solution set as $\varnothing$ or $\{\ \}$.

Because the last statement, $9 = 5$, is false, the equation is a contradiction. The original equation is a contradiction and has no solution. The solution set is $\varnothing$ or $\{\ \}$. ●

EXAMPLE 8 **Classifying a Linear Equation**

Solve the linear equation $-4(x - 2) + 3(4x + 2) = 2(4x + 7)$. Then state whether the equation is an identity, a contradiction, or a conditional equation.

Solution

$$-4(x - 2) + 3(4x + 2) = 2(4x + 7)$$

Use the Distributive Property: $\quad -4x + 8 + 12x + 6 = 8x + 14$

Combine like terms: $\quad 8x + 14 = 8x + 14$

The equation $8x + 14 = 8x + 14$ is true for all real numbers x. So the original equation is an identity, and its solution set is $\{x \mid x \text{ is any real number}\}$ or $\mathbb{R}$.

If we continued to solve the equation in Example 8, we could write

$$8x + 14 - 8x = 8x + 14 - 8x$$
$$14 = 14$$

The statement $14 = 14$ is true for all real numbers x, so the solution set of the original equation is all real numbers.

Quick ✓

25. A(n) _____ _____ is an equation that is true for some values of the variable and false for others.

26. A(n) _____ is an equation that is false for every value of the variable. A(n) _____ is an equation that is satisfied by every allowed choice of the variable.

In Problems 27–30, solve the equation and state whether it is an identity, a contradiction, or a conditional equation.

27. $4(x + 2) = 4x + 2$

28. $3(x - 2) = 2x - 6 + x$

29. $-4x + 2 + x + 1 = -4(x + 2) + 11$

30. $-3(z + 1) + 2(z - 3) = z + 6 - 2z - 15$

2.IR4 Exercises   MyLab Statistics
Underlined exercises have complete video solutions in MyLab.

Problems 1–30 are the Quick ✓ s that follow the EXAMPLES.

Building Skills

In Problems 31–36, determine which of the numbers are solutions to the given equation. See Objective 1.

31. $8x - 10 = 6$; $x = -2, x = 1, x = 2$

32. $-4x - 3 = -15$; $x = -2, x = 1, x = 3$

33. $5m - 3 = -3m + 5$; $m = -2, m = 1, m = 3$

34. $6x + 1 = -2x + 9$; $x = -2, x = 1, x = 4$

35. $4(x - 1) = 3x + 1$; $x = -1, x = 2, x = 5$

36. $3(t + 1) - t = 4t + 9$; $t = -3, t = -1, t = 2$

In Problems 37–56, solve each linear equation, and be sure to verify your solution. See Objective 2.

37. $3x + 1 = 7$

38. $8x - 6 = 18$

39. $4z + 3 = 2$

40. $-6x - 5 = 13$

41. $-3w + 2w + 5 = -4$

42. $-7t - 3 + 5t = 11$

43. $5x + 2 - 2x + 3 = 7x + 2 - x + 5$

44. $-6x + 2 + 2x + 9 + x = 5x + 10 - 6x + 11$

45. $3(x + 2) = -6$

46. $4(z - 2) = 12$

47. $\dfrac{4y}{5} - \dfrac{14}{15} = \dfrac{y}{3}$

48. $\dfrac{3x}{2} + \dfrac{x}{6} = -\dfrac{5}{3}$

49. $\dfrac{4x + 3}{9} - \dfrac{2x + 1}{2} = \dfrac{1}{6}$

50. $\dfrac{2x + 1}{3} - \dfrac{6x - 1}{4} = -\dfrac{5}{12}$

51. $\dfrac{y}{10} + 6 = \dfrac{y}{4} + 12$

52. $\dfrac{n}{5} + 3 = \dfrac{n}{2} + 6$

53. $0.5x - 3.2 = -1.7$

54. $0.3z + 0.8 = -0.1$

55. $0.14x + 2.23 = 0.09x + 1.98$

56. $0.12y - 5.26 = 0.05y + 1.25$

In Problems 57–72, solve the equation. Identify each equation as an identity, a contradiction, or a conditional equation. Be sure to verify your solution. See Objective 3.

57. $4(x + 1) = 4x$

58. $5(s + 3) = 3s + 2s$

59. $4m + 1 - 6m = 2(m + 3) - 4m$

60. $10(x - 1) - 4x = 2x - 1 + 4(x + 1)$

61. $2(y + 1) - 3(y - 2) = 5y + 8 - 6y$

62. $8(w + 2) - 3w = 7(w + 2) + 2(1 - w)$

63. $\dfrac{x}{4} + \dfrac{3x}{10} = -\dfrac{33}{20}$

64. $\dfrac{z - 2}{4} + \dfrac{2z - 3}{6} = 7$

65. $3p - \dfrac{p}{4} = \dfrac{11p}{4} + 1$

66. $\dfrac{r}{2} + 2(r - 1) = \dfrac{5r}{2} + 4$

67. $\dfrac{2x + 1}{2} - \dfrac{x + 1}{5} = \dfrac{23}{10}$

68. $\dfrac{3x + 1}{4} - \dfrac{7x - 4}{2} = \dfrac{26}{3}$

69. $0.4(z + 1) - 0.7z = -0.1z + 0.7 - 0.2z - 0.3$

70. $0.9(z - 3) - 0.2(z - 5) = 0.4(z + 1) + 0.3z - 2.1$

71. $\dfrac{1}{3}(2x - 3) + 2 = \dfrac{5}{6}(x + 3) - \dfrac{11}{12}$

72. $\dfrac{4}{5}(y - 4) + 3 = \dfrac{2}{3}(y + 1) + \dfrac{4}{15}$

Mixed Practice

In Problems 73–90, solve each linear equation and verify the solution. State whether the equation is an identity, a contradiction, or a conditional equation.

73. $7y - 8 = -7$

74. $4y + 5 = 7$

75. $4a + 3 - 2a + 4 = 5a - 7 + a$

76. $-5x + 5 + 3x + 7 = 5x - 6 + x + 12$

77. $4(p + 3) = 3(p - 2) + p + 18$

78. $7(x + 2) = 5(x - 2) + 2(x + 12)$

79. $4b - 3(b + 1) - b = 5(b - 1) - 5b$

80. $13z - 8(z + 1) = 2(z - 3) + 3z$

81. $8(4x + 6) = 11 - (x + 7)$

82. $5(2x + 3) = 9 - (x + 5)$

83. $\dfrac{m + 1}{4} + \dfrac{5}{6} = \dfrac{2m - 1}{12}$

84. $\dfrac{z - 4}{6} - \dfrac{2z + 1}{9} = \dfrac{1}{3}$

85. $0.3x - 1.3 = 0.5x - 0.7$

86. $-0.8y + 0.3 = 0.2y - 3.7$

87. $-0.8(x + 1) = 0.2(x + 4)$

88. $0.5(x + 3) = 0.2(x - 6)$

89. $\dfrac{1}{4}(x - 4) + 3 = \dfrac{1}{3}(2x + 6) - \dfrac{5}{6}$

90. $\dfrac{1}{5}(2a - 5) - 4 = \dfrac{1}{2}(a + 4) - \dfrac{7}{10}$

Applying the Concepts

91. Find a such that the solution set of $ax + 3 = 15$ is $\{-3\}$.

92. Find a such that the solution set of $ax + 6 = 20$ is $\{7\}$.

93. Find a such that the solution set of $a(x - 1) = 3(x - 1)$ is the set of all real numbers.

94. Find a such that the solution set of $ax + 3 = 2x + 3(x + 1)$ is the set of all real numbers.

95. **Interest** Suppose you have a credit card debt of $2000. Last month, the bank charged you $25 interest on the debt. The solution to the equation $25 = 2000 \cdot \dfrac{r}{12}$ represents the annual interest rate r on the credit card. Find the annual interest rate on the credit card.

96. **How Much Do I Make?** Last week, before taxes, you earned $539 after working 26 hours at your regular hourly rate and 6 hours at time-and-a-half. The solution to the equation $26x + 9x = 539$ represents your regular hourly rate, x. Determine your regular hourly rate.

97. **Paying Your Taxes** You are single and have just determined that you paid $4412.50 in federal income tax in 2015. The solution to the equation $4412.5 = 0.15(I - 9225) + 922.50$ represents your annual adjusted income I for 2015. Determine your annual adjusted income for 2015. (*Source: Internal Revenue Service*)

98. **Paying Your Taxes** You are married and have just determined that you paid $11,037.50 in federal income tax in 2015. The solution to the equation $11,037.5 = 0.25(I - 74,900) + 10,312.50$ represents your annual adjusted income I for 2015. Determine your annual adjusted income for 2015. (*Source: Internal Revenue Service*)

Explaining the Concepts

99. Explain the difference between $4(x + 1) - 2$ and $4(x + 1) = 2$. In general, what is the difference between an algebraic expression and an equation?

100. Explain the difference between the directions "solve" and "simplify."

101. Make up a linear equation that has one solution. Make up a linear equation that has no solution. Make up a linear equation that is an identity. Comment on the differences and similarities in making up these equations.

2.IR5 Using Linear Equations to Solve Problems

Objectives

1. Translate English Sentences into Mathematical Statements
2. Model and Solve Direct Translation Problems
3. Solve for a Variable in a Formula

Work Smart

We learned in the last section that an equation is a statement in which two algebraic expressions, separated by an equal sign, are equal.

1 Translate English Sentences into Mathematical Statements

In English a complete sentence must contain a subject and a verb, so expressions or "phrases" are not complete sentences. For example, the expression "5 more than a number x" does not contain a verb and therefore is not a complete sentence. The statement "5 more than a number x is 18" is a complete sentence because it contains a subject and a verb, so we can translate it into a mathematical statement. Mathematical statements are represented symbolically through equations. In English, statements can be true or false. Mathematical statements can be true or false as well.

You may want to look back at Section 1.IR8 for a review of key words that translate into mathematical symbols. Table 1 provides a summary of words that typically translate into an equal sign.

Table 1 Words That Translate into an Equal Sign

is	yields	are	equals	is equivalent to
was	gives	results in	is equal to	

Let's translate some English sentences into mathematical statements.

EXAMPLE 1 Translating English Sentences into Mathematical Statements

Translate each sentence into a mathematical statement. Do not solve the equation.

(a) Five more than a number x is 20.
(b) Four times the sum of a number z and 3 is 15.
(c) The difference of x and 5 equals the quotient of x and 2.

Solution

(a) 5 more than a number x is 20
$$x + 5 = 20$$

(b) The expression "Four times the sum of" tells us to find the sum first and then multiply this result by 4.

Four times the sum of a number z and 3 is 15
$$4(z + 3) = 15$$

Work Smart

The English statement "The sum of four times a number z and 3 is 15" would be expressed mathematically as $4z + 3 = 15$. Do you see the difference between this statement and the one in Example 1(b)?

(c) The difference of x and 5 equals the quotient of x and 2
$$x - 5 = \frac{x}{2}$$

Quick ✓

1. Mathematical statements are represented symbolically through _____.

In Problems 2–6, translate each English statement into a mathematical statement. Do not solve the equation.

2. The sum of x and 7 results in 12.
3. The product of 3 and y is equal to 21.
4. Two times the sum of n and 3 is 5.
5. The difference of x and 10 equals the quotient of x and 2.
6. The sum of two times n and 3 is 5.

An Introduction to Problem Solving and Mathematical Models

Every day we encounter various types of problems that must be solved. **Problem solving** is the ability to use information, tools, and our own skills to achieve a goal. For example, suppose Kevin wants a glass of water, but he is too short to reach the sink. Kevin has a problem. To solve the problem, he finds a step stool and pulls it over to the sink. He uses the step stool to climb on the counter, opens the kitchen cabinet, and pulls out a cup. He then crawls along the counter top, turns on the faucet, and proceeds to fill his cup with water. Problem solved!

Of course, Kevin could solve the problem in other ways. Just as there are various ways to solve life's everyday problems, there are many ways to solve problems using mathematics. However, regardless of the approach, there are always some common aspects in solving any problem. For example, regardless of how Kevin ultimately ends up with his cup of water, someone must get a cup from the cabinet and someone must turn on the faucet.

One of the purposes of learning algebra (and statistics) is to be able to solve certain types of problems. To solve these problems, translate the verbal description of the problem into equations that can be solved. The process of turning a verbal description into a mathematical equation is **mathematical modeling.** The equation that is developed is called the **mathematical model.**

Not all models are mathematical. In general, a **model** is a way of using graphs, pictures, equations, or even verbal descriptions to represent a real-life situation. Because the world is a complex place, we need to simplify information when developing a model. For example, a map is a model of our road system. Maps don't show all the details of the system (such as trees, buildings, or potholes), but they do a good job of describing how to get from point A to point B. Mathematical models are similar in that we often make assumptions regarding our world in order to make the mathematics less complicated.

Every problem is unique in some way. However, the following guidelines will help you solve many problems.

Work Smart

When you solve a problem by making a mathematical model, check your work to make sure you have the right answer. Typically, errors can happen in two ways.

- One type of error occurs if you correctly translate the problem into a model but then make an error solving the equation. This type of error is usually easy to find.

- However, if you misinterpret the problem and develop an incorrect model, then the solution you obtain may still satisfy your model, but it probably will not be the correct solution to the original problem. We can check for this type of error by determining whether the solution is reasonable. Does your answer make sense? Always be sure that you are answering the question that is being asked.

> **Solving Problems with Mathematical Models**
>
> **Step 1: Identify What You Are Looking For** Read the problem very carefully, perhaps two or three times. Identify the type of problem and the information you wish to learn. It is fairly typical for the last sentence in the problem to indicate what you need to solve for.
>
> **Step 2: Give Names to the Unknowns** Assign variables to the unknown quantities. Choose a variable that is representative of the unknown quantity. For example, use t for time.
>
> **Step 3: Translate the Problem into Mathematics** Read the problem again. This time, after each sentence is read, determine whether the sentence can be translated into a mathematical statement or expression in terms of the variables identified in Step 2. It is often helpful to create a table, chart, or figure. Combine the mathematical statements or expressions into an equation that can be solved.
>
> **Step 4: Solve the Equation(s) Found in Step 3** Solve the equation for the variable.
>
> **Step 5: Check the Reasonableness of Your Answer** Check your answer to be sure that it makes sense. If it does not, go back and try again.
>
> **Step 6: Answer the Question** Write your answer in a complete sentence.

Model and Solve Direct Translation Problems

Let's solve some **direct translation** problems, which can be set up by reading the problem and translating the verbal description into an equation.

EXAMPLE 2 Finding Consecutive Integers

The sum of three consecutive odd integers results in 45. Find the integers.

Solution

Step 1: Identify This is a direct translation problem. We are looking for three odd integers. The odd integers are 1, 3, 5, and so on.

Step 2: Name Let x represent the first odd integer, then $x + 2$ is the next odd integer, and $x + 4$ is the third odd integer.

Step 3: Translate Since we know that their sum is 45, we have

$$\underbrace{x}_{\text{First Integer}} + \underbrace{x + 2}_{\text{Second Integer}} + \underbrace{x + 4}_{\text{Third Integer}} = 45 \quad \text{The Model}$$

Step 4: Solve Solve the equation.

$$x + x + 2 + x + 4 = 45$$

Combine like terms: $\quad 3x + 6 = 45$

Subtract 6 from both sides: $\quad 3x = 39$

Divide both sides by 3: $\quad x = 13$

Step 5: Check Because x represents the first odd integer, the remaining two odd integers are $13 + 2 = 15$ and $13 + 4 = 17$. It is always a good idea to make sure your answer is reasonable. Since $13 + 15 + 17 = 45$, we know we have the right answer!

Step 6: Answer the Question The three consecutive odd integers are 13, 15, and 17.

Quick ✓

7. *True or False* If n represents the first of two consecutive odd integers, $n + 1$ represents the next consecutive odd integer.

In Problems 8 and 9, translate each English statement into an equation and solve the equation.

8. The sum of three consecutive even integers is 60. Find the integers.

9. The sum of three consecutive integers is 78. Find the integers.

EXAMPLE 3 Finding Hourly Wage

Before taxes, Marissa earned $725 one week after working 52 hours. Her employer pays time-and-a-half for all hours worked in excess of 40 hours. What is Marissa's hourly wage?

Solution

Step 1: Identify In this direct translation problem, we want to find Marissa's hourly wage.

Step 2: Name Let w represent Marissa's hourly wage.

Step 3: Translate Marissa earned $725 by working 40 hours at her regular wage and 12 hours at 1.5 times her regular wage. If her regular wage is w dollars, then her overtime wage is $1.5w$ dollars. Therefore, her total salary is

$$\underbrace{40w}_{\text{Regular Earnings}} + \underbrace{12(1.5w)}_{\text{Overtime Earnings}} = \underbrace{725}_{\text{Total Earnings}} \quad \text{The Model}$$

Step 4: Solve

$$40w + 12(1.5w) = 725$$

Simplify: $\quad 40w + 18w = 725$

Combine like terms: $\quad 58w = 725$

Divide both sides by 58: $\quad \dfrac{58w}{58} = \dfrac{725}{58}$

$$w = 12.50$$

Step 5: Check If Marissa's hourly wage is $12.50, then for the first 40 hours she earned $12.50(40) = $500 and for the next 12 hours she earned 1.5($12.50)(12) = $225. Her total salary was $500 + $225 = $725. This checks with the information presented in the problem.

Step 6: Answer the Question Marissa's hourly wage is $12.50.

Quick ✓

10. Before taxes, Melody earned $735 one week after working 46 hours. Her employer pays time-and-a-half for all hours worked in excess of 40 hours. What is Melody's hourly wage?

11. Before taxes, Jim earned $564 one week after working 30 hours at his regular wage, 6 hours at time-and-a-half on Saturday, and 4 hours at double-time on Sunday. What is Jim's hourly wage?

EXAMPLE 4 **Choosing Take-Out Pizza**

Domino's Pizza charges $13.25 plus $1.95 per topping for their large 14-inch pizza. Extreme Pizza charges $14.45 plus $1.70 per topping for the same size pizza. For how many toppings will the two orders cost the same?
(*Source: Domino's and Extreme Pizza*)

Solution

Step 1: Identify This is a direct translation problem. Find the number of toppings for which the two orders cost the same.

Step 2: Name Let t represent the number of toppings on a large 14-inch pizza.

Step 3: Translate The cost of a 14-inch pizza at Domino's Pizza is $13.25 plus $1.95 for each topping. So if one topping is picked, the cost is $13.25 + $1.95(1) = $15.20. For 2 toppings, the cost is $13.25 + $1.95(2) = $17.15. In general, for t toppings, the cost is $13.25 + 1.95t$ dollars. Similar logic reveals that the cost of a 14-inch pizza at Extreme Pizza is $14.45 + 1.70t$ dollars. To find the number of toppings for which the cost for the two pizzas will be the same, solve:

Cost for Domino's pizza = Cost for Extreme Pizza pizza

$$13.25 + 1.95t = 14.45 + 1.70t \quad \text{The Model}$$

Step 4: Solve $\quad 13.25 + 1.95t = 14.45 + 1.70t$

Subtract 13.25 from both sides: $\quad 1.95t = 1.20 + 1.70t$

Subtract 1.70t from both sides: $\quad 0.25t = 1.20$

Divide both sides by 0.25: $\quad t = 4.8$

Step 5: Check For $t = 4.8$ toppings, the cost of a Domino's pizza will be $13.25 + $1.95(4.8) = $22.61, and the cost of an Extreme Pizza pizza will be $14.45 + $1.70(4.8) = $22.61. They are the same! Of course, because you cannot order a partial topping, round 4.8 topping to 5 toppings.

Step 6: Answer the Question The two orders will cost roughly the same when 5 toppings are ordered.

Quick ✓

12. You need to rent a moving truck. EZ-Rental charges $35 per day plus $0.15 per mile. Do It Yourself Rental charges $20 per day plus $0.25 per mile. For how many miles will the cost of renting be the same?

13. You are comparing cell phone plans. Verizon offers two plans you are considering. Plan A charges $79.99 per month for 900 minutes of talk and text, with a $0.40 per-minute rate after the allowance. Plan B charges $89.99 per month for unlimited minutes of talk and text. For how many minutes will the monthly cost be the same? (*Source: Verizon*)

There are many types of direct translation problems. One type of direct translation problem is a "percent problem." Typically, these problems involve discounts or mark-ups that businesses use in determining their prices.

Percent means *divided by* 100 or *per hundred*. The symbol % denotes percent, so 45% means 45 out of 100 or $\frac{45}{100}$ or 0.45. In applications involving percents, we often see the word "of," as in 20% of 60. Remember from Section 1.IR8 that the word "of" translates into multiplication in mathematics, so 20% of 60 means

$$20\% \cdot 60 \quad \text{or} \quad 0.20 \cdot 60$$

When dealing with percents and the price of goods, it is helpful to remember the following:

$$\text{Original Price} - \text{Discount} = \text{Sale Price}$$

$$\text{Wholesale Price} + \text{Markup} = \text{Selling Price}$$

▶ **EXAMPLE 5** **Finding Discounted Price**

Everything in your favorite clothing store is marked at a discount of 40% off. If the sale price of a suit is $144, what was its original price?

Solution

Step 1: Identify This is a direct translation problem involving percents. We wish to find the original price of the suit.

Step 2: Name Let p represent the original price.

Step 3: Translate The original price minus the discount is the sale price, $144, so

$$p - \text{discount} = 144$$

The discount was 40% off of the original price, so the discount is $0.40p$. Substitute this expression into the equation $p - \text{discount} = 144$:

$$p - 0.40p = 144 \quad \text{The Model}$$

Step 4: Solve
$$p - 0.40p = 144$$
Combine like terms: $\quad 0.60p = 144$
Divide both sides by 0.60: $\quad \dfrac{0.60p}{0.60} = \dfrac{144}{0.60}$
Simplify: $\quad p = 240$

Work Smart
You could eliminate the decimals by multiplying both sides of the equation by 10.

Step 5: Check If p, the original price of the suit, was $240, then the discount would be $0.4(\$240) = \$96$. Subtracting $96 from the original price gives $144, the sale price. This agrees with the information in the problem.

Step 6: Answer the Question The original price of the suit was $240.

Quick ✓

14. Percent means "divided by _____."
15. What is 40% of 100?
16. 8 is 5% of what number?
17. 15 is what percent of 20?
18. Suppose you have just entered your favorite clothing store and find that everything in the store is marked at a discount of 30% off. If the sale price of a shirt is $21, what was its original price?
19. A Milex Tune-Up automotive facility marks up its parts 35%. Suppose that Milex charges its customers $1.62 for each spark plug it installs. What is Milex's cost for each spark plug?

Interest is money paid for the use of money. The total amount borrowed is called the **principal.** The principal can be in the form of a loan (an individual borrows from the bank) or a deposit (the bank borrows from the individual). The **rate of interest,** expressed as a percent, is the amount charged for the use of the principal for a given period of time, usually for a year (that is, on a per annum basis).

Simple Interest Formula

If a principal of P dollars is borrowed for a period of t years at an annual interest rate r, expressed as a decimal, then the interest I charged is

$$I = Prt$$

Interest charged according to this formula is called **simple interest.**

EXAMPLE 6   Computing Credit Card Interest

Yolanda has a credit card balance of $2800. Each month, the credit card company charges 14% annual simple interest on any outstanding balances. How much interest will Yolanda be charged on this loan after one month? What is her credit card balance after one month? Round answers to the nearest penny.

Solution
We want to find the interest I charged on the loan. Because the interest rate is given as an annual rate we must express the length of time the money is borrowed in years. One month is $\frac{1}{12}$ of a year, so $t = \frac{1}{12}$. The outstanding balance, or principal, is $P = \$2800$. The annual interest rate is $r = 14\% = 0.14$. Substituting into $I = Prt$, we obtain

$$I = (\$2800)(0.14)\left(\frac{1}{12}\right) = \$32.67$$

Yolanda will owe the amount borrowed, $2800, plus accrued interest, $32.67, for a total of $2800 + $32.67 = $2832.67.

Quick ✓

20. _____ is money paid for the use of money. The total amount borrowed is called the _____.
21. Suppose that Dave has a car loan of $6500. The bank charges 6% annual simple interest. What is the interest charge on Dave's car loan after 1 month?
22. Suppose that you have $1400 in a savings account. The bank pays 1.5% annual simple interest. What would be the interest paid after 6 months? What is the balance in the account?

❸ Solve for a Variable in a Formula

A **formula** is an equation that describes how two or more variables are related. To "solve for a variable" means to isolate the variable on one side of the equation, with all the other variables and constants, if any, on the other side, by forming equivalent equations. For example, the formula for the area of a rectangle, $A = lw$, is solved for A because A is by itself on one side of the equation. The steps we follow when solving a formula for a certain variable are identical to those we follow when solving an equation.

EXAMPLE 7

Solving for a Variable in a Formula

Figure 2

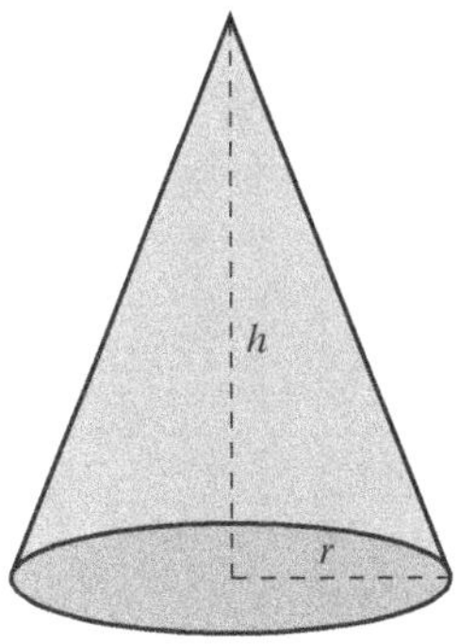

The volume V of a cone is given by the formula $V = \frac{1}{3}\pi r^2 h$, where r is the radius and h is the height of the cone. See Figure 2.

(a) Solve the formula for h.

(b) Use the result from part (a) to find the height of a cone if its volume is 50π cubic feet and its radius is 5 feet.

Solution

(a) To solve the formula for h, isolate h on one side of the equation.

$$V = \frac{1}{3}\pi r^2 h$$

Multiply both sides by 3: $$3 \cdot V = \left(\frac{1}{3}\pi r^2 h\right) \cdot 3$$

Divide out common factors: $$3V = \pi r^2 h$$

Divide both sides by πr^2: $$\frac{3V}{\pi r^2} = \frac{\pi r^2 h}{\pi r^2}$$

Divide out common factors: $$\frac{3V}{\pi r^2} = h$$

If $a = b$, then $b = a$: $$h = \frac{3V}{\pi r^2}$$

(b) Substitute $V = 50\pi$ ft^3 and $r = 5$ ft into $h = \frac{3V}{\pi r^2}$.

$$h = \frac{3(50\pi \text{ ft}^3)}{\pi(5 \text{ ft})^2}$$

$$h = \frac{150\pi \text{ ft}^3}{25\pi \text{ ft}^2}$$

$$h = 6 \text{ feet}$$

Work Smart

Solving for a variable is just like solving an equation with one unknown. When solving for a variable, treat all the other variables as constants.

Work Smart

When working with formulas, keep track of the units. Verify that the units in your answer are reasonable.

Quick ✓

23. The area A of a triangle is given by the formula $A = \frac{1}{2}bh$, where b is the base of the triangle and h is the height.

(a) Solve the formula for h.

(b) Find the height of the triangle whose area is 10 square inches and whose base is 4 inches.

24. The perimeter P of a parallelogram is given by the formula $P = 2a + 2b$, where a is the length of one side of the parallelogram and b is the length of the adjacent side.

(a) Solve the formula for b.

(b) Find the length of one side of a parallelogram whose perimeter is 60 cm and length of the other side is 20 cm.

EXAMPLE 8 Solving for a Variable in a Formula

The formula $Y = C + bY + I + G + N$ is a model used in economics to describe the total income of an economy. In the model, Y is income, C is consumption, I is investment in capital, G is government spending, N is net exports, and b is a constant. Solve the formula for Y.

Solution

We want to get all terms with Y on the same side of the equal sign.

$$Y = C + bY + I + G + N$$

Subtract bY from both sides: $Y - bY = C + bY + I + G + N - bY$

Combine like terms: $Y - bY = C + I + G + N$

Use the Distributive Property "in reverse" to isolate Y: $Y(1 - b) = C + I + G + N$

Divide both sides by $1 - b$: $\dfrac{Y(1 - b)}{1 - b} = \dfrac{C + I + G + N}{1 - b}$

Simplify: $Y = \dfrac{C + I + G + N}{1 - b}$

Quick ✓

In Problems 25–28, solve for the indicated variable.

25. $I = Prt$ for P
26. $Ax + By = C$ for y
27. $2xh - 4x = 3h - 3$ for h
28. $S = na + (n - 1)d$ for n

2.IR5 Exercises MyLab Statistics

Underlined exercises have complete video solutions in MyLab.

Problems 1–28 are the Quick ✓ s that follow the EXAMPLES.

Building Skills

29. What is 25% of 40?
30. What is 150% of 70?
31. 12 is 30% of what?
32. 50 is 90% of what?
33. 30 is what percent of 80?
34. 90 is what percent of 120?

In Problems 35–44, translate each English statement into a mathematical statement. Then solve the equation. See Objective 1.

35. The sum of a number x and 12 is 20.
36. The difference between 10 and a number z is 6.
37. Twice the sum of y and 3 is 16.
38. The sum of two times y and 3 is 16.
39. The difference between w and 22 equals three times w.
40. The sum of x and 4 results in twice x.
41. Four times a number x is equivalent to the sum of two times x and 14.
42. Five times a number x is equivalent to the difference of three times x and 10.
43. 80% of a number is equivalent to the sum of the number and 5.
44. 40% of a number equals the difference between the number and 10.

In Problems 45–56, solve the formula for the indicated variable. See Objective 1.

45. **Uniform Motion** Solve $d = rt$ for r.
46. **Direct Variation** Solve $y = kx$ for k.
47. **Algebra** Solve $y - y_1 = m(x - x_1)$ for m.
48. **Algebra** Solve $y = mx + b$ for m.
49. **Statistics** Solve $Z = \dfrac{x - \mu}{\sigma}$ for x.
50. **Statistics** Solve $E = \dfrac{Z \cdot \sigma}{\sqrt{n}}$ for $\sqrt{n}$.
51. **Newton's Law of Gravitation** Solve $F = G\dfrac{m_1 m_2}{r^2}$ for m_1.

52. Sequences Solve $S - rS = a - ar^5$ for S.

53. Finance Solve $A = P + Prt$ for P.

54. Bernoulli's Equation Solve $p + \frac{1}{2}\rho v^2 + \rho g y = a$ for ρ.

55. Temperature Conversion Solve $C = \frac{5}{9}(F - 32)$ for F.

56. Trapezoid Solve $A = \frac{1}{2}h(B + b)$ for b.

In Problems 57–64, solve for y.

57. $2x + y = 13$

58. $-4x + y = 12$

59. $9x - 3y = 15$

60. $4x + 2y = 20$

61. $4x + 3y = 13$

62. $5x - 6y = 18$

63. $\frac{1}{2}x + \frac{1}{6}y = 2$

64. $\frac{2}{3}x - \frac{5}{2}y = 5$

Applying the Concepts

65. Number Sense Grant is thinking of two numbers. He says that one of the numbers is twice the other number and the sum of the numbers is 39. What are the numbers?

66. Number Sense Pattie is thinking of two numbers. She says that one of the numbers is 8 more than the other number and the sum of the numbers is 56. What are the numbers?

67. Consecutive Integers The sum of three consecutive integers is 75. Find the integers.

68. Consecutive Integers The sum of four consecutive odd integers is 104. Find the integers.

69. Computing Grades Going into the final exam, which counts as two grades, Kendra has test scores of 84, 78, 64, and 88. What score does Kendra need on the final exam in order to have an average of 80?

70. Computing Grades Going into the final exam, which counts as three grades, Mark has test scores of 65, 79, 83, and 68. What score does Mark need on the final exam in order to have an average of 70?

71. Comparing Printers Jacob is trying to decide between two laser printers, one manufactured by Hewlett-Packard, the other by Brother. They have similar features and warranties, so price is the determining factor. The Hewlett-Packard costs $180 and printing costs are approximately $0.03 per page. The Brother costs $230 and printing costs are approximately $0.01 per page. How many pages need to be printed for the two printers to cost the same?

72. Comparing Job Offers Maria has just been offered two sales jobs. The first job offer has a base monthly salary of $2500 plus a commission of 3% of total sales. The second job offer has a base monthly salary of $1500 plus a commission of 3.5% of total sales. For what level of monthly sales are the salaries offered by these two jobs equivalent?

73. Finance An inheritance of $800,000 is to be divided among Avery, Connor, and Olivia in the following manner: Olivia is to receive $3/4$ of what Connor gets, while Avery gets $1/4$ of what Connor gets. How much does each receive?

74. Sharing the Cost of a Pizza Judy and Linda agree to share the cost of a $21 pizza based on how much each ate. If Judy ate $3/4$ of the amount that Linda ate, how much should each pay?

75. Sales Tax In the state of Connecticut there is a state sales tax of 6.35% on all goods purchased. If Jan buys a television for $599, what will be the final bill, including state sales tax?

76. Sales Tax In Austin, Texas, there is a sales tax of 8.25% on all goods purchased. If Megan buys a sofa for $450, what will be the final bill, including the sales tax?

77. Markups A new Honda Accord has a list price of $23,950. Suppose that the dealer mark-up on this car is 15%. What is the dealer's cost?

78. Markups Suppose that the price of a new Statistics text is $95. The bookstore has a policy of marking texts up 30%. What is the cost of the text to the bookstore?

79. Discount Pricing Suppose that you just received an e-mail alert indicating that 32-gigabyte flash drives have just been discounted by 60%. If the sale price of the flash drive is now $27.55, what was the original price?

80. Discount Pricing Suppose that you just received an e-mail alert from Kohls indicating that the fall line of clothing has just been discounted by 30% and knit polo shirts are now $28. What was the original price of a polo shirt?

81. Cars A Mazda 6s weighs 136 pounds more than a Nissan Altima. A Honda Accord EX weighs 119 pounds more than a Nissan Altima. The total weight of all three cars is 9834 pounds. How much does each car weigh? (*Source: Road and Track magazine*)

82. Cars The Honda Accord EX has 1.9 cubic feet less cargo space than a Mazda 6s. Together the cars have 31.3 cubic feet of cargo space. How much cargo space does each car have? (*Source: Road and Track magazine*)

83. Finance A total of $20,000 is going to be split between Adam and Krissy with Adam receiving $3000 less than Krissy. How much will each get?

84. Finance A total of $40,000 is going to be invested in stocks and bonds. A financial advisor recommends that $6000 more be invested in stocks than in bonds. How much is invested in stocks? How much is invested in bonds?

85. **Investments** Suppose that your long-lost Aunt Sara has left you an unexpected inheritance of $24,000. You have decided to invest the money rather than spend it on frivolous purchases. Your financial advisor has recommended that you diversify by placing some of the money in stocks and some in bonds. Based upon current market conditions, she has recommended that the amount in bonds equal three-fifths of the amount invested in stocks. How much should be invested in stocks? How much should be invested in bonds?

86. **Investments** Jack and Diane have $60,000 to invest. Their financial advisor has recommended that they diversify by placing some of the money in stocks and some in bonds. Based upon current market conditions, he has recommended that the amount in bonds equal two-thirds of the amount invested in stocks. How much should be invested in stocks? How much should be invested in bonds?

87. **Simple Interest** Elena has a credit card balance of $2500. The credit card company charges 14% per annum simple interest. What is the interest charge on Elena's credit card after 1 month?

88. **Simple Interest** Faye has a home equity loan of $70,000. The bank charges Faye 6% per annum simple interest. What is the interest charge on Faye's loan after 1 month?

89. **Cylinders** The volume V of a right circular cylinder is given by the formula $V = \pi r^2 h$, where r is the radius and h is the height.
    (a) Solve the formula for h.
    (b) Find the height of a right circular cylinder whose volume is 32π cubic inches and whose radius is 2 inches.

90. **Cylinders** The surface area S of a right circular cylinder is given by the formula $S = 2\pi rh + 2\pi r^2$, where r is the radius and h is the height.
    (a) Solve the formula for h.
    (b) Determine the height of a right circular cylinder whose surface area is 72π square centimeters and whose radius is 4 centimeters.

91. **Maximum Heart Rate** The model $M = -0.711A + 206.3$ was developed by Londeree and Moeschberger to determine the maximum heart rate M of an individual who is age A. (*Source:* Londeree and Moeschberger, "Effect of Age and Other Factors on HR max," *Research Quarterly for Exercise and Sport*, 53(4), 297–304)
    (a) Solve the model for A.
    (b) According to this model, what is the age of an individual whose maximum heart rate is 160?

92. **Maximum Heart Rate** The model $M = -0.85A + 217$ was developed by Miller to determine the maximum heart rate M of an individual who is age A. (*Source:* Miller et al., "Predicting Max HR," *Medicine and Science in Sports and Exercise*, 25(9), 1077–1081)
    (a) Solve the model for A.
    (b) According to this model, what is the age of an individual whose maximum heart rate is 160?

93. **Finance** The formula $A = P(1 + r)^t$ can be used to relate the future value A of a deposit of P dollars in an account that earns an annual interest rate r (expressed as a decimal) after t years.
    (a) Solve the formula for P.
    (b) How much would you have to deposit today in order to have $5000 in 5 years in a bank account that pays 4% annual interest?

94. **Federal Income Taxes** For a single filer with an annual adjusted income in 2015 over $37,450 but less than $90,751, the federal income tax T for an annual adjusted income I is found using the formula $T = 0.25(I - 37,450) + \$5156.25$.
    (a) Solve the formula for I.
    (b) Determine the adjusted income of a single filer whose tax bill is $14,780.

Explaining the Concepts

95. How is mathematical modeling related to problem solving?

96. Why do we make assumptions when creating mathematical models?

CHAPTER 3
Integrated Review: Getting Ready for Least-Squares Regression

Outline

3.IR1 The Rectangular Coordinate System and Equations in Two Variables
3.IR2 Graphing Equations in Two Variables
3.IR3 Slope
3.IR4 Slope-Intercept Form of a Line
3.IR5 Point-Slope Form of a Line

3.IR1 The Rectangular Coordinate System and Equations in Two Variables

Objectives

1. Plot Points in the Rectangular Coordinate System
2. Determine Whether an Ordered Pair Satisfies an Equation
3. Create a Table of Values That Satisfy an Equation

1 Plot Points in the Rectangular Coordinate System

Because pictures allow individuals to visualize ideas, they are typically more powerful than other forms of printed communication. Figure 1, which shows the results of the Manhattan Project from the test conducted on July 16, 1945, illustrates the power of the atom in a way that words never could.

Although the pictures that we use in mathematics might not deliver as powerful a message as Figure 1, they are powerful nonetheless. To draw pictures of mathematical relationships, we need a "canvas." The "canvas" that we use in this chapter is the *rectangular coordinate system*.

In Section 1.IR3, we learned how to plot points on a real number line. We can think of plotting points on the real number line as plotting in one dimension. In this chapter, we use the *rectangular coordinate system* to plot points in two dimensions.

Begin by drawing two real number lines, one horizontal and one vertical, that intersect at right (90°) angles. Call the horizontal real number line the ***x*-axis** and the vertical real number line the ***y*-axis**. The point where the *x*-axis and *y*-axis intersect is called the **origin**, *O*. See Figure 2.

Figure 1

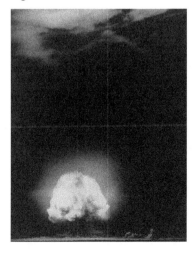

Figure 2
The rectangular coordinate system.

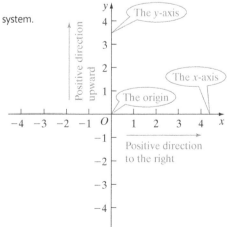

IR-111

The origin O has a value of 0 on each axis. Points on the x-axis to the right of O represent positive real numbers, and points on the x-axis to the left of O represent negative real numbers. On the y-axis, points above O represent positive real numbers, and points below O represent negative real numbers. Notice in Figure 2 that we label the x-axis "x" and the y-axis "y." An arrow at the end of each axis shows the positive direction.

The coordinate system in Figure 2 is called a **rectangular,** or **Cartesian, coordinate system,** named after René Descartes (1596–1650), a French mathematician, philosopher, and theologian. The plane formed by the x-axis and y-axis is often called the ***xy*-plane**, and the x-axis and y-axis are called the **coordinate axes**.

We can represent any point P in the rectangular coordinate system by using an **ordered pair (x, y)** of real numbers. We say that x represents the distance that P is from the y-axis. If $x > 0$ (that is, if x is positive), then P is x units to the right of the y-axis. If $x < 0$, then P is $|x|$ units to the left of the y-axis. We say that y represents the distance that P is from the x-axis. If $y > 0$, then P is y units above the x-axis. If $y < 0$, then P is $|y|$ units below the x-axis. The ordered pair (x, y) is also called the **coordinates** of P. To plot the point with coordinates $(2, 5)$, begin at the origin, move 2 units to the right, and then move 5 units up, as shown in Figure 3(a).

Work Smart

Be careful: The order in which numbers appear in the ordered pairs matters. For example, (3, 2) represents a different point from (2, 3).

Figure 3

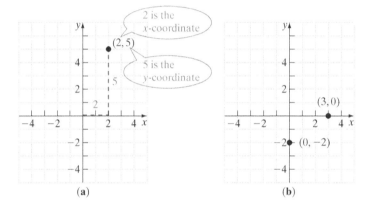

The origin O has coordinates $(0, 0)$. Any point on the x-axis has coordinates of the form $(x, 0)$, and any point on the y-axis has coordinates of the form $(0, y)$. For example, the points with coordinates $(3, 0)$ and $(0, -2)$, respectively, are shown in Figure 3(b).

If point P has coordinates (x, y), then x is called the ***x*-coordinate** of P, and y is called the ***y*-coordinate** of P.

The x- and y-axes divide the plane into four separate regions, or **quadrants**. In quadrant I, both the x-coordinate and the y-coordinate are positive. In quadrant II, x is negative and y is positive. In quadrant III, both x and y are negative. In quadrant IV, x is positive and y is negative. Points on the coordinate axes do not belong to a quadrant. See Figure 4.

Figure 4

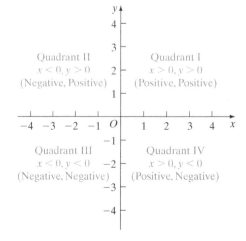

The Rectangular Coordinate System

- Composed of two real number lines–one horizontal (the *x*-axis) and one vertical (the *y*-axis). The *x*- and *y*-axes intersect at the origin.
- Also called the Cartesian coordinate system or *xy*-plane.
- Points in the rectangular coordinate system are denoted (x, y) and are called the coordinates of the point. We call *x* the *x*-coordinate and *y* the *y*-coordinate.
- If both *x* and *y* are positive, the point lies in quadrant I; if *x* is negative and *y* is positive, the point lies in quadrant II; if *x* is negative and *y* is negative, the point lies in quadrant III; if *x* is positive and *y* is negative, the point lies in quadrant IV.
- Points on the *x*-axis have a *y*-coordinate of 0; points on the *y*-axis have an *x*-coordinate of 0. Points on the *x*- or *y*-axis do not lie in a quadrant.

EXAMPLE 1 Plotting Points in the Rectangular Coordinate System

Plot the following ordered pairs in the rectangular coordinate system. Tell which quadrant each point lies in or state that the point lies on the *x*- or *y*-axis.

(a) $A(3, 1)$ (b) $B(-4, 2)$ (c) $C(3, -5)$

(d) $D(4, 0)$ (e) $E(0, -3)$ (f) $F\left(-\dfrac{5}{2}, -\dfrac{7}{2}\right)$

Solution

Before we can plot the points, draw a rectangular, or Cartesian, coordinate system. See Figure 5(a). Now plot the points.

Figure 5

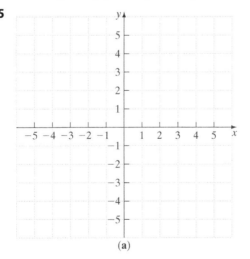
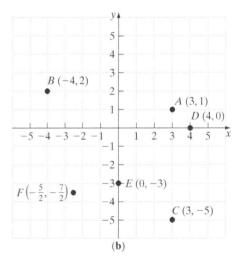

(a)   (b)

(a) To plot the point whose coordinates are $A(3, 1)$, begin at the origin *O* and travel 3 units to the right and then 1 unit up. Label the point *A*. Point *A* lies in quadrant I because both *x* and *y* are positive. See Figure 5(b).

(b) To plot $B(-4, 2)$, begin at the origin *O* and travel 4 units to the left and 2 units up. Label the point *B*. See Figure 5(b). Point *B* lies in quadrant II.

(c) See Figure 5(b). Point *C* lies in quadrant IV.

(d) See Figure 5(b). Point *D* lies on the *x*-axis, not in a quadrant.

(e) See Figure 5(b). Point *E* lies on the *y*-axis, not in a quadrant.

(f) It helps to convert the fractions into decimals, so $F\left(-\dfrac{5}{2}, -\dfrac{7}{2}\right) = F(-2.5, -3.5)$.

The *x*-coordinate of the point is halfway between -3 and -2; the *y*-coordinate is halfway between -4 and -3. See Figure 5(b). Point *F* lies in quadrant III.

Quick ✓

1. In the rectangular coordinate system, we call the horizontal real number line the _____ and we call the vertical real number line the _____. The point where these two axes intersect is called the _____.
2. If (x, y) are the coordinates of a point P, then x is called the _____ of P and y is called the _____ of P.
3. *True or False* The point whose ordered pair is $(-2, 4)$ is located in quadrant IV.
4. *True or False* The ordered pairs $(7, 4)$ and $(4, 7)$ represent the same point in the Cartesian plane.

In Problems 5 and 6, plot the ordered pairs in the rectangular coordinate system. Tell which quadrant each point lies in or state that the point lies on the x-axis or y-axis.

5. **(a)** $(5, 2)$ **(b)** $(-4, -3)$ **(c)** $(1, -3)$ **(d)** $(-2, 0)$ **(e)** $(0, 6)$ **(f)** $\left(-\dfrac{3}{2}, \dfrac{5}{2}\right)$

6. **(a)** $(-6, 2)$ **(b)** $(1, 7)$ **(c)** $(-3, -2)$ **(d)** $(4, 0)$ **(e)** $(0, -1)$ **(f)** $\left(\dfrac{3}{2}, -\dfrac{7}{2}\right)$

EXAMPLE 2 **Identifying Points in the Rectangular Coordinate System**

Identify the coordinates of each point labeled in Figure 6.

Figure 6

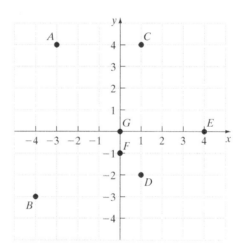

Solution

In an ordered pair (x, y), remember that x represents the position of the point left or right of the y-axis, while y represents the position of the point above or below the x-axis.

Point A is 3 units left of the y-axis so it has an x-coordinate of -3; point A is 4 units above the x-axis, so its y-coordinate is 4. The ordered pair $(-3, 4)$ corresponds to point A. We find the remaining coordinates in a similar fashion.

Point	Position	Ordered Pair
A	3 units left of the y-axis, 4 units above the x-axis	$(-3, 4)$
B	4 units left of the y-axis, 3 units below the x-axis	$(-4, -3)$
C	1 unit right of the y-axis, 4 units above the x-axis	$(1, 4)$
D	1 unit right of the y-axis, 2 units below the x-axis	$(1, -2)$
E	4 units right of the y-axis, on the x-axis	$(4, 0)$
F	On the y-axis, 1 unit below the x-axis	$(0, -1)$
G	On the x-axis, on the y-axis	$(0, 0)$

Quick ✓

7. Identify the coordinates of each point labeled in the figure below.

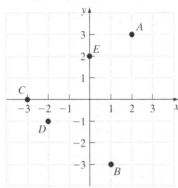

(a) A (b) B (c) C (d) D (e) E

▶ ❷ Determine Whether an Ordered Pair Satisfies an Equation

Recall from Section 2.IR4, that the solution of a linear equation in one variable is either a single value of the variable (conditional equation), the empty set (contradiction), or all real numbers (identity). See Table 1.

Table 1 Categories of Linear Equations in One Variable

Conditional Equation	Contradiction	Identity
$3x + 2 = 11$	$4(x - 2) - x = 2(x + 1) + x$	$-2(x + 3) + 4x = 2(x - 3)$
$3x = 9$	$4x - 8 - x = 2x + 2 + x$	$-2x - 6 + 4x = 2x - 6$
$x = 3$	$3x - 8 = 3x + 2$	$2x - 6 = 2x - 6$
	$-8 = 2$	$-6 = -6$
	no solution	solution is all real numbers

We will now look at equations in two variables.

Definition

An **equation in two variables**, x and y, is a statement in which the algebraic expressions involving x and y are equal. The expressions are called **sides** of the equation.

For example, the following statements are all equations in two variables.
$$x + 2y = 6 \qquad y = -3x + 7 \qquad y = x^2 + 3$$

Since an equation is a statement, it may be true or false, depending on the values of the variables. Any values of the variables that make the equation a true statement **satisfy** the equation.

The first equation, $x + 2y = 6$, is satisfied when $x = 4$ and $y = 1$.

$$x + 2y = 6$$
Substitute $x = 4$ and $y = 1$: $4 + 2(1) \stackrel{?}{=} 6$
$$6 = 6 \quad \text{True}$$

We can also say that the ordered pair $(4, 1)$ satisfies the equation. Does $x = 8$ and $y = -1$ satisfy the equation $x + 2y = 6$?

$$x + 2y = 6$$
$x = 8, y = -1$: $8 + 2(-1) \stackrel{?}{=} 6$
$$6 = 6 \quad \text{True}$$

So the ordered pair $(8, -1)$ satisfies the equation as well. In fact, infinitely many choices of x and y satisfy the equation $x + 2y = 6$, but some choices of x and y do not satisfy the equation $x + 2y = 6$. For example, $x = 3$ and $y = 4$ do not satisfy the equation $x + 2y = 6$.

$$x + 2y = 6$$
$$x = 3, y = 4: \quad 3 + 2(4) \stackrel{?}{=} 6$$
$$11 = 6 \quad \text{False}$$

EXAMPLE 3 Determining Whether an Ordered Pair Satisfies an Equation

Determine whether the following ordered pairs satisfy the equation $x + 2y = 8$.

(a) $(2, 3)$ (b) $(6, -1)$ (c) $(-2, 5)$

Solution

(a) Check to see whether $x = 2$, $y = 3$ satisfies the equation $x + 2y = 8$.

$$x + 2y = 8$$
$$\text{Let } x = 2, y = 3: \quad 2 + 2(3) \stackrel{?}{=} 8$$
$$2 + 6 \stackrel{?}{=} 8$$
$$8 = 8 \quad \text{True}$$

The statement is true, so $(2, 3)$ satisfies the equation $x + 2y = 8$.

(b) For the ordered pair $(6, -1)$, we have

$$x + 2y = 8$$
$$\text{Let } x = 6, y = -1: \quad 6 + 2(-1) \stackrel{?}{=} 8$$
$$6 - 2 \stackrel{?}{=} 8$$
$$4 = 8 \quad \text{False}$$

The statement $4 = 8$ is false, so $(6, -1)$ does not satisfy the equation.

(c) For the ordered pair $(-2, 5)$, we have

$$x + 2y = 8$$
$$\text{Let } x = -2, y = 5: \quad -2 + 2(5) \stackrel{?}{=} 8$$
$$-2 + 10 \stackrel{?}{=} 8$$
$$8 = 8 \quad \text{True}$$

The statement is true, so $(-2, 5)$ satisfies the equation. ●

Quick ✓

8. *True or False* An equation in two variables can have more than one solution.
9. Determine whether the following ordered pairs satisfy the equation $x + 4y = 12$.

    (a) $(4, 2)$ (b) $(-2, 4)$ (c) $(1, 8)$

10. Determine whether the following ordered pairs satisfy the equation $y = 4x + 3$.

    (a) $(1, 3)$ (b) $(-2, -5)$ (c) $\left(-\dfrac{3}{2}, -3\right)$

3 Create a Table of Values That Satisfy an Equation

In Example 3 we learned how to determine whether a given ordered pair satisfies an equation. Now, we learn how to find an ordered pair that satisfies an equation.

EXAMPLE 4 **How to Find an Ordered Pair That Satisfies an Equation**

Find an ordered pair that satisfies the equation $3x + y = 5$.

Step-by-Step Solution

Step 1: Choose any value for one of the variables in the equation.

Choose any value of x or y that you want. We will let $x = 2$.

Step 2: Substitute the chosen value of the variable into the equation. Solve for the remaining variable.

Substitute 2 for x in $3x + y = 5$ and then solve for y.

$$3x + y = 5$$
$$\text{Let } x = 2: \quad 3(2) + y = 5$$
$$\text{Simplify:} \quad 6 + y = 5$$
$$\text{Subtract 6 from both sides:} \quad 6 - 6 + y = 5 - 6$$
$$y = -1$$

One ordered pair that satisfies the equation is $(2, -1)$. ●

Quick ✓

In Problems 11–13, determine an ordered pair that satisfies the given equation by substituting the given value of the variable into the equation.

11. $2x + y = 10; x = 3$ **12.** $-3x + 2y = 11; y = 1$ **13.** $4x + 3y = 0; x = \dfrac{1}{2}$

Look again at Example 3. Did you notice that two different ordered pairs satisfy the equation? In fact, an infinite number of ordered pairs satisfy the equation $x + 2y = 8$ because for any real number y, we can find a value of x that makes the equation a true statement. One way to find some of the solutions of an equation in two variables is to create a table of values that satisfy the equation. The table is created by choosing values of x and using the equation to find the corresponding value of y (as we did in Example 4) or by choosing a value of y and using the equation to find the corresponding value of x.

EXAMPLE 5 **Creating a Table of Values That Satisfy an Equation**

Use the equation $y = -2x + 5$ to complete Table 2, and then use the table to list some of the ordered pairs that satisfy the equation.

Table 2

x	y	(x, y)
−2		
0		
1		

Solution

The first entry in the table is $x = -2$. Substitute -2 for x and use the equation $y = -2x + 5$ to find y.

$$y = -2x + 5$$
$$x = -2: \quad y = -2(-2) + 5$$
$$y = 4 + 5$$
$$y = 9$$

Now substitute 0 for x in the equation $y = -2x + 5$.

$$y = -2x + 5$$
$$x = 0: \quad y = -2(0) + 5$$
$$y = 0 + 5$$
$$y = 5$$

Finally, substitute 1 for x in the equation $y = -2x + 5$.

$$y = -2x + 5$$
$$x = 1: \quad y = -2(1) + 5$$
$$y = -2 + 5$$
$$y = 3$$

Table 3

x	y	(x, y)
-2	9	(-2, 9)
0	5	(0, 5)
1	3	(1, 3)

The completed table is shown as Table 3. Three ordered pairs that satisfy the equation are $(-2, 9)$, $(0, 5)$, and $(1, 3)$. ●

Quick ✓

In Problems 14 and 15, use the equation to complete the table. Use the table to list some of the ordered pairs that satisfy the equation.

14. $y = 5x - 2$

x	y	(x, y)
-2		
0		
1		

15. $y = -3x + 4$

x	y	(x, y)
-1		
2		
5		

EXAMPLE 6   **Creating a Table of Values That Satisfy an Equation**

Use the equation $2x - 3y = 12$ to complete Table 4, and then use the table to list some of the ordered pairs that satisfy the equation.

Solution

Table 4

x	y	(x, y)
-3		
	-4	
6		

The first entry in the table is $x = -3$. Substitute -3 for x and use the equation $2x - 3y = 12$ to find y.

$$2x - 3y = 12$$
$$x = -3: \quad 2(-3) - 3y = 12$$
$$-6 - 3y = 12$$

Add 6 to both sides: $\quad -3y = 18$

Divide both sides by -3: $\quad y = -6$

Substitute -4 for y in the equation $2x - 3y = 12$.

$$2x - 3y = 12$$
$$y = -4: \quad 2x - 3(-4) = 12$$
$$2x + 12 = 12$$

Subtract 12 from both sides: $\quad 2x = 0$

Divide both sides by 2: $\quad x = 0$

Substitute 6 for x in the equation $2x - 3y = 12$.

$$2x - 3y = 12$$
$$x = 6: \quad 2(6) - 3y = 12$$
$$12 - 3y = 12$$

Subtract 12 from both sides: $\quad -3y = 0$

Divide both sides by -3: $\quad y = 0$

Table 5 shows that three ordered pairs that satisfy the equation are $(-3, -6)$, $(0, -4)$, and $(6, 0)$.

Table 5

x	y	(x, y)
−3	−6	(−3, −6)
0	−4	(0, −4)
6	0	(6, 0)

Quick ✓

In Problems 16 and 17, use the equation to complete the table. Use the table to list some of the ordered pairs that satisfy the equation.

16. $2x + y = -8$

x	y	(x, y)
−5		
	−4	
2		

17. $2x - 5y = 18$

x	y	(x, y)
−6		
	−4	
2		

When modeling a situation, we can use variables other than x and y. For example, we might use the equation $C = 1.20m + 3$ to represent the cost of taking a taxi where C represents the cost (in dollars) and m represents the number of miles driven.

Recall from Section 1.IR3 that the scale of a number line refers to the distance between tick marks on the number line. In Figures 2 through 6, we used a scale of 1 on both the x-axis and the y-axis. In applications, a different scale is often used on each axis.

EXAMPLE 7 **An Electric Bill**

In North Carolina, Duke Power determines that the monthly electric bill for a household will be C dollars for using x kilowatt-hours (kWh) of electricity using the formula

$$C = 9.90 + 0.092896x$$

Source: Duke Power

(a) Complete Table 6, and use the table to list ordered pairs that satisfy the equation. Express answers rounded to the nearest penny.

Table 6

x(kWh)	50 kWh	100 kWh	200 kWh
C($)			

(b) Plot in a rectangular coordinate system the ordered pairs (x, C) found in part (a).

Solution

(a) The first entry in the table is $x = 50$. Substitute 50 for x into the equation $C = 9.90 + 0.092896x$ to find C. Round C to two decimal places because C represents the cost, so our answer is rounded to the nearest penny.

$$C = 9.90 + 0.092896x$$
$$x = 50: \quad C = 9.90 + 0.092896(50)$$
$$\text{Use a calculator:} \quad C = 14.54$$

Now substitute 100 for x into the equation to find C.

$$C = 9.90 + 0.092896x$$
$$x = 100: \quad C = 9.90 + 0.092896(100)$$
$$\text{Use a calculator:} \quad C = 19.19$$

Figure 7

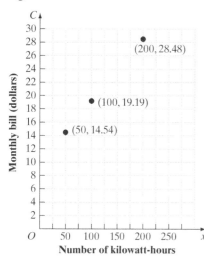

Now substitute 200 for x into the equation to find C.

$$C = 9.90 + 0.092896x$$

$x = 200$: $\quad C = 9.90 + 0.092896(200)$

Use a calculator: $\quad C = 28.48$

Table 7 shows the completed table. Three ordered pairs that satisfy the equation are $(50, 14.54)$, $(100, 19.19)$, and $(200, 28.48)$.

Table 7

x(kWh)	50 kWh	100 kWh	200 kWh
C($)	$14.54	$19.19	$28.48

(b) Because x represents the number of kilowatt-hours used, label the horizontal axis x. Because C represents the bill, label the vertical axis C. The ordered pairs found in part (a) are plotted in Figure 7. Note that we used different scales on the horizontal and vertical axes.

Notice in Figure 7 that we labeled the horizontal and vertical axes so that it is clear what they represent. Labeling the axes is a good practice to follow whenever you draw a graph.

Quick ✓

18. Piedmont Natural Gas charges its customers in North Carolina a monthly fee of C dollars for using x therms of natural gas using the formula

$$C = 10 + 0.86676x$$

Source: Piedmont Natural Gas

(a) Complete the table, and use the results to list ordered pairs (x, C) that satisfy the equation. Express answers rounded to the nearest penny.

x (therms)	50 therms	100 therms	150 therms
C ($)			

(b) Plot in a rectangular coordinate system the ordered pairs found in part (a).

3.IR1 Exercises MyLab Statistics

Underlined exercises have complete video solutions in MyLab.

Problems 1–18 are the Quick ✓s that follow the EXAMPLES.

Building Skills

In Problems 19–22, plot the following ordered pairs in the rectangular coordinate system. Tell which quadrant each point lies in or state that the point lies on the x-axis or y-axis. See Objective 1.

19. $A(-3, 2)$; $B(4, 1)$; $C(-2, -4)$; $D(5, -4)$; $E(-1, 3)$; $F(2, -4)$

20. $P(-3, -2)$; $Q(2, -4)$; $R(4, 3)$; $S(-1, 4)$; $T(-2, -4)$; $U(3, -3)$

21. $A\left(\frac{1}{2}, 0\right)$; $B\left(\frac{3}{2}, -\frac{1}{2}\right)$; $C\left(4, \frac{7}{2}\right)$; $D\left(0, -\frac{5}{2}\right)$; $E\left(\frac{9}{2}, 2\right)$; $F\left(-\frac{5}{2}, -\frac{3}{2}\right)$; $G(0, 0)$

22. $P\left(\frac{3}{2}, -2\right)$; $Q\left(0, \frac{5}{2}\right)$; $R\left(-\frac{9}{2}, 0\right)$; $S(0, 0)$; $T\left(-\frac{3}{2}, -\frac{9}{2}\right)$; $U\left(3, \frac{1}{2}\right)$; $V\left(\frac{5}{2}, -\frac{7}{2}\right)$

In Problems 23 and 24, plot the following ordered pairs in the rectangular coordinate system. Tell the location of each point: positive x-axis, negative x-axis, positive y-axis, or negative y-axis. See Objective 1.

23. $A(3, 0);\quad B(0, -1);\quad C(0, 3);\quad D(-4, 0)$

24. $P(0, -1);\quad Q(-2, 0);\quad R(0, 3);\quad S(1, 0)$

In Problems 25 and 26, identify the coordinates of each point labeled in the figure. Name the quadrant in which each point lies or state that the point lies on the x- or y-axis. See Objective 1.

25.
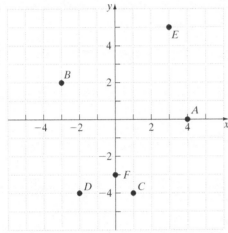

26.
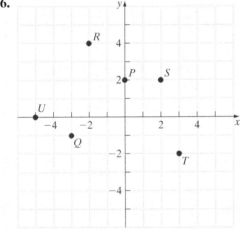

In Problems 27–32, determine whether or not the ordered pair satisfies the equation. See Objective 2.

27. $y = -3x + 5;\quad A(-2, -1)\quad B(2, -1)\quad C\left(\dfrac{1}{3}, 4\right)$

28. $y - 2x - 3;\quad A(-1, -5)\quad B(4, -5)\quad C(-2, -7)$

29. $3x + 2y = 4;\quad A(0, 2)\quad B(1, 0)\quad C(4, -4)$

30. $5x - y = 12;\quad A(2, 0)\quad B(0, 12)\quad C(-2, -22)$

31. $\dfrac{4}{3}x + y - 1 = 0;\quad A(3, -3)\quad B(-6, -9)\quad C\left(\dfrac{3}{4}, 0\right)$

32. $\dfrac{3}{4}x + 2y = 0;\quad A\left(-4, \dfrac{3}{2}\right)\quad B(0, 0)\quad C\left(1, -\dfrac{3}{2}\right)$

For Problems 33–38, see Objective 3.

33. Find an ordered pair that satisfies the equation $x + y = 5$ by letting $x = 4$.

34. Find an ordered pair that satisfies the equation $x + y = 7$ by letting $x = 2$.

35. Find an ordered pair that satisfies the equation $2x + y = 9$ by letting $y = -1$.

36. Find an ordered pair that satisfies the equation $-4x - y = 5$ by letting $y = 7$.

37. Find an ordered pair that satisfies the equation $-3x + 2y = 15$ by letting $x = -3$.

38. Find an ordered pair that satisfies the equation $5x - 3y = 11$ by letting $y = 3$.

In Problems 39–52, use the equation to complete the table. Use the table to list some of the ordered pairs that satisfy the equation. See Objective 3.

39. $y = -x$

x	y	(x, y)
-3		
0		
1		

40. $y = x$

x	y	(x, y)
-4		
0		
2		

41. $y = -3x + 1$

x	y	(x, y)
-2		
-1		
4		

42. $y = 4x - 5$

x	y	(x, y)
-3		
1		
2		

43. $2x + y = 6$

x	y	(x, y)
-1		
2		
3		

44. $3x + 4y = 2$

x	y	(x, y)
-2		
2		
4		

45. $y = 6$

x	y	(x, y)
-4		
1		
12		

46. $x = 2$

x	y	(x, y)
	-4	
	0	
	8	

47. $x - 2y + 6 = 0$

x	y	(x, y)
1		
	1	
-2		

48. $2x + y - 4 = 0$

x	y	(x, y)
-1		
	-4	
	2	

49. $y = 5 + \dfrac{1}{2}x$

x	y	(x, y)
	7	
-4		
	2	

50. $y = 8 - \dfrac{1}{3}x$

x	y	(x, y)
	10	
9		
	27	

51. $\dfrac{x}{2} + \dfrac{y}{3} = -1$

x	y	(x, y)
0		
	0	
	-6	

52. $\dfrac{x}{5} - \dfrac{y}{2} = 1$

x	y	(x, y)
	0	
0		
	-5	

In Problems 53–64, for each equation find the missing value in the ordered pair. See Objective 3.

53. $y = -3x - 10$ $A(\_\_, -16)$ $B(-3, \_\_)$
    $C(\_\_, -9)$

54. $y = 5x - 4$ $A(-1, \_\_)$ $B(\_\_, 31)$
    $C\left(-\dfrac{2}{5}, \_\_\right)$

55. $x = -\dfrac{1}{3}y$ $A(2, \_\_)$ $B(\_\_, 0)$
    $C\left(\_\_, -\dfrac{1}{2}\right)$

56. $y = \dfrac{2}{3}x$ $A\left(\_\_, -\dfrac{8}{3}\right)$ $B(0, \_\_)$
    $C\left(-\dfrac{5}{6}, \_\_\right)$

57. $x = 4$ $A(\_\_, -8)$ $B(\_\_, -19)$ $C(\_\_, 5)$

58. $y = -1$ $A(6, \_\_)$ $B(-1, \_\_)$ $C(0, \_\_)$

59. $y = \dfrac{2}{3}x + 2$ $A(\_\_, 4)$ $B(-6, \_\_)$
    $C\left(\dfrac{1}{2}, \_\_\right)$

60. $y = -\dfrac{5}{4}x - 1$ $A(\_\_, -6)$ $B(-8, \_\_)$
    $C\left(\_\_, -\dfrac{11}{6}\right)$

61. $\dfrac{1}{2}x - 3y = 2$ $A(-4, \_\_)$ $B(\_\_, -1)$
    $C\left(-\dfrac{2}{3}, \_\_\right)$

62. $\dfrac{1}{3}x + 2y = -1$ $A(-4, \_\_)$ $B\left(\_\_, -\dfrac{3}{4}\right)$
    $C(0, \_\_)$

63. $0.5x - 0.3y = 3.1$ $A(20, \_\_)$ $B(\_\_, -17)$
    $C(2.6, \_\_)$

64. $-1.7x + 0.2y = -5$ $A(\_\_, -110)$
    $B(40, \_\_)$ $C(2.4, \_\_)$

Applying the Concepts

65. **Book Value** Residential investment property such as apartment buildings may be depreciated over 28.5 years, according to the Internal Revenue Service (IRS). The IRS allows an investor to depreciate an apartment building whose value (excluding the land) is $285,000 by $10,000 per year.
    The equation $V = -10{,}000x + 285{,}000$ represents the book value, V, of an apartment after x years.
    (a) What will be the book value of the apartment building after 2 years?
    (b) What will be the book value of the apartment building after 5 years?
    (c) After how many years will the book value of the apartment building be $205,000?
    (d) If (x, V) represents any ordered pair that satisfies $V = -10{,}000x + 285{,}000$, interpret the meaning of $(3, 255000)$.

66. **Taxi Ride** The cost to take a taxi is $1.70 plus $2.00 per mile for each mile driven. The total cost, C, is given by the equation $C = 1.7 + 2m$, where m represents the total miles driven.
    (a) How much will it cost to take a taxi 5 miles?
    (b) How much will it cost to take a taxi 20 miles?
    (c) If you spent $32.70 on cab fare, how far was your trip?
    (d) If (m, C) represents any ordered pair that satisfies $C = 1.7 + 2m$, interpret the meaning of $(14, 29.7)$ in the context of this problem.

67. **College Graduates** An equation to approximate the percentage of the U.S. population 25 years of age or older with a bachelor's degree can be given by the model $P = 0.38n + 30.5$, where n is the number of years after 2009.
    (a) According to the model, what percentage of the U.S. population 25 years of age or older had a bachelor's degree in 2009 $(n = 0)$?
    (b) According to the model, what percentage of the U.S. population 25 years of age or older had a bachelor's degree in 2015 $(n = 6)$?
    (c) According to the model, what percentage of the U.S. population 25 years of age or older will have a bachelor's degree in 2020 $(n = 11)$?
    (d) In which year will 50% of the U.S. population 25 years of age or older have a bachelor's degree? Round your answer to the nearest year.

(e) According to the model, 100% of the U.S. population 25 years of age or older will have a bachelor's degree in 2192. Do you think this is reasonable? Why or why not?

68. Life Expectancy The model $A = 0.183n + 67.895$ is used to estimate the life expectancy A of residents of the United States born n years after 1950.

(a) According to the model, what is the life expectancy for a person born in 1950?

(b) According to the model, what is the life expectancy for a person born in 1980 ($n = 30$)?

(c) If the model holds true for future generations, what is the life expectancy for a person born in 2020?

(d) If a person has a life expectancy of 77 years, to the nearest year, when was the person born?

(e) Do you think life expectancy will continue to increase in the future? What could happen that would change this model?

In Problems 69–72, use the equation to complete the table. Use the table to list some of the ordered pairs that satisfy the equation.

69. $4a + 2b = -8$

a	b	(a, b)
2		
	-4	
	6	

70. $2r - 3s = 3$

r	s	(r, s)
	3	
	-1	
-3		

71. $\dfrac{2p}{5} + \dfrac{3q}{10} = 1$

p	q	(p, q)
0		
	0	
-10		

72. $\dfrac{4a}{3} + \dfrac{2b}{5} = -1$

a	b	(a, b)
	0	
0		
	10	

Extending the Concepts

In Problems 73–78, determine the value of k so that the given ordered pair is a solution to the equation.

73. Find the value of k for which $(1, 2)$ satisfies $y = -2x + k$.

74. Find the value of k for which $(-1, 10)$ satisfies $y = 3x + k$.

75. Find the value of k for which $(2, 9)$ satisfies $7x - ky = -4$.

76. Find the value of k for which $(3, -1)$ satisfies $4x + ky = 9$.

77. Find the value of k for which $\left(-8, -\dfrac{5}{2}\right)$ satisfies $kx - 4y = 6$.

78. Find the value of k for which $\left(-9, \dfrac{1}{2}\right)$ satisfies $kx + 2y = -2$.

In Problems 79 and 80, use the equation to complete the table. Choose any value for x and solve the resulting equation to find the corresponding value for y. Then plot these ordered pairs in a rectangular coordinate system. Connect the points and describe the figure.

79. $3x - 2y = -6$

x	y	(x, y)

80. $-x + y = 4$

x	y	(x, y)

In Problems 81–84, use the equation to complete the table. Then plot the points in a rectangular coordinate system.

81. $y = x^2 - 4$

x	y	(x, y)
-2		
-1		
0		
1		
2		

82. $y = -x^2 + 3$

x	y	(x, y)
-2		
-1		
0		
1		
2		

83. $y = -x^3 + 2$

x	y	(x, y)
-2		
-1		
0		
1		
2		

84. $y = 2x^3 - 1$

x	y	(x, y)
-2		
-1		
0		
1		
2		

Explaining the Concepts

85. Describe how the quadrants in the rectangular coordinate system are labeled and how you can determine the quadrant in which a point lies. Describe the characteristics of a point that lies on either the x- or y-axis.

86. Describe how to plot the point whose coordinates are $(3, -5)$.

The Graphing Calculator

Graphing calculators can also create tables of values that satisfy an equation. To do this, first solve the equation for y. For example, to obtain a table of values that satisfy the equation $2x - 3y = 12$ (Example 6), solve for y as follows:

$$2x - 3y = 12$$

Subtract 2x from both sides: $\quad -3y = -2x + 12$

Divide both sides by -3: $\quad y = \dfrac{-2x + 12}{-3}$

Divide -3 into each term in the numerator: $\quad y = \dfrac{-2x}{-3} + \dfrac{12}{-3}$

Simplify: $\quad y = \dfrac{2}{3}x - 4$

Now enter the equation $y = \dfrac{2}{3}x - 4$ into the calculator and create the table shown.

In Problems 87–94, use a graphing calculator to create a table of values that satisfy each equation. Have the table begin at -3 and increase by 1.

87. $y = 2x - 9$
88. $y = -3x + 8$
89. $y = -x + 8$
90. $y = 2x - 4$
91. $y + 2x = 13$
92. $y - x = -15$
93. $y = -6x^2 + 1$
94. $y = -x^2 + 3x$

3.IR2 Graphing Equations in Two Variables

Objectives

1. Graph a Line by Plotting Points
2. Graph a Line Using Intercepts
3. Graph Vertical and Horizontal Lines

▶ 1 Graph a Line by Plotting Points

In the previous section, we found values of x and y that satisfy an equation. What does this mean? Well, it means that the ordered pair (x, y) is a point on the graph of the equation.

In Words
The graph of an equation is a geometric way of representing the set of all ordered pairs that make the equation a true statement. Think of the graph as a picture of the solution set.

Definition
The **graph of an equation in two variables** x and y is the set of points whose coordinates, (x, y), in the xy-plane satisfy the equation.

But how do we obtain the graph of an equation? One method for graphing an equation is the **point-plotting method**.

EXAMPLE 1 How to Graph an Equation Using the Point-Plotting Method

Graph the equation $y = 2x - 3$ using the point-plotting method.

Step-by-Step Solution

Step 1: Find ordered pairs that satisfy the equation by choosing some values of x and using the equation to find the corresponding values of y. See Table 8.

Table 8

x	y	(x, y)
-2	$y = 2(-2) - 3$ $= -4 - 3$ $= -7$	$(-2, -7)$
-1	$y = 2(-1) - 3$ $= -5$	$(-1, -5)$
0	$y = 2(0) - 3$ $= -3$	$(0, -3)$
1	$y = 2(1) - 3$ $= -1$	$(1, -1)$
2	$y = 2(2) - 3$ $= 1$	$(2, 1)$

Step 2: Plot in a rectangular coordinate system the points whose coordinates, (x, y), were found in Step 1. See Figure 8.

Figure 8

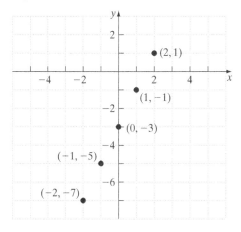

Step 3: Connect the points with a straight line. See Figure 9.

Figure 9
$y = 2x - 3$

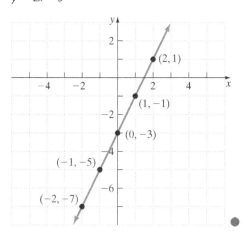

Work Smart

Remember, all coordinates of the points on the graph shown in Figure 9 satisfy the equation $y = 2x - 3$.

The graph of the equation in Figure 9 does not show all the points that satisfy the equation. For example, the point $(5, 7)$ is a part of the graph of $y = 2x - 3$ but is not shown in Figure 9. Because the graph of $y = 2x - 3$ could be extended as far as we please, use arrows to indicate that the pattern shown continues. It is important to show enough of the graph so that anyone can "see" how it continues. This is called a **complete graph**.

For the remainder of the text, we will say "the point (x, y)" for short rather than "the point whose coordinates are (x, y)."

> **Graphing an Equation Using the Point-Plotting Method**
>
> **Step 1:** Find several ordered pairs that satisfy the equation.
>
> **Step 2:** Plot the points found in Step 1 in a rectangular coordinate system.
>
> **Step 3:** Connect the points with a smooth curve or line.

Quick ✓

1. The _____ of an equation in two variables is the set of points whose coordinates, (x, y), in the xy-plane satisfy the equation.

In Problems 2 and 3, draw a complete graph of each equation using point plotting.

2. $y = 3x - 2$

3. $y = -4x + 8$

A question you may be asking yourself is "How many points do I need to find before I can be sure that I have a complete graph?" It depends on the type of equation you are graphing. The equation that we graphed in Example 1 is called a *linear equation*.

Definition

A **linear equation in two variables** is an equation that can be written in the form
$$Ax + By = C$$
where A, B, and C are real numbers. A and B cannot both be 0. A linear equation written in the form $Ax + By = C$ is said to be in **standard form.***

EXAMPLE 2 Identifying Linear Equations in Two Variables

Determine whether the equation is a linear equation in two variables.

(a) $3x - 4y = 9$

(b) $\frac{1}{2}x + \frac{2}{3}y = 4$

(c) $x^2 + 5y = 10$

(d) $-2y = 5$

Solution

(a) The equation $3x - 4y = 9$ is a linear equation in two variables because it is written in the form $Ax + By = C$ with $A = 3$, $B = -4$, and $C = 9$.

(b) The equation $\frac{1}{2}x + \frac{2}{3}y = 4$ is a linear equation in two variables because it is written in the form $Ax + By = C$ with $A = \frac{1}{2}$, $B = \frac{2}{3}$, and $C = 4$.

(c) The equation $x^2 + 5y = 10$ is not a linear equation because x is squared.

(d) The equation $-2y = 5$ is a linear equation in two variables because it is written in the form $Ax + By = C$ with $A = 0$, $B = -2$, and $C = 5$.

Quick ✓

4. A(n) _____ equation is an equation that can be written in the form $Ax + By = C$, where A, B, and C are real numbers, and A and B are not both zero. Equations written in this form are said to be in _____ _____.

In Problems 5–7, determine whether or not the equation is a linear equation in two variables.

5. $4x - y = 12$

6. $5x - y^2 = 10$

7. $5x = 20$

Work Smart

When graphing a line, be sure to find three points—just to be safe!

We will refer to linear equations in two variables as **linear equations.** The graph of a linear equation is a **line.** To graph a linear equation requires only two points; however, we recommend finding a third point as a check.

*Some texts call $Ax + By = C$ the **general form** of a line.

EXAMPLE 3 Graphing a Linear Equation Using the Point-Plotting Method

Graph the linear equation $2x + y = 4$.

Solution

Because the coefficient of y is 1, it is easier to choose values of x and find the corresponding values of y. We will determine the value of y for $x = -2, 0,$ and 2. There is nothing magical about these choices. Any three different values of x will give us the results we want.

Work Smart

Choose values of x (or y) that make the algebra easy.

$x = -2:$
$$2x + y = 4$$
Let $x = -2$: $2(-2) + y = 4$
$$-4 + y = 4$$
$$y = 8$$

$x = 0:$
$$2x + y = 4$$
Let $x = 0$: $2(0) + y = 4$
$$y = 4$$

$x = 2:$
$$2x + y = 4$$
Let $x = 2$: $2(2) + y = 4$
$$4 + y = 4$$
$$y = 0$$

We plot the three points from Table 9 and connect them with a straight line. See Figure 10.

Table 9

x	y	(x, y)
-2	8	(-2, 8)
0	4	(0, 4)
2	0	(2, 0)

Figure 10

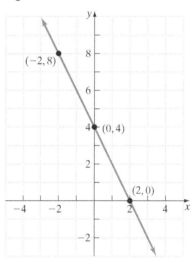

Quick

8. The graph of a linear equation is a _____.

In Problems 9 and 10, graph each linear equation using the point-plotting method.

9. $-3x + y = -6$

10. $2x + 3y = 12$

EXAMPLE 4 Cost of Renting a Car

A car-rental agency quotes you the cost of renting a car in Washington, D.C., as $30 per day plus $0.20 per mile. The linear equation $C = 0.20m + 30$ models the cost, where C represents total cost and m represents the number of miles that were traveled in one day.

(a) Complete Table 10 to find ordered pairs that satisfy the equation. Round answers to the nearest penny.

(b) Graph the linear equation $C = 0.20m + 30$ using the points obtained in part (a).

Table 10

m	C	(m, C)
0		
50		
100		

Solution

(a) The first entry in the table is $m = 0$. Substitute 0 for m in the equation $C = 0.20m + 30$ to find C.

$m = 0$: $\quad C = 0.20m + 30$
$\quad\quad\quad C = 0.20(0) + 30$
$\quad\quad\quad C = 30$

Now substitute 50 for m in the equation $C = 0.20m + 30$ to find C.

$m = 50$: $\quad C = 0.20m + 30$
$\quad\quad\quad\; C = 0.20(50) + 30$
$\quad\quad\quad\; C = 40$

Now substitute 100 for m in the equation to find C.

$m = 100$: $\quad C = 0.20m + 30$
$\quad\quad\quad\quad C = 0.20(100) + 30$
$\quad\quad\quad\quad C = 50$

Table 11

m	C	(m, C)
0	30	(0, 30)
50	40	(50, 40)
100	50	(100, 50)

Table 11 shows the completed table. The ordered pair (50, 40) means that if a car is driven 50 miles in a day, the rental cost will be $40.

(b) Because m represents the number of miles driven, label the horizontal axis m. Label the vertical axis C, the cost of renting the car. Let the scale of the horizontal axis be 10, so each tick mark represents 10 miles. Scale the vertical axis to 5, so each tick mark represents 5 dollars. Scaling in this way makes it easier to plot the ordered pairs. In Figure 11, plot the points found in part (a) and then draw the line.

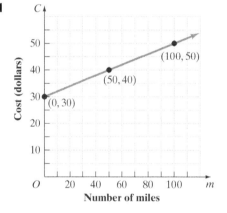

Figure 11

Quick ✓

11. Michelle sells computers. Her monthly salary is $3000 plus 8% of total sales. The linear equation $S = 0.08x + 3000$ models Michelle's monthly salary, S, where x represents her total sales in the month.

(a) Complete the table and use the results to list ordered pairs that satisfy the equation.

x	S	(x, S)
0		
10,000		
25,000		

(b) Graph the linear equation $S = 0.08x + 3000$ using the points obtained in part (a).

Figure 12

 ❷ **Graph a Line Using Intercepts**

Intercepts should always be displayed in a complete graph.

Definitions

The **intercepts** are the points, if any, where a graph crosses or touches a coordinate axis. A point at which the graph crosses or touches the x-axis is an **x-intercept,** and a point at which the graph crosses or touches the y-axis is a **y-intercept.**

See Figure 12 for an illustration. The graph in Figure 12 is the graph obtained in Example 3.

EXAMPLE 5   **Finding Intercepts from a Graph**

Find the intercepts of the graphs shown in Figures 13(a) and 13(b). What are the x-intercepts? What are the y-intercepts?

Figure 13

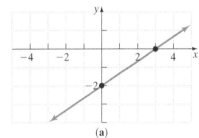
(a)

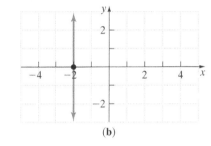
(b)

> **In Words**
> An x-intercept exists when $y = 0$.
> A y-intercept exists when $x = 0$.

Solution

(a) The intercepts of the graph in Figure 13(a) are the points $(0, -2)$ and $(3, 0)$. The x-intercept is $(3, 0)$. The y-intercept is $(0, -2)$.

(b) The intercept of the graph in Figure 13(b) is the point $(-2, 0)$. The x-intercept is $(-2, 0)$. There are no y-intercepts.

Quick ✓

12. The _____ are the points, if any, where a graph crosses or touches a coordinate axis.

In Problems 13 and 14, find the intercepts of the graphs shown in the figures. What are the x-intercepts? What are the y-intercepts?

13.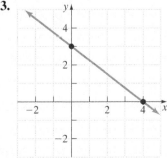

14.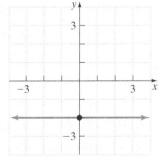

From Figure 12, it should be clear that an x-intercept exists when the value of y is 0 and that a y-intercept exists when the value of x is 0. This leads to the following procedure for finding the intercepts algebraically.

Work Smart

Every point on the x-axis has a y-coordinate of 0. That's why we set $y = 0$ to find an x-intercept. Likewise, every point on the y-axis has an x-coordinate of 0. That's why we set $x = 0$ to find a y-intercept.

Procedure for Finding Intercepts

1. To find the x-intercept(s), if any, of the graph of an equation, let $y = 0$ in the equation and solve for x.
2. To find the y-intercept(s), if any, of the graph of an equation, let $x = 0$ in the equation and solve for y.

We can use the intercepts to graph a line. Because the intercepts represent only two points, find a third point so that we can check our work.

EXAMPLE 6   **How to Graph a Linear Equation by Finding Its Intercepts**

Graph the linear equation $4x - 3y = 24$ by finding its intercepts.

Step-by-Step Solution

Step 1: Find the y-intercept by letting $x = 0$ and solving the equation for y.

$$4x - 3y = 24$$
$$\text{Let } x = 0: \quad 4(0) - 3y = 24$$
$$0 - 3y = 24$$
$$-3y = 24$$
Divide both sides by -3: $y = -8$

The y-intercept is $(0, -8)$.

Step 2: Find the x-intercept by letting $y = 0$ and solving the equation for x.

$$4x - 3y = 24$$
$$\text{Let } y = 0: \quad 4x - 3(0) = 24$$
$$4x - 0 = 24$$
$$4x = 24$$
Divide both sides by 4: $x = 6$

The x-intercept is $(6, 0)$.

Step 3: Find one additional point on the graph by choosing any value of x that is convenient and solving the equation for y.

Let $x = 3$ and solve the equation $4x - 3y = 24$ for y.

$$\text{Let } x = 3: \quad 4(3) - 3y = 24$$
$$12 - 3y = 24$$
Subtract 12 from both sides: $-3y = 12$
Divide both sides by -3: $y = -4$

The point $(3, -4)$ is on the graph of the equation.

Step 4: Plot the points found in Steps 1–3 and draw in the line.

Plot the points $(0, -8)$, $(6, 0)$, and $(3, -4)$. Connect the points with a straight line and obtain the graph in Figure 14.

Figure 14
$4x - 3y = 24$

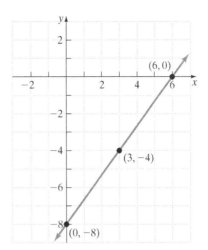

EXAMPLE 7 Graphing a Linear Equation by Finding Its Intercepts

Graph the linear equation $\frac{1}{2}x - 2y = 3$ by finding its intercepts.

Solution

x-intercept:

$$\frac{1}{2}x - 2y = 3$$

Let $y = 0$: $\frac{1}{2}x - 2(0) = 3$

$$\frac{1}{2}x = 3$$

$$x = 6$$

The x-intercept is $(6, 0)$.

y-intercept:

$$\frac{1}{2}x - 2y = 3$$

Let $x = 0$: $\frac{1}{2}(0) - 2y = 3$

$$-2y = 3$$

$$y = -\frac{3}{2}$$

The y-intercept is $\left(0, -\frac{3}{2}\right)$.

Additional point (choose $x = 2$):

$$\frac{1}{2}x - 2y = 3$$

Let $x = 2$: $\frac{1}{2}(2) - 2y = 3$

$$1 - 2y = 3$$

$$-2y = 2$$

$$y = -1$$

The additional point is $(2, -1)$.

Plot the points $(6, 0), \left(0, -\frac{3}{2}\right)$, and $(2, -1)$ and connect them with a straight line. See Figure 15.

Figure 15
$\frac{1}{2}x - 2y = 3$

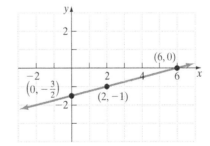

Quick ✓

15. *True or False* To find the y-intercept(s), if any, of the graph of an equation, let $y = 0$ in the equation and solve for x.

In Problems 16–18, graph each linear equation by finding its intercepts.

16. $x + y = 3$

17. $2x - 5y = 20$

18. $\frac{3}{2}x - 2y = 9$

EXAMPLE 8 Graphing a Linear Equation of the Form $Ax + By = 0$

Graph the linear equation $2x + 3y = 0$ by finding its intercepts.

Solution

x-intercept:

$$2x + 3y = 0$$
Let $y = 0$: $\quad 2x + 3(0) = 0$
$$2x + 0 = 0$$
Divide both sides by 2: $\quad x = 0$

The x-intercept is $(0, 0)$.

y-intercept:

$$2x + 3y = 0$$
Let $x = 0$: $\quad 2(0) + 3y = 0$
$$3y = 0$$
Divide both sides by 3: $\quad y = 0$

The y-intercept is $(0, 0)$.

Work Smart

Linear equations of the form $Ax + By = 0$, where $A \neq 0$ and $B \neq 0$, have only one intercept at $(0, 0)$, so two additional points should be plotted to obtain the graph.

Because both the x- and y-intercepts are $(0, 0)$, we find *two* additional points.

Additional point (choose $x = 3$):

$$2x + 3y = 0$$
Let $x = 3$: $\quad 2(3) + 3y = 0$
$$6 + 3y = 0$$
Subtract 6 from both sides: $\quad 3y = -6$
Divide both sides by 3: $\quad y = -2$

The additional point is $(3, -2)$.

Additional point (choose $x = -3$):

$$2x + 3y = 0$$
Let $x = -3$: $\quad 2(-3) + 3y = 0$
$$-6 + 3y = 0$$
Add 6 to both sides: $\quad 3y = 6$
Divide both sides by 2: $\quad y = 2$

The additional point is $(-3, 2)$.

Plot the points $(0, 0)$, $(-3, 2)$, and $(3, -2)$ and connect them with a straight line. See Figure 16.

Figure 16
$2x + 3y = 0$

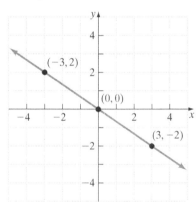

Quick ✓

In Problems 19 and 20, graph the equation by finding its intercepts.

19. $y = \dfrac{1}{2}x$

20. $4x + y = 0$

▶ ❸ Graph Vertical and Horizontal Lines

In the equation of a line, $Ax + By = C$, we said that A and B cannot both be zero. But what if $A = 0$ or $B = 0$? We find that this leads to special types of lines called *vertical lines* (when $B = 0$) and *horizontal lines* (when $A = 0$).

EXAMPLE 9 Graphing a Vertical Line

Graph the equation $x = 3$ using the point-plotting method.

Solution

Because the equation $x = 3$ can be written as $1x + 0y = 3$, the graph is a line. For the equation $x = 3$, any value of y has a corresponding x-value of 3. For example, if $y = -1$, then

$$1x + 0(-1) = 3$$
$$x = 3$$

See Table 12 for other choices for y. The points $(3, -2)$, $(3, -1)$, $(3, 0)$, $(3, 1)$, and $(3, 2)$ are all on the line. See Figure 17.

Table 12

x	y	(x, y)
3	−2	(3, −2)
3	−1	(3, −1)
3	0	(3, 0)
3	1	(3, 1)
3	2	(3, 2)

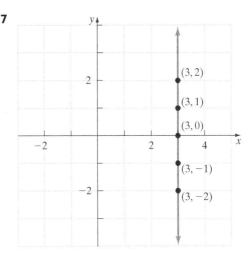

Figure 17 $x = 3$

The results of Example 9 lead to a definition of a *vertical line*.

Definition Equation of A Vertical Line

A **vertical line** is given by an equation of the form

$$x = a$$

where $(a, 0)$ is the x-intercept.

EXAMPLE 10 **Graphing a Horizontal Line**

Graph the equation $y = -2$ using the point-plotting method.

Solution

Because the equation $y = -2$ can be written as $0x + 1y = -2$, the graph is a line. For $y = -2$, any value of x has a corresponding y-value of -2. For example, if $x = -2$, then

$$0(-2) + 1y = -2$$
$$y = -2$$

See Table 13 for other choices of x. The points $(-2, -2)$, $(-1, -2)$, $(0, -2)$, $(1, -2)$, and $(2, -2)$ are all on the line. See Figure 18.

Table 13

x	y	(x, y)
−2	−2	(−2, −2)
−1	−2	(−1, −2)
0	−2	(0, −2)
1	−2	(1, 2)
2	−2	(2, −2)

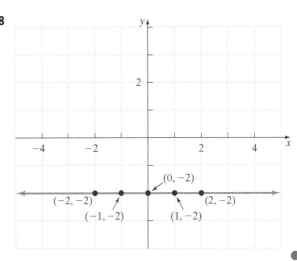

Figure 18 $y = -2$

The results of Example 10 lead to a definition of a *horizontal line*.

Definition Equation of a Horizontal Line

A **horizontal line** is given by an equation of the form

$$y = b$$

where $(0, b)$ is the y-intercept.

Quick ✓

21. A _____ line is given by an equation of the form $x = a$, where _____ is the x-intercept.

22. A _____ line is given by an equation of the form $y = b$, where _____ is the y-intercept.

In Problems 23 and 24, graph each equation.

23. $x = -5$ 24. $y = -4$

We present a summary below to help you organize the information presented in this section.

Summary Intercepts and Equations of Lines

Topic	Comments
Intercepts: Points where the graph crosses or touches a coordinate axis.	Intercepts need to be shown for a graph to be complete.
x-intercept: Point where the graph crosses or touches the x-axis. Found by letting $y = 0$ in the equation.	
y-intercept: Point where the graph crosses or touches the y-axis. Found by letting $x = 0$ in the equation.	
Standard Form of an Equation of a Line: $Ax + By = C$, where A and B are not both zero.	Can be graphed using point-plotting or intercepts.
Equation of a Vertical Line: $x = a$	Graph is a vertical line whose x-intercept is $(a, 0)$.
Equation of a Horizontal Line: $y = b$	Graph is a horizontal line whose y-intercept is $(0, b)$.

3.IR2 Exercises MyLab Statistics

Underlined exercises have complete video solutions in MyLab.

Problems 1–24 are the Quick ✓s that follow the EXAMPLES.

Building Skills

In Problems 25–32, determine whether or not the equation is a linear equation in two variables. See Objective 1.

25. $2x - 5y = 10$
26. $y^2 = 2x + 3$
27. $x^2 + y = 1$
28. $y - 2x = 9$
29. $y = \dfrac{4}{x}$
30. $x - 8 = 0$
31. $y - 1 = 0$
32. $y = -\dfrac{2}{x}$

In Problems 33–50, graph each linear equation using the point-plotting method. See Objective 1.

33. $y = 2x$
34. $y = 3x$
35. $y = 4x - 2$
36. $y = -3x - 1$
37. $y = -2x + 5$
38. $y = x - 6$
39. $x + y = 5$
40. $x - y = 6$
41. $-2x + y = 6$
42. $5x - 2y = -10$
43. $4x - 2y = -8$
44. $x + 3y = 6$
45. $x = -4y$
46. $x = \dfrac{1}{2}y$
47. $y + 7 = 0$
48. $x - 6 = 0$
49. $y - 2 = 3(x + 1)$
50. $y + 3 = -2(x - 2)$

In Problems 51–58, find the intercepts of each graph. See Objective 2.

51.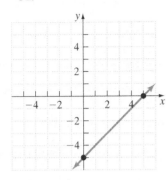

52.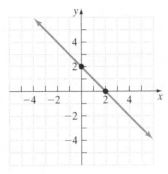

53.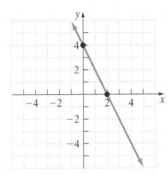

54.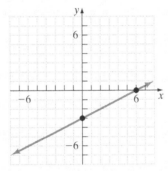

55.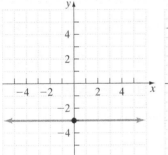

56.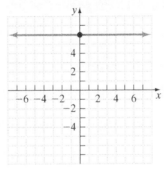

57.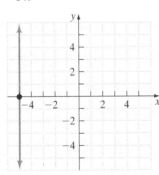

58.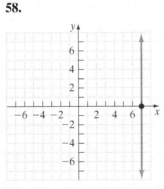

In Problems 59–70, find the intercepts of each equation. See Objective 2.

59. $2x + 3y = -12$ **60.** $3x - 5y = 30$

61. $x = -6y$ **62.** $y = 10x$

63. $y = x - 5$ **64.** $y = -x + 7$

65. $\dfrac{x}{6} + \dfrac{y}{8} = 1$ **66.** $\dfrac{x}{2} - \dfrac{y}{8} = 1$

67. $x = 4$ **68.** $y = 6$

69. $y + 2 = 0$ **70.** $x + 8 = 0$

In Problems 71–86, graph each linear equation by finding its intercepts. See Objective 2.

71. $3x + 6y = 18$ **72.** $3x - 5y = 15$

73. $-x + 5y = 15$ **74.** $-2x + y = 14$

75. $\dfrac{1}{2}x = y + 3$ **76.** $\dfrac{4}{3}x = -y + 1$

77. $9x - 2y = 0$ **78.** $\dfrac{1}{3}x - y = 0$

79. $y = -\dfrac{1}{2}x + 3$ **80.** $y = \dfrac{2}{3}x - 3$

81. $\dfrac{1}{3}y + 2 = 2x$ **82.** $\dfrac{1}{2}x - 3 = 3y$

83. $\dfrac{x}{2} + \dfrac{y}{3} = 1$ **84.** $\dfrac{y}{4} - \dfrac{x}{3} = 1$

85. $4y - 2x + 1 = 0$ **86.** $2y - 3x + 2 = 0$

In Problems 87–94, graph each horizontal or vertical line. See Objective 3.

87. $x = 5$ **88.** $x = -7$

89. $y = -6$ **90.** $y = 2$

91. $y - 12 = 0$ **92.** $y + 3 = 0$

93. $3x - 5 = 0$ **94.** $2x - 7 = 0$

Mixed Practice

In Problems 95–106, graph each linear equation by the point-plotting method or by finding intercepts.

95. $y = 2x - 5$ **96.** $y = -3x + 2$

97. $y = -5$ **98.** $x = 2$

99. $2x + 5y = -20$ **100.** $3x - 4y = 12$

101. $2x = -6y + 4$ **102.** $5x = 3y - 10$

103. $x - 3 = 0$ **104.** $y + 4 = 0$

105. $3y - 12 = 0$ **106.** $-4x + 8 = 0$

107. If $(3, y)$ is a point on the graph of $4x + 3y = 18$, find y.

108. If $(-4, y)$ is a point on the graph of $3x - 2y = 10$, find y.

109. If $(x, -2)$ is a point on the graph of $3x + 5y = 11$, find x.

110. If $(x, -3)$ is a point on the graph of $4x - 7y = 19$, find x.

Applying the Concepts

111. Plot the points $(3, 5)$ and $(-2, 5)$ and draw a line through the points. What is the equation of this line?

112. Plot the points $(-1, 2)$ and $(5, 2)$ and draw a line through the points. What is the equation of this line?

113. Plot the points $(-2, -4)$ and $(-2, 1)$ and draw a line through the points. What is the equation of this line?

114. Plot the points $(3, -1)$ and $(3, 2)$ and draw a line through the points. What is the equation of this line?

In Problems 115–118, find the equation of each line.

115. **116.** **117.** **118.**

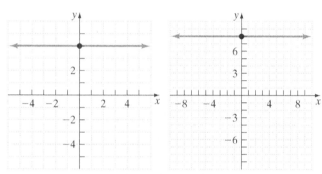

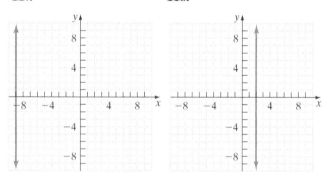

119. Create a set of ordered pairs in which the x-coordinates are twice the y-coordinates. What is the equation of this line?

120. Create a set of ordered pairs in which the y-coordinates are twice the x-coordinates. What is the equation of this line?

121. Create a set of ordered pairs in which the y-coordinates are 2 more than the x-coordinates. What is the equation of this line?

122. Create a set of ordered pairs in which the x-coordinates are 3 less than the y-coordinates. What is the equation of this line?

123. Calculating Wages Marta earns $500 per week plus $100 in commission for every car she sells. The linear equation that calculates her weekly earnings is $E = 100n + 500$, where E represents her weekly earnings in dollars and n represents the number of cars she sold during the week.

(a) Create a set of ordered pairs (n, E) if, in three consecutive weeks, she sold 0 cars, 4 cars, and 10 cars.
(b) Graph the linear equation $E = 100n + 500$ using the ordered pairs obtained in part (a). Be sure to label the axes appropriately.
(c) Explain the meaning of the E-intercept.

124. Carpet Cleaning Harry's Carpet Cleaning charges a $50 service charge plus $0.10 for each square foot of carpeting to be cleaned. The linear equation that calculates the total cost to clean a carpet is $C = 0.1f + 50$, where C is the total cost in dollars and f is the number of square feet of carpet.

(a) Create a set of ordered pairs (f, C) for the following number of square feet to be cleaned: 1000 sq ft, 2000 sq ft, 2500 sq ft.
(b) Graph the linear equation $C = 0.1f + 50$ using the ordered pairs obtained in part (a). Be sure to label the axes appropriately.
(c) Explain the meaning of the C-intercept.

Extending the Concepts

125. Graph each of the following linear equations in the same xy-plane. What do you notice about each of these graphs?

$$y = 2x - 1 \quad y = 2x + 3 \quad 2x - y = 5$$

126. Graph each of the following linear equations in the same xy-plane. What do you notice about each of these graphs?

$$y = 3x + 2 \quad 6x - 2y = -4 \quad x = \frac{1}{3}y - \frac{2}{3}$$

127. Graph each of the following linear equations in the same xy-plane. What statement can you make about the steepness of the line as the coefficient of x gets larger?

$$y = x \quad y = 2x \quad y = 10x$$

128. Graph each of the following linear equations in the same xy-plane. What statement can you make about the steepness of the line as the coefficient of x gets smaller?

$$y = x + 2 \quad y = \frac{1}{2}x + 2 \quad y = \frac{1}{8}x + 2$$

In Problems 129–132, find the intercepts of each graph.

129. **130.**

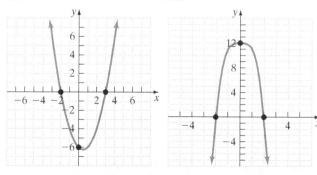

131.

132.

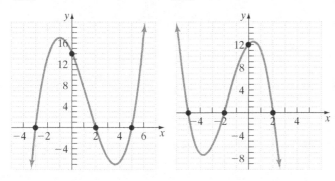

The Graphing Calculator

Graphing calculators can graph equations. In fact, graphing calculators also use the point-plotting method to obtain the graph by choosing 95 values of x and using the equation to find the corresponding values of y. As with creating tables, first solve the equation for y. For example, to obtain a table of values that satisfy the equation $2x - 3y = 12$ (Example 6 from Section 3.IR1), solve for y, and obtain $y = \frac{2}{3}x - 4$. Enter the equation $y = \frac{2}{3}x - 4$ into the calculator and create the graph shown below.

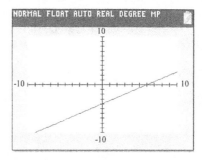

Explaining the Concepts

133. Explain what the graph of an equation represents.

134. What is meant by a complete graph?

135. How many points are required to graph a line? Explain your reasoning and why you might include additional point(s) when graphing a line.

136. Explain how to use the intercepts to graph the equation $Ax + By = C$, where A, B, and C are not equal to zero. Explain how to graph the same equation when C is equal to zero. Can you use the same techniques for both equations? Why or why not?

In Problems 137–142, use a graphing calculator to graph each equation.

137. $y = 2x - 9$ **138.** $y = -3x + 8$

139. $y + 2x = 13$ **140.** $y - x = -15$

141. $y = -6x^2 + 1$ **142.** $y = -x^2 + 3x$

3.IR3 Slope

Objectives

1. Find the Slope of a Line Given Two Points
2. Find the Slope of Vertical and Horizontal Lines
3. Graph a Line Using Its Slope and a Point on the Line
4. Work with Applications of Slope

Figure 19

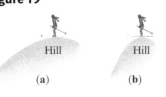

(a)   (b)

Pretend you are on snow skis for the first time in your life. The ski resort that you are visiting has two "beginner" hills, as shown in Figure 19. Which hill would you prefer to go down? Why?

It is clear that the hill in Figure 19(a) is not as steep as the hill in Figure 19(b). Mathematicians like to numerically describe situations such as the steepness of a hill. Measuring the steepness of each hill allows for an easy comparison. The numerical measure that describes the steepness of a hill is its *slope*.

▶ ① Find the Slope of a Line Given Two Points

Consider the staircase drawn in Figure 20(a) below. If we draw a line through the top of each riser on the staircase, we can see that each step contains exactly the same horizontal change (or **run**) and the same vertical change (or **rise**). We define *slope* in terms of the rise and run.

Figure 20

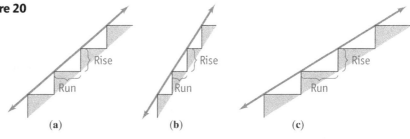

(a)   (b)   (c)

Definition

The **slope** of a line, denoted by the letter m, is the ratio of the rise to the run. That is,

$$\text{Slope} = m = \frac{\text{rise}}{\text{run}}$$

The symbol for the slope of a line is m, which comes from the French word *monter*, which means "to go up, ascend, or climb." Slope is a numerical measure of the steepness of the line. For example, if the run is decreased and the rise remains the same, then the staircase becomes steeper. See Figure 20(b). If the run is increased and the rise remains the same, then the staircase becomes less steep. See Figure 20(c).

If the staircase in Figure 20(a) has a rise of 6 inches and a run of 6 inches, then the slope of the line is

$$m = \frac{\text{rise}}{\text{run}} = \frac{6 \text{ inches}}{6 \text{ inches}} = 1$$

If the rise of the staircase is increased to 9 inches, then the slope of the line is

$$m = \frac{\text{rise}}{\text{run}} = \frac{9 \text{ inches}}{6 \text{ inches}} = \frac{3}{2}$$

The main idea is the steeper the line, the larger the slope. We can define the slope of a line using rectangular coordinates.

Work Smart

The subscripts 1 and 2 on $x_1, x_2, y_1,$ and y_2 do not represent a computation (as superscripts do in x^2). Instead, they are used to indicate that the values of the variable x_1 may be different from x_2, and y_1 may be different from y_2.

Definition

If $x_1 \neq x_2$, the **slope** m of the line containing the points (x_1, y_1) and (x_2, y_2) is defined by the formula

$$m = \frac{y_2 - y_1}{x_2 - x_1} \quad x_1 \neq x_2$$

Figure 21 illustrates the slope of a line.

Figure 21

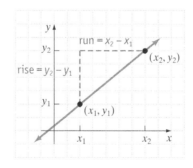

Notice that the slope m of a line may be viewed as

$$m = \frac{\text{rise}}{\text{run}} = \frac{y_2 - y_1}{x_2 - x_1}$$

We can also write the slope m of a line as

In Words

Slope is rise over run, or the change in y divided by the change in x.

$$m = \frac{y_2 - y_1}{x_2 - x_1} = \frac{\text{change in } y}{\text{change in } x} = \frac{\Delta y}{\Delta x}$$

The symbol Δ is the Greek letter delta. In mathematics, we read the symbol Δ as "change in." Thus $\frac{\Delta y}{\Delta x}$ is read "change in y divided by change in x." The symbol Δ comes from the first letter of the Greek word for "difference," *diaphora*.

EXAMPLE 1   How to Find the Slope of a Line

Find the slope of the line drawn in Figure 22.

Figure 22

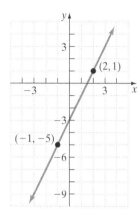

Step-by-Step Solution

Step 1: Let one of the points be (x_1, y_1) and the other point be (x_2, y_2).

Let's say that $(x_1, y_1) = (-1, -5)$ and $(x_2, y_2) = (2, 1)$.

Step 2: Find the slope by evaluating

$$m = \frac{y_2 - y_1}{x_2 - x_1}$$

$$m = \frac{y_2 - y_1}{x_2 - x_1}$$

$x_1 = -1, y_1 = -5;$
$x_2 = 2, y_2 = 1:$

$$= \frac{1 - (-5)}{2 - (-1)}$$

$$= \frac{6}{3}$$

$$m = 2$$

Figure 23

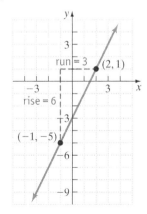

Work Smart

It doesn't matter which point is called (x_1, y_1) and which is called (x_2, y_2). The answer will be the same. In Example 1, if we let $(x_1, y_1) = (2, 1)$ and $(x_2, y_2) = (-1, -5)$, we get

$$m = \frac{y_2 - y_1}{x_2 - x_1} = \frac{-5 - 1}{-1 - 2} = \frac{-6}{-3} = 2$$

Remember that slope can be thought of as "rise divided by run." This description of the slope of a line is illustrated in Figure 23. Interpret the slope of the line drawn in Figure 23 as follows: "The value of y will increase by 6 units whenever x increases by 3 units." Or, "If x increases by 3 units, then y will increase by 6 units."

Because $\frac{6}{3} = 2 = \frac{2}{1}$, "the value of y will increase by 2 units whenever x increases by 1 unit." Or, "If x increases by 1 unit, then y will increase by 2 units."

EXAMPLE 2   Finding and Interpreting the Slope of a Line

Plot the points $(-1, 3)$ and $(2, -2)$ and draw a line through the two points. Find and interpret the slope of the line.

Solution

We plot the points $(x_1, y_1) = (-1, 3)$ and $(x_2, y_2) = (2, -2)$ in the rectangular coordinate system and draw a line through the two points. See Figure 24. The slope

Figure 24

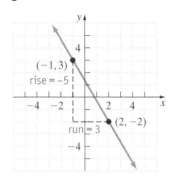

Work Smart

A third interpretation is: If x increases by 3 units, y will decrease by 5 units.

of the line drawn in Figure 24 is

$$m = \frac{y_2 - y_1}{x_2 - x_1} = \frac{-2 - 3}{2 - (-1)}$$

$$= \frac{-5}{3}$$

$$= -\frac{5}{3}$$

You can interpret a slope of $-\frac{5}{3} = \frac{-5}{3}$ this way: The value of y will go down 5 units whenever x increases by 3 units. Because $-\frac{5}{3} = \frac{5}{-3} = \frac{\text{rise}}{\text{run}}$, a second interpretation is as follows: The value of y will increase by 5 units whenever x decreases by 3 units.

Notice that the line drawn in Figure 23 goes up and to the right and the slope is positive, while the line drawn in Figure 24 goes down and to the right and slope is negative. In general, lines that go down and to the right have negative slope, and lines that go up and to the right have positive slope. See Figure 25.

Figure 25

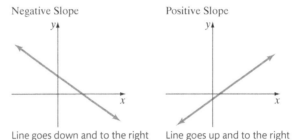

Negative Slope — Line goes down and to the right
Positive Slope — Line goes up and to the right

Quick ✓

1. If the run of a line is 10 and its rise is 6, then its slope is _____.
2. *True or False* If $P = (x_1, y_1)$ and $Q = (x_2, y_2)$ are two distinct points with $y_1 \neq y_2$, the slope m of the line that contains points P and Q is defined by the formula $m = \frac{x_2 - x_1}{y_2 - y_1}$.
3. *True or False* If the slope of a line is $\frac{3}{2}$, then y will increase by 3 units when x increases by 2 units.
4. If the graph of a line goes up as you move to the right, then the slope of this line must be _____.

In Problems 5 and 6, plot the points in a rectangular coordinate system. Then draw a line through the two points. Find and interpret the slope of the line containing the points.

5. $(0, 2)$ and $(4, 10)$
6. $(-2, 2)$ and $(3, -7)$

▶ ❷ Find the Slope of Vertical and Horizontal Lines

Did you notice that in the definition of slope, $m = \frac{y_2 - y_1}{x_2 - x_1}$, there is a restriction that $x_1 \neq x_2$? This means the formula does not apply if the x-coordinates of the two points are the same. Why? See Example 3.

EXAMPLE 3 **The Slope of a Vertical Line**

Plot the points $(2, -1)$ and $(2, 3)$ in a rectangular coordinate system. Then draw a line through the two points. Find and interpret the slope of the line.

Solution

Plot the points $(x_1, y_1) = (2, -1)$ and $(x_2, y_2) = (2, 3)$ in the rectangular coordinate system and draw a line through the two points. See Figure 26. The slope of the line drawn in Figure 26 is

$$m = \frac{y_2 - y_1}{x_2 - x_1} = \frac{3 - (-1)}{2 - 2}$$

$$= \frac{4}{0}$$

Figure 26

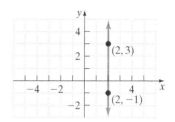

Because division by 0 is undefined, the slope of the line is undefined. When y increases by 4, there is no change in x.

Let's generalize the results of Example 3. Let (x_1, y_1) and (x_2, y_2) be two distinct points. If $x_1 = x_2$, then we have a **vertical line** whose slope m is **undefined** (since this results in division by 0). See Figure 27.

Figure 27
Vertical line

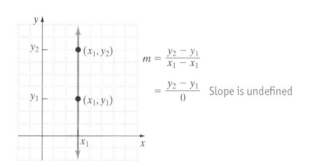

Okay, but what if $y_1 = y_2$?

EXAMPLE 4 **The Slope of a Horizontal Line**

Plot the points $(-2, 4)$ and $(3, 4)$ in a rectangular coordinate system. Then draw a line through the two points. Find and interpret the slope of the line.

Solution

Plot the points $(x_1, y_1) = (-2, 4)$ and $(x_2, y_2) = (3, 4)$ in the rectangular coordinate system and draw a line through the two points. See Figure 28. The slope of the line drawn in Figure 28 is

$$m = \frac{y_2 - y_1}{x_2 - x_1} = \frac{4 - 4}{3 - (-2)}$$

$$= \frac{0}{5}$$

$$= 0$$

Figure 28

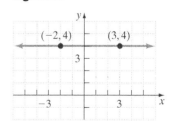

The slope of the line is 0. A slope of 0 can be interpreted as: When x increases by 1 unit, there is no change in y.

Let's generalize the results of Example 4. Let (x_1, y_1) and (x_2, y_2) be two distinct points. If $y_1 = y_2$, then we have a **horizontal line** whose slope m is 0. See Figure 29 on the next page.

Figure 29
Horizontal line

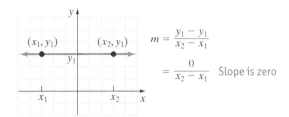

Quick ✓

7. The slope of a horizontal line is _____, while the slope of a vertical line is _____.

In Problems 8 and 9, plot the given points in a rectangular coordinate system. Then draw a line through the two points. Find and interpret the slope of the line.

8. $(2, 5)$ and $(2, -1)$ **9.** $(2, 5)$ and $(6, 5)$

Summary The Slope of a Line

Figure 30 illustrates the four possibilities for the slope of a line. Remember, just as we read a text from left to right, we also read graphs from left to right.

Figure 30

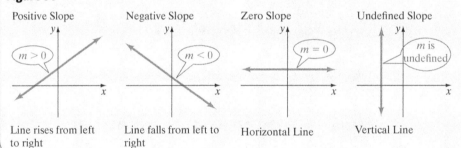

Positive Slope — Line rises from left to right
Negative Slope — Line falls from left to right
Zero Slope — Horizontal Line
Undefined Slope — Vertical Line

▶ ❸ Graph a Line Using Its Slope and a Point on the Line

We now illustrate how to use slope to graph lines.

EXAMPLE 5 **Graphing a Line Given a Point and Its Slope**

Draw a graph of the line that contains the point $(1, 3)$ and has a slope of 2.

Solution

We know that $m = 2 = \dfrac{2}{1} = \dfrac{\text{rise}}{\text{run}}$. This means that y will increase by 2 units (the rise), when x increases by 1 unit (the run). So if we start at $(1, 3)$ and move 2 units up and then 1 unit to the right, we end up at the point $(2, 5)$. We then draw a line through the points $(1, 3)$ and $(2, 5)$ to obtain the graph of the line. See Figure 31.

Figure 31

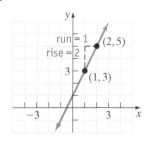

Work Smart

Because $m = \dfrac{2}{1}$ also means that if x increases by 1 unit, then y will increase by 2 units, we could start at $(1, 3)$, move 1 unit right and then 2 units up to end at $(2, 5)$.

EXAMPLE 6 — Graphing a Line Given a Point and Its Slope

Draw a graph of the line that contains the point $(1, 3)$ and has a slope of $-\dfrac{2}{3}$.

Solution

Because slope $= -\dfrac{2}{3} = \dfrac{-2}{3} = \dfrac{\text{rise}}{\text{run}}$, y will decrease by 2 units when x increases by 3 units. If we start at $(1, 3)$ and move 2 units down and then 3 units to the right, we end up at the point $(4, 1)$. We then draw a line through the points $(1, 3)$ and $(4, 1)$ to obtain the graph of the line. See Figure 32.

Figure 32

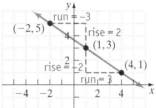

It is perfectly acceptable to set $\dfrac{\text{rise}}{\text{run}} = -\dfrac{2}{3} = \dfrac{2}{-3}$ so that we move 2 units up from $(1, 3)$ and then 3 units to the left. We would then end up at $(-2, 5)$, which is also on the graph of the line, as indicated in Figure 32.

Quick ✓

10. Draw a graph of the line that contains the point $(1, 2)$ and has a slope of

(a) $\dfrac{1}{2}$ (b) -3 (c) 0

▶ ④ Work with Applications of Slope

Work Smart

If the "rise" is positive, we go up. If the "rise" is negative, we go down. Similarly, if the "run" is positive, then we go to the right. If the "run" is negative, then we go to the left.

In its simplest form, slope is a ratio of rise over run. For example, a hill whose grade is $5\%\left(= 0.05 = \dfrac{5}{100}\right)$ goes up 5 feet (the rise) for every 100 feet it goes horizontally (the run). See Figure 33.

Figure 33

100 feet — 5 feet

Work Smart

The pitch of a roof or grade of a road is always represented as a positive number.

Consider the pitch of a roof. If a roof's pitch is $\dfrac{5}{12}$, then every 5-foot measurement upward results in a horizontal measurement of 12 feet. See Figure 34.

Figure 34

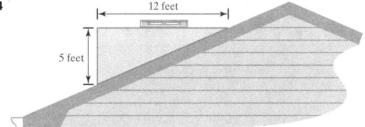

EXAMPLE 7 — Finding the Grade of a Road

In Heckman Pass, British Columbia, there is a road that rises 9 feet for every 50 feet of horizontal distance covered. What is the grade of the road?

Solution

The grade of the road is given by $\dfrac{\text{rise}}{\text{run}}$. Since a rise of 9 feet is accompanied by a run of 50 feet, the grade of the road is $\dfrac{9 \text{ feet}}{50 \text{ feet}} = 0.18 = 18\%$.

The slope m of a line measures the amount that y changes as x changes from x_1 to x_2. The slope of a line is also called the **average rate of change** of y with respect to x.

In applications, we are often interested in knowing how the change in one variable might affect some other variable. For example, if your income increases by $1000, how much will your spending (on average) change? Or, if the speed of your car increases by 10 miles per hour, how much (on average) will your car's gas mileage change?

EXAMPLE 8   Slope as an Average Rate of Change

In Naples, Florida, the price of a new three-bedroom house that is 2100 square feet is $239,000. The price of a new three-bedroom house that is 2500 square feet is $295,000. Find and interpret the slope of the line joining the points (2100, 239,000) and (2500, 295,000). (*Source: www.fiddlerscreek.com*)

Solution

Let x represent the square footage of the house and y represent the price. Let $(x_1, y_1) = (2100, 239{,}000)$ and $(x_2, y_2) = (2500, 295{,}000)$ and compute the slope as

$$m = \frac{y_2 - y_1}{x_2 - x_1} = \frac{295{,}000 - 239{,}000}{2500 - 2100}$$

$$= \frac{56{,}000}{400}$$

$$= 140$$

Work Smart

Notice how the words "on average" are included in the interpretation. This is important because we cannot say each additional square foot will definitely increase the price $140. Instead, over the data observed, this was the "average" price change.

The unit of measure for y is dollars, and the unit of measure for x is square feet. So, the slope can be interpreted as follows: Between 2100 and 2500 square feet, the price increases by $140 per square foot, on average.

Quick ✓

11. A road rises 4 feet for every 50 feet of horizontal distance covered. What is the grade of the road?

12. The average annual cost of operating a Chevy Cobalt is $1370 when it is driven 10,000 miles. The average annual cost of operating a Chevy Cobalt is $1850 when it is driven 14,000 miles. Find and interpret the slope of the line joining (10,000, 1370) and (14,000, 1850).

3.IR3 Exercises   MyLab Statistics

Underlined exercises have complete video solutions in MyLab.

Problems 1–12 are the Quick ✓s that follow the EXAMPLES.

Building Skills

In Problems 13–18, find the slope of the line whose graph is given. See Objective 1.

13.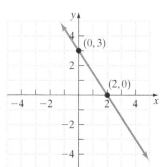

14.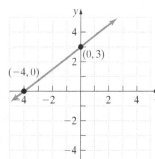

15.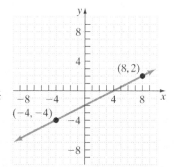

16.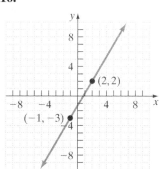

17.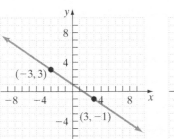

18.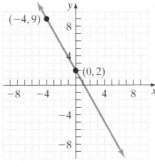

39.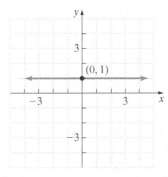

40.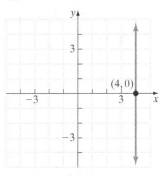

In Problems 19–22, (a) plot the points in a rectangular coordinate system, (b) draw a line through the points, (c) and find and interpret the slope of the line. See Objective 1.

19. $(-3, 2)$ and $(3, 5)$ **20.** $(2, 6)$ and $(-2, -4)$

21. $(2, -9)$ and $(-2, -1)$ **22.** $(4, -5)$ and $(-2, -4)$

In Problems 23–36, find and interpret the slope of the line containing the given points. See Objective 1.

23. $(10, 4)$ and $(6, 12)$ **24.** $(7, 3)$ and $(0, -11)$

25. $(4, -4)$ and $(12, -12)$ **26.** $(-3, 2)$ and $(2, -3)$

27. $(7, -2)$ and $(4, 3)$ **28.** $(-8, -1)$ and $(2, 3)$

29. $(-4, -1)$ and $(2, 3)$ **30.** $(5, 1)$ and $(-1, -1)$

31. $(0.4, 0.7)$ and $(1.2, 2.3)$

32. $(-0.2, 1.3)$ and $(0.2, 0.1)$

33. $\left(\dfrac{1}{2}, \dfrac{3}{4}\right)$ and $\left(-\dfrac{5}{2}, -\dfrac{1}{4}\right)$ **34.** $\left(-\dfrac{1}{3}, \dfrac{2}{5}\right)$ and $\left(\dfrac{2}{3}, -\dfrac{3}{5}\right)$

35. $(0.21, 1.38)$ and $(0.61, -0.60)$

36. $(1.72, 0.35)$ and $(-0.28, 0.05)$

In Problems 37–40, find the slope of the line whose graph is given. See Objective 2.

37.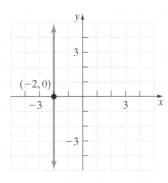

38.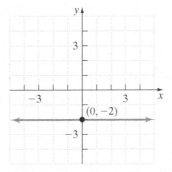

In Problems 41–44, find and interpret the slope of the line containing the given points. See Objective 2.

41. $(4, -6)$ and $(-1, -6)$ **42.** $(-1, -3)$ and $(-1, 2)$

43. $(3, 9)$ and $(3, -2)$ **44.** $(5, 1)$ and $(-2, 1)$

In Problems 45–62, draw a graph of the line that contains the given point and has the given slope. See Objective 3.

45. $(4, 2); m = 1$ **46.** $(3, -1); m = -1$

47. $(0, 6); m = -2$ **48.** $(-1, 3); m = 3$

49. $(-1, 0); m = \dfrac{1}{4}$ **50.** $(5, 2); m = -\dfrac{1}{2}$

51. $(2, -3); m = 0$ **52.** $(1, 4); m$ is undefined

53. $(2, 1); m = \dfrac{2}{3}$ **54.** $(-2, -3); m = \dfrac{5}{2}$

55. $(-1, 4); m = -\dfrac{5}{4}$ **56.** $(0, -2); m = -\dfrac{3}{2}$

57. $(0, 0); m$ is undefined   **58.** $(3, -1); m = 0$

59. $(0, 2); m = -4$ **60.** $(0, 0); m = \dfrac{1}{5}$

61. $(2, -3); m = \dfrac{3}{4}$ **62.** $(-3, 0); m = -3$

Applying the Concepts

In Problems 63–66, draw the graph of the two lines with the given properties on the same rectangular coordinate system.

63. Both lines pass through the point $(2, -1)$. One has slope of 2 and the other has slope of $-\dfrac{1}{2}$.

64. Both lines pass through the point $(3, 0)$. One has slope of $\dfrac{2}{3}$ and the other has slope of $-\dfrac{3}{2}$.

65. Both lines have a slope of $\dfrac{3}{4}$. One passes through the point $(-1, -2)$, and the other passes through the point $(2, 1)$.

66. Both lines have a slope of -1. One passes through the point $(0, -3)$, and the other passes through the point $(2, -1)$.

67. **Roof Pitch** A carpenter was installing a new roof on a garage and noticed that for every 1-foot horizontal run, the roof was elevated by 4 inches. What is the pitch of this roof?

68. **Roof Pitch** A canopy is set up on the football field. On the 45-yard line, the height of the canopy is 68 inches. The peak of the canopy is at the 50-yard marker, where the height is 84 inches. What is the pitch of the roof of the canopy?

69. **Building a Roof** To build a shed in his backyard, Moises has decided to use a pitch of $\frac{2}{5}$ for his roof. The shed measures 30 in. from the side to the center. How much height should he add to the roof to get the desired pitch?

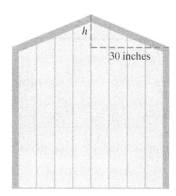

70. **Building a Roof** The design for the bedroom of a house requires a roof pitch of $\frac{7}{20}$. If the room measures 5 feet from the wall to the center, how high above the ceiling is the peak of the roof?

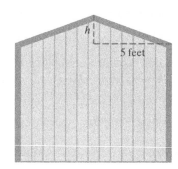

71. **Road Grade** Fall River Road was completed in 1920 and was the first road built through the Rocky Mountains in Colorado. It was so steep that sometimes the early model cars had to drive up the hill in reverse to maximize the output of their weak engines and fuel systems. If the road rises 200 feet for every 1250 feet of horizontal change, what is the grade of this road in percent?

72. **Road Grade** Barbara decided to take a bicycle trip up to the observatory on Mauna Kea on the island of Hawaii. The road has a vertical rise of 120 feet for every 800 feet of horizontal change. In percent, what is the grade of this road?

73. **Population Growth** The population of the United States was 123,202,624 in 1930 and 309,330,219 in 2010. Use the ordered pairs $(0, 123$ million$)$ and $(80, 309$ million$)$ to find and interpret the slope of the line representing the average rate of change in the population of the United States.

74. **Earning Potential** On average, a person who graduates from high school can expect to have lifetime earnings of 1.2 million dollars. It takes four years to earn a bachelor's degree, but the lifetime earnings will increase to 2.1 million dollars. Use the ordered pairs $(0, 1.2$ million$)$ and $(4, 2.1$ million$)$ to find and interpret the slope of the line representing the increase in earnings due to finishing college.

Extending the Concepts

In Problems 75–78, find any two ordered pairs that lie on the given line. Graph the line and then determine the slope of the line.

75. $3x + y = -5$

76. $2x + 5y = 12$

77. $y = 3x + 4$

78. $y = -x - 6$

In Problems 79–84, find the slope of the line containing the given points.

79. $(2a, a)$ and $(3a, -a)$

80. $(4p, 2p)$ and $(-2p, 5p)$

81. $(2p + 1, q - 4)$ and $(3p + 1, 2q - 4)$

82. $(3p + 1, 4q - 7)$ and $(5p + 1, 2q - 7)$

83. $(a + 1, b - 1)$ and $(2a - 5, b + 5)$

84. $(2a - 3, b + 4)$ and $(4a + 7, 5b - 1)$

In economics, **marginal revenue** is a rate of change defined as the change in total revenue divided by the change in output. If Q_1 represents the number of units sold, then the total revenue from selling these goods is represented by R_1. If Q_2 represents a different number of units sold, then the total revenue from this sale is represented by R_2. We compute marginal revenue as

$$MR = \frac{R_2 - R_1}{Q_2 - Q_1}.$$

So marginal revenue is a rate of change, or slope. Marginal revenue is important in economics because it is used to determine the level of output that maximizes profits for a company. Use the marginal revenue formula to solve Problems 85 and 86.

85. Determine and interpret marginal revenue if total revenue is $1000 when 400 hot dogs are sold at a baseball game and total revenue is $1200 when 500 hot dogs are sold.

86. Determine and interpret marginal revenue if total revenue is $300 when 30 flash drives are sold and total revenue is $400 when 50 flash drives are sold.

Explaining the Concepts

87. Describe a line that has one x-intercept but no y-intercept. Give two ordered pairs that could lie on this line and then describe how to find its slope.

88. Describe a line that has one y-intercept but no x-intercept. Give two ordered pairs that could lie on this line and then describe how to find its slope.

3.IR4 Slope-Intercept Form of a Line

Objectives

1. Use the Slope-Intercept Form to Identify the Slope and y-Intercept of a Line
2. Graph a Line Whose Equation Is in Slope-Intercept Form
3. Graph a Line Whose Equation Is in the Form $Ax + By = C$
4. Find the Equation of a Line Given Its Slope and y-Intercept
5. Work with Linear Models in Slope-Intercept Form

1 Use the Slope-Intercept Form to Identify the Slope and y-Intercept of a Line

In this section, we use the slope and y-intercept to graph a line. This method for graphing will be more efficient than plotting points. Why? Well, suppose we want to graph the equation $-2x + y = 5$ by plotting points. To do this, we might first solve the equation for y (get y by itself) by adding $2x$ to both sides of the equation.

$$-2x + y = 5$$

Add 2x to both sides: $y = 2x + 5$

Now create Table 14, which gives points on the graph of the equation. Figure 35 shows the graph of the line.

Notice two things about the line in Figure 35. First, the slope is $m = 2$. Second, the y-intercept is $(0, 5)$. If you look at the equation after we solved for y, $y = 2x + 5$, you should notice that the coefficient of the variable x is 2 and the constant is 5. This is no coincidence!

> **Slope-Intercept Form of an Equation of a Line**
>
> An equation of a line with slope m and y-intercept $(0, b)$ is
>
> $$y = mx + b$$

Table 14

x	$y = 2x + 5$	(x, y)
-2	$2(-2) + 5 = 1$	$(-2, 1)$
-1	$2(-1) + 5 = 3$	$(-1, 3)$
0	$2(0) + 5 = 5$	$(0, 5)$

Figure 35
$-2x + y = 5$

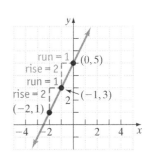

EXAMPLE 1 Finding the Slope and y-Intercept of a Line

Find the slope and y-intercept of the line whose equation is $y = -3x + 1$.

Solution

Compare the equation $y = -3x + 1$ to the slope-intercept form of a line $y = mx + b$. The coefficient of x, -3, is the slope, and the constant is 1, so the y-intercept is $(0, 1)$.

EXAMPLE 2 Finding the Slope and y-Intercept of a Line Whose Equation Is in Standard Form

Find the slope and y-intercept of the line whose equation is $3x + 2y = 6$.

Solution

Rewrite the equation $3x + 2y = 6$ so that it is in the form $y = mx + b$.

$$3x + 2y = 6$$

Subtract 3x from both sides: $2y = -3x + 6$

Divide both sides by 2: $y = \dfrac{-3x + 6}{2}$

$\dfrac{a+b}{c} = \dfrac{a}{c} + \dfrac{b}{c}$: $y = \dfrac{-3}{2}x + \dfrac{6}{2}$

Simplify: $y = -\dfrac{3}{2}x + 3$

Work Smart

Notice that after we subtracted 3x from both sides, we wrote the equation as $2y = -3x + 6$ rather than $2y = 6 - 3x$. This is because we want to get the equation in the form $y = mx + b$, so the term involving x should be first.

Now compare the equation $y = -\dfrac{3}{2}x + 3$ to the slope-intercept form of a line $y = mx + b$. The coefficient of x, $-\dfrac{3}{2}$, is the slope, and the constant is 3, so the y-intercept is $(0, 3)$.

Any equation of the form $y = b$ can be written $y = 0x + b$. Thus the slope of the line whose equation is $y = b$ is 0, and the y-intercept is $(0, b)$.

Any equation of the form $x = a$ cannot be written in the form $y = mx + b$. The line whose equation is $x = a$ has an undefined slope and no y-intercept.

Quick ✓

1. An equation of a line with slope m and y-intercept $(0, b)$ is _____.

In Problems 2–6, find the slope and y-intercept of the line whose equation is given.

2. $y = 4x - 3$
3. $3x + y = 7$
4. $2x + 5y = 15$
5. $y = 8$
6. $x = 3$

▶ ❷ Graph a Line Whose Equation Is in Slope-Intercept Form

If an equation is in slope-intercept form, we can graph the line by plotting the y-intercept and using the slope to find another point on the line.

EXAMPLE 3 How to Graph a Line Whose Equation Is in Slope-Intercept Form

Graph the line $y = 3x - 1$ using the slope and y-intercept.

Step-by-Step Solution

Step 1: Identify the slope and y-intercept of the line.

$$y = 3x - 1$$
$$y = 3x + (-1)$$
$$m = 3 \qquad b = -1$$

The slope is $m = 3$ and the y-intercept is $(0, -1)$.

Step 2: Plot the y-intercept and then use the slope to find a second point on the graph. Draw a line through the points.

Plot the y-intercept at $(0, -1)$. Use the slope $m = \dfrac{3}{1} = \dfrac{\text{rise}}{\text{run}}$ to find a second point on the graph. See Figure 36.

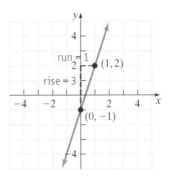

Figure 36
$y = 3x - 1$

Quick ✓

In Problems 7 and 8, graph the line using the slope and y-intercept.

7. $y = 2x - 5$ **8.** $y = \dfrac{1}{2}x - 5$

EXAMPLE 4 Graphing a Line Whose Equation Is in Slope-Intercept Form

Graph the line $y = -\dfrac{4}{3}x + 2$ using the slope and y-intercept.

Solution

First, determine the slope and y-intercept.

$$y = -\dfrac{4}{3}x + 2$$
$$m = -\dfrac{4}{3} \qquad b = 2$$

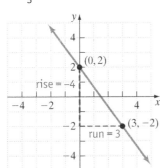

Figure 37
$y = -\dfrac{4}{3}x + 2$

Plot the y-intercept at $(0, 2)$. Now use the slope $m = -\dfrac{4}{3} = \dfrac{-4}{3} = \dfrac{\text{rise}}{\text{run}}$ to find a second point on the graph. Then draw a line through these two points. See Figure 37.

Quick ✓

In Problems 9 and 10, graph the line using the slope and y-intercept.

9. $y = -3x + 1$ 10. $y = -\dfrac{3}{2}x + 4$

③ Graph a Line Whose Equation Is in the Form $Ax + By = C$

If a linear equation is written in standard form $Ax + By = C$, we can still use the slope and y-intercept to obtain the graph of the equation. Let's see how.

EXAMPLE 5 How to Graph a Line Whose Equation Is in the Form $Ax + By = C$

Graph the line $8x + 2y = 10$ using the slope and y-intercept.

Step-by-Step Solution

Step 1: Solve the equation for y to put it in the form $y = mx + b$.

$$8x + 2y = 10$$

Subtract 8x from both sides: $\quad 2y = -8x + 10$

Divide both sides by 2: $\quad y = \dfrac{-8x + 10}{2}$

Simplify: $\quad y = -4x + 5$

Step 2: Identify the slope and y-intercept of the line.

The slope is $m = -4$ and the y-intercept is $(0, 5)$.

Step 3: Plot the y-intercept and then use the slope to find a second point on the graph. Draw a line through the points.

Plot the point $(0, 5)$ and use the slope $m = -4 = \dfrac{-4}{1} = \dfrac{\text{rise}}{\text{run}}$ to find a second point on the graph. See Figure 38.

Figure 38
$8x + 2y = 10$

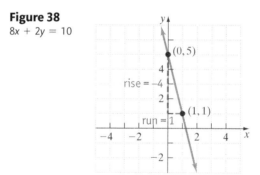

Work Smart

An alternative to graphing the equation in Example 5 using the slope and y-intercept would be to graph the line using intercepts.

Quick ✓

In Problems 11–13, graph each line using the slope and y-intercept.

11. $-2x + y = -3$ 12. $6x - 2y = 2$ 13. $3x + 5y = 0$

14. List three techniques that can be used to graph a line.

④ Find the Equation of a Line Given Its Slope and y-Intercept

Up to now, we have identified the slope and y-intercept of a line from an equation. We will now reverse the process and find the equation of a line given its slope and y-intercept. This is a straightforward process—replace m with the slope and b with the y-intercept.

EXAMPLE 6 **Finding the Equation of a Line Given Its Slope and y-Intercept**

Find the equation of a line whose slope is $\frac{3}{8}$ and whose y-intercept is $(0, -4)$. Graph the line.

Solution

The slope is $m = \frac{3}{8}$ and the y-intercept is $(0, b) = (0, -4)$. Substitute $\frac{3}{8}$ for m and -4 for b in the slope-intercept form of a line, $y = mx + b$, to obtain

$$y = \frac{3}{8}x - 4$$

Figure 39 shows the graph of the equation.

Figure 39
$y = \frac{3}{8}x - 4.$

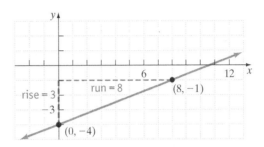

Quick ✓

In Problems 15–18, find the equation of the line whose slope and y-intercept are given. Graph the line.

15. $m = 3, (0, -2)$ **16.** $m = -\frac{1}{4}, (0, 3)$

17. $m = 0, (0, -1)$ **18.** $m = 1, (0, 0)$

⑤ Work with Linear Models in Slope-Intercept Form

In many situations (especially in Statistics), linear equations are used to describe the relationship between two variables. For example, your long-distance phone bill depends linearly on the number of minutes used, or the cost of renting a moving truck depends linearly on the number of miles driven.

EXAMPLE 7 **A Model for Total Cholesterol**

When you have a physical exam, your doctor draws blood for your cholesterol test. Your total cholesterol count is measured in milligrams per deciliter (mg/dL). It is the sum of low-density lipoprotein cholesterol (LDL)—sometimes called "bad cholesterol"—and high-density lipoprotein cholesterol (HDL)—sometimes called "good cholesterol." Based on data from the National Center for Health Statistics, a woman's total cholesterol y is related to her age x by the following linear equation:

$$y = 1.1x + 157$$

(a) Use the equation to predict the total cholesterol of a 40-year-old woman.

(b) Determine and interpret the slope of the equation.

(c) Determine and interpret the y-intercept of the equation.

(d) Graph the equation in a rectangular coordinate system.

Solution

(a) Substitute 40 for x, the women's age, in the equation $y = 1.1x + 157$ to find the total cholesterol y.

$$y = 1.1x + 157$$
$$x = 40: \quad y = 1.1(40) + 157$$
$$= 201$$

We predict that a 40-year-old woman will have a total cholesterol of 201 mg/dL.

(b) The slope of the equation $y = 1.1x + 157$ is 1.1. Because slope equals $\dfrac{\text{rise}}{\text{run}} = \dfrac{1.1 \text{ mg/dL}}{1 \text{ year}}$, we interpret the slope as follows: "The total cholesterol of a female increases by 1.1 mg/dL as age increases by 1 year."

(c) The y-intercept of the equation $y = 1.1x + 157$ is $(0, 157)$. The y-intercept is the value of total cholesterol, y, when $x = 0$. Since x represents age, we interpret the y-intercept as follows: "The total cholesterol of a newborn girl is 157 mg/dL."

(d) Figure 40 shows the graph of the equation. Because it does not make sense for x to be less than 0, we graph the equation only in quadrant I.

Work Smart

The slope in Example 7 may also be interpreted as, "If age increases by 1 year, the total cholesterol of a female increases by 1.1 mg/dL.

Work Smart

Notice that we did not use the slope to obtain an additional point on the graph of the equation. It would be difficult to find an additional point with a slope of 1.1. For example, from the y-intercept, we would go up 1.1 mg/dL and right 1 year and end up at $(1, 158.1)$. It would be hard to draw the line through these two points!

Figure 40
$y = 1.1x + 157$

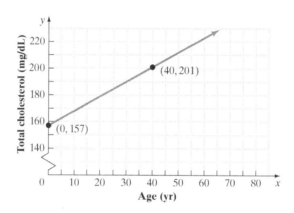

In Figure 40, notice the "broken line" ($\lessgtr$) on the y-axis near the origin. This symbol indicates that a portion of the graph has been removed. This is done to avoid a lot of white space in the graph. Whenever you are reading a graph, always look carefully at how the axes are labeled and at the units on each axis.

Quick ✓

19. Based on data obtained from the National Center for Health Statistics, the birth weight y of a baby, measured in grams, is linearly related to gestation period x (in weeks) according to the equation

$$y = 143x - 2215$$

(a) Use the equation to predict the birth weight of a baby if the gestation period is 30 weeks.

(b) Use the equation to predict the birth weight of a baby if the gestation period is 36 weeks.

(c) Determine and interpret the slope of the equation.

(d) Explain why it does not make sense to interpret the y-intercept of the equation.

(e) Graph the equation in a rectangular coordinate system for $28 \leq x \leq 43$.

We know that slope can be interpreted as a rate of change. For this reason, when information in a problem is given as a rate of change (as in miles per gallon or dollars per pound), the rate of change will represent the slope in a linear model.

EXAMPLE 8 **Cost of Owning and Operating a Car**

Some costs involved in owning a car are affected by the number of miles driven (gas and maintenance), while others are not (comprehensive insurance, license plates, depreciation). Suppose the annual cost of operating a Chevy Cobalt is $0.25 per mile plus $3000.

(a) Write a linear equation that relates the annual cost of operating the car y to the number of miles driven in a year x.

(b) What is the annual cost of driving 11,000 miles?

(c) Graph the equation in a rectangular coordinate system.

Solution

(a) The rate of change in the problem is $0.25 per mile. We can express this as $\frac{\$0.25}{1 \text{ mile}}$, which is the slope m of the linear equation. The cost of $3000 is a cost that does not change with the number of miles driven. Put another way, if we drive 0 miles, the cost will be $3000, so this value represents the y-intercept, $(0, b)$. The linear equation that relates cost y to number of miles driven x is

$$y = 0.25x + 3000$$

(b) Let $x = 11{,}000$ in the equation $y = 0.25x + 3000$.

$$y = 0.25(11{,}000) + 3000$$
$$= 2750 + 3000$$
$$= \$5750$$

The cost of driving 11,000 miles in a year is $5750. This cost includes gas, insurance, maintenance, and depreciation in the value of the vehicle.

(c) See Figure 41.

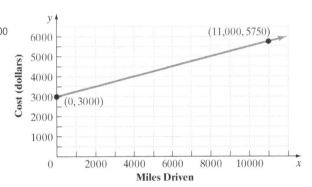

Figure 41
$y = 0.25x + 3000$

Quick ✓

20. The daily cost, y, of renting a 16-foot moving truck for a day is $50 plus $0.38 per mile driven, x.

(a) Write a linear equation relating the daily cost y to the number of miles driven, x.

(b) Determine the cost of renting the truck if the truck is driven 75 miles.

(c) If the cost of renting the truck is $84.20, how many miles were driven?

(d) Graph the linear equation.

3.IR4 Exercises — MyLab Statistics
Underlined exercises have complete video solutions in MyLab.

Problems 1–20 are the Quick ✓s that follow the EXAMPLES.

Building Skills

In Problems 21–40, find the slope and y-intercept of the line whose equation is given. See Objective 1.

21. $y = 5x + 2$
22. $y = 7x + 1$
23. $y = x - 9$
24. $y = x - 7$
25. $y = -10x + 7$
26. $y = -6x + 2$
27. $y = -x - 9$
28. $y = -x - 12$
29. $2x + y = 4$
30. $3x + y = 9$
31. $2x + 3y = 24$
32. $6x - 8y = -24$
33. $5x - 3y = 9$
34. $10x + 6y = 24$
35. $x - 2y = 5$
36. $-x - 5y = 3$
37. $y = -5$
38. $y = 3$
39. $x = 6$
40. $x = -2$

In Problems 41–48, use the slope and y-intercept to graph each line whose equation is given. See Objective 2.

41. $y = x + 3$
42. $y = x + 4$
43. $y = -2x - 3$
44. $y = -4x - 1$
45. $y = \frac{2}{3}x + 2$
46. $y = \frac{4}{3}x - 3$
47. $y = -\frac{5}{2}x - 2$
48. $y = -\frac{2}{5}x + 3$

In Problems 49–56, graph each line using the slope and y-intercept. See Objective 3.

49. $4x + y = 5$
50. $3x + y = 2$
51. $x + 2y = -6$
52. $x - 2y = -4$
53. $3x - 2y = 10$
54. $4x + 3y = -6$
55. $6x + 3y = -15$
56. $5x - 2y = 6$

In Problems 57–68, find the equation of the line with the given slope and intercept. See Objective 4.

57. slope is -1; y-intercept is $(0, 8)$
58. slope is 1; y-intercept is $(0, 10)$
59. slope is $\frac{6}{7}$; y-intercept is $(0, -6)$
60. slope is $\frac{4}{7}$; y-intercept is $(0, -9)$
61. slope is $-\frac{1}{3}$; y-intercept is $\left(0, \frac{2}{3}\right)$
62. slope is $\frac{1}{4}$; y-intercept is $\left(0, \frac{3}{8}\right)$
63. slope is undefined; x-intercept is $(-5, 0)$
64. slope is 0; y-intercept is $(0, -2)$
65. slope is 0; y-intercept is $(0, 3)$
66. slope is undefined; x-intercept is $(4, 0)$
67. slope is 5; y-intercept is $(0, 0)$
68. slope is -3; y-intercept is $(0, 0)$

Mixed Practice

In Problems 69–92, graph each equation using any method you wish.

69. $y = 2x - 7$
70. $y = -4x + 1$
71. $3x - 2y = 24$
72. $2x + 5y = 30$
73. $y = -5$
74. $y = 4$
75. $x = -6$
76. $x = -3$
77. $6x - 4y = 0$
78. $3x + 8y = 0$
79. $y = -\frac{5}{3}x + 6$
80. $y = -\frac{3}{5}x + 4$
81. $2y = x + 4$
82. $3y = x - 9$
83. $y = \frac{x}{3}$
84. $y = -\frac{x}{4}$
85. $2x = -8y$
86. $-3x = 5y$
87. $y = -\frac{2}{3}x + 1$
88. $y = \frac{3}{2}x - 4$
89. $x + 2 = -7$
90. $y - 4 = -1$
91. $5x + y + 1 = 0$
92. $2x - y + 4 = 0$

Applying the Concepts

93. Mobile Device Usage In 2008, Americans spent an average of 19 minutes daily on their mobile devices. By 2015 that figure was up to 171 minutes daily. Assuming the increase in usage was constant, the number of minutes used can be calculated by the equation $y = 21.7x + 19$, where x represents the number of years after 2008 and y represents the number of daily minutes spent on mobile devices.
(*Source: CTIA, The Wireless Association*)

(a) To the nearest whole minute, how many minutes daily did Americans spend on their mobile devices in 2010?

(b) In which year did an average mobile device user spend approximately 106 minutes on her or his mobile devices?

(c) Interpret the slope of the equation $y = 21.7x + 19$.

(d) Can this trend continue indefinitely?

(e) Graph the equation in a rectangular coordinate system. Label the axes appropriately.

94. Counting Calories According to a National Academy of Sciences report, the recommended daily intake of calories for males between the ages of 7 and 15 can be calculated by the equation $y = 125x + 1125$, where x represents the boy's age and y represents the recommended caloric intake.

(a) What is the recommended caloric intake for a 12-year-old boy?

(b) What is the age of a boy whose recommended caloric intake is 2250 calories?

(c) Interpret the slope of $y = 125x + 1125$.

(d) Why would this equation not be accurate for a 3-year-old male?

(e) Graph the equation in a rectangular coordinate system. Label the axes appropriately.

95. Weekly Salary Dien is paid a salary of $400 per week plus an 8% commission on all sales he makes during the week.

(a) Write a linear equation that calculates his weekly income, where y represents his income and x represents the amount of sales.

(b) What is Dien's weekly income if he sold $1200 worth of merchandise?

(c) Graph the equation in a rectangular coordinate system. Label the axes appropriately.

96. Car Rental To rent a car for a day, Gloria pays $75 plus $0.10 per mile.

(a) Write a linear equation that calculates the daily cost, y, to rent a car that will be driven x miles.

(b) What is the cost to drive this car for 200 miles?

(c) If Gloria paid $87.50, how many miles did she drive?

(d) Graph the equation in a rectangular coordinate system. Label the axes appropriately.

Extending the Concepts

In Problems 97–102, find the value of the missing coefficient so that the line will have the given property.

97. $2x + By = 12$; slope is $\frac{1}{2}$

98. $Ax + 2y = 5$; slope is $\frac{3}{2}$

99. $Ax - 2y = 10$; slope is -2

100. $12x + By = -1$; slope is -4

101. $x + By = \frac{1}{2}$; y-intercept is $-\frac{1}{6}$

102. $4x + By = \frac{4}{3}$; y-intercept is $\frac{2}{3}$

In Problems 103 and 104, use the following information. In business, a cost equation relates the total cost of producing a product or good such as a refrigerator, rug, or blender to the number of goods produced. The simplest cost model is the linear cost model. In the linear cost model, the slope of the linear equation represents the cost of producing one additional unit of a product. Variable cost is reported as a rate of change, such as $40 per calculator. Examples of variable costs include labor costs and materials. The y-intercept of the linear equation represents the fixed costs of production—these are costs that exist regardless of the level of production. Fixed costs would include the cost of the manufacturing facility and insurance.

103. Manufacturing Costs Suppose the variable cost of manufacturing a graphing calculator is $40 per calculator, and the daily fixed cost is $4000.

(a) Write a linear equation that relates cost y to the number of calculators manufactured x.

(b) What is the daily cost of manufacturing 500 calculators?

(c) One day, the total cost was $19,000. How many calculators were manufactured?

(d) Graph the equation relating cost and number of calculators manufactured.

104. Cost Equations Manufacturing Costs Suppose the variable cost of manufacturing a cell phone is $35 per phone, and the daily fixed cost is $3600.

(a) Write a linear equation that relates the daily cost y to the number of cell phones manufactured x.

(b) What is the daily cost of manufacturing 400 cell phones?

(c) One day, the total cost was $13,225. How many cellular telephones were manufactured?

(d) Graph the equation relating cost and number of cellular telephones manufactured.

Explaining the Concepts

105. Describe the line whose graph is shown. Which of the following equations could have the graph that is shown?

(a) $y = 3x - 2$
(b) $y = -2x + 5$
(c) $y = 3$
(d) $2x + 3y = 6$
(e) $3x - 2y = 8$
(f) $4x - y = -4$
(g) $-5x + 2y = 12$
(h) $x - y = -3$

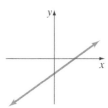

106. Without graphing, describe the orientation of each line (rises from left to right, and so on). Explain how you came to this conclusion.

(a) $y = 4x - 3$
(b) $y = -2x + 5$
(c) $y = x$
(d) $y = 4$

3.IR5 Point-Slope Form of a Line

Objectives

① Find the Equation of a Line Given a Point and a Slope

② Find the Equation of a Line Given Two Points

③ Build Linear Models from Data

Figure 42

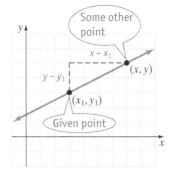

① Find the Equation of a Line Given a Point and a Slope

We have discussed two forms for the equation of a line: the standard form of a line, $Ax + By = C$, where A and B are not both zero, and the slope-intercept form of a line, $y = mx + b$, where m is the slope and b is the y-intercept. We now introduce another form for the equation of a line.

Suppose we have a nonvertical line with slope m containing the point (x_1, y_1). See Figure 42. For any other point (x, y) on the line, the slope of the line is

$$m = \frac{y - y_1}{x - x_1}$$

Multiplying both sides by $x - x_1$ gives us

$$m(x - x_1) = y - y_1 \quad \text{or} \quad y - y_1 = m(x - x_1)$$

Point-Slope Form of an Equation of a Line

An equation of a nonvertical line with slope m containing the point (x_1, y_1) is

$$y - y_1 = m(x - x_1)$$

where m is the slope and (x_1, y_1) is the given point.

The point-slope form of a line can be used to write an equation in either slope-intercept form ($y = mx + b$) or standard form ($Ax + By = C$).

EXAMPLE 1 Using the Point-Slope Form of an Equation of a Line–Positive Slope

Find the equation of a line that has a slope of 3 and contains the point $(-1, 4)$. Write the equation in slope-intercept form. Graph the line.

Section 3.IR5 Point-Slope Form of a Line IR-157

Figure 43
$y = 3x + 7$

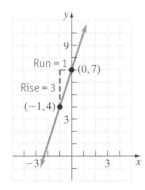

Solution
Because we are given the slope and a point on the line, use the point-slope form of a line with $m = 3$ and $(x_1, y_1) = (-1, 4)$.

$$y - y_1 = m(x - x_1)$$
$m = 3, x_1 = -1, y_1 = 4$: $\quad y - 4 = 3(x - (-1))$
$$y - 4 = 3(x + 1)$$

To put the equation in slope-intercept form, $y = mx + b$, solve the equation for y.

Distribute: $\quad y - 4 = 3x + 3$
Add 4 to both sides: $\quad y = 3x + 7$

See Figure 43 for a graph of the line.

EXAMPLE 2 Using the Point-Slope Form of an Equation of a Line–Negative Slope

Find the equation of a line that has a slope of $-\frac{3}{4}$ and contains the point $(-4, 3)$. Write the equation in slope-intercept form. Graph the line.

Solution
Because we are given the slope and a point on the line, use the point-slope form of a line with $m = -\frac{3}{4}$ and $(x_1, y_1) = (-4, 3)$.

Figure 44
$y = -\frac{3}{4}x$

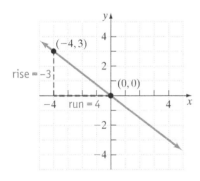

$$y - y_1 = m(x - x_1)$$
$m = -\frac{3}{4}, x_1 = -4, y_1 = 3$: $\quad y - 3 = -\frac{3}{4}(x - (-4))$

Simplify: $\quad y - 3 = -\frac{3}{4}(x + 4)$

Distribute: $\quad y - 3 = -\frac{3}{4}x - 3$

Add 3 to both sides: $\quad y = -\frac{3}{4}x$

See Figure 44 for a graph of the line.

Quick ✓
1. The point-slope form of a nonvertical line that has a slope of m and contains the point (x_1, y_1) is _____.
2. *True or False* The slope of the line $y - 3 = 4(x - 1)$ is 4.

In Problems 3–6, find the equation of the line with the given properties. Write the equation in slope-intercept form. Graph the line.

3. $m = 3$ containing $(x_1, y_1) = (2, 1)$
4. $m = \frac{1}{3}$ containing $(x_1, y_1) = (3, -4)$
5. $m = -4$ containing $(x_1, y_1) = (-2, 5)$
6. $m = -\frac{5}{2}$ containing $(x_1, y_1) = (-4, 5)$

EXAMPLE 3 Finding the Equation of a Horizontal Line

Find the equation of a horizontal line that contains the point $(-4, 2)$. Write the equation in slope-intercept form. Graph the line.

Solution

The line is horizontal, so its slope is 0. Because we know the slope and a point on the line, use the point-slope form with $m = 0$, $x_1 = -4$, and $y_1 = 2$.

$$y - y_1 = m(x - x_1)$$

$m = 0, x_1 = -4, y_1 = 2:$ $y - 2 = 0(x - (-4))$

$$y - 2 = 0$$

Add 2 to both sides: $y = 2$ Slope-intercept form, $y = 0x + 2$

See Figure 45 for a graph of the line.

Figure 45
$y = 2$

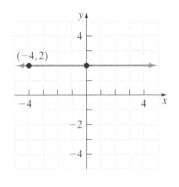

Quick ✓

7. Find the equation of a horizontal line that contains the point $(-2, 3)$. Write the equation of the line in slope-intercept form. Graph the line.

▶ ❷ Find the Equation of a Line Given Two Points

From Section 3.IR2, we know that only two points are needed to graph a line. Given two points, we can find the equation of the line through the points by first finding the slope of the line and then using the point-slope form of a line.

EXAMPLE 4 How to Find the Equation of a Line from Two Points

Find the equation of a line through the points $(1, 3)$ and $(4, 9)$. Write the equation in slope-intercept form. Graph the line.

Step-by-Step Solution

Step 1: Find the slope of the line containing the points.

Let $(x_1, y_1) = (1, 3)$ and $(x_2, y_2) = (4, 9)$. Substitute these values into the formula for the slope of a line.

$$m = \frac{y_2 - y_1}{x_2 - x_1} = \frac{9 - 3}{4 - 1} = \frac{6}{3} = 2$$

Step 2: Substitute the slope found in Step 1 and either point into the point-slope form of a line to find the equation.

$$y - y_1 = m(x - x_1)$$

Use $m = 2, x_1 = 1, y_1 = 3:$ $y - 3 = 2(x - 1)$

Step 3: Solve the equation for y.

Distribute the 2: $y - 3 = 2x - 2$

Add 3 to both sides: $y = 2x + 1$

Figure 46
$y = 2x + 1$

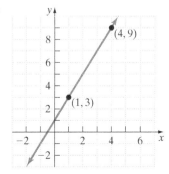

The slope-intercept form of the equation is $y = 2x + 1$. The slope of the line is 2, and the y-intercept is $(0, 1)$. See Figure 46 for the graph.

Quick ✓

In Problems 8 and 9, find the equation of the line containing the given points. Write the equation in slope-intercept form. Graph the line.

8. $(0, 2); (3, 5)$ **9.** $(-1, 4); (1, -2)$

> **Work Smart: Study Skills**
> To write the equation of a nonvertical line, we must know either the slope of the line and a point on the line or two points on the line.
> - If the **slope** and the **y-intercept** are known, use the slope-intercept form, $y = mx + b$.
> - If the **slope** and a **point** that is not the y-intercept are known, use the **point-slope** form, $y - y_1 = m(x - x_1)$.
> - If **two points** are known, first find the **slope** and then use that slope and one of the **points** in the **point-slope** formula, $y - y_1 = m(x - x_1)$.

EXAMPLE 5 Finding the Equation of a Vertical Line from Two Points

Find the equation of a line through the points $(-3, 2)$ and $(-3, -4)$. Write the equation in slope-intercept form, if possible. Graph the line.

Solution

Let $(x_1, y_1) = (-3, 2)$ and $(x_2, y_2) = (-3, -4)$ in the formula for the slope of a line.

$$m = \frac{y_2 - y_1}{x_2 - x_1} = \frac{-4 - 2}{-3 - (-3)} = \frac{-6}{0}$$

Work Smart
The equation of a vertical line cannot be written in slope-intercept form.

The slope is undefined, so the line is vertical. No matter what value of y we choose, the x-coordinate of the point on the line will be -3. The equation of the line is $x = -3$, which cannot be written in slope-intercept form. See Figure 47 for the graph.

Figure 47
$x = -3$

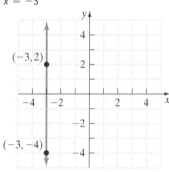

Quick ✓

10. Find the equation of the line containing the points $(3, 2)$ and $(3, -4)$. Write the equation in slope-intercept form, if possible. Graph the line.

Summary Equations of Lines

Form of Line	Formula	Comments
Horizontal line	$y = b$	Graph is a horizontal line (slope is 0) with y-intercept $(0, b)$.
Vertical line	$x = a$	Graph is a vertical line (undefined slope) with x-intercept $(a, 0)$.
Point-slope	$y - y_1 = m(x - x_1)$	Useful for finding the equation of a line, given a point and the slope, or two points.
Slope-intercept	$y = mx + b$	Useful for finding the equation of a line, given the slope and y-intercept, or for quickly determining the slope and y-intercept of the line, given the equation of the line.
Standard form	$Ax + By = C$	Useful for finding the x- and y-intercepts.

Quick ✓

11. List the five forms of the equation of a line: Horizontal line: _____; Vertical line: _____; Point-slope: _____; Slope-intercept: _____; Standard form: _____

③ Build Linear Models from Data

If we have a set of data with more than two points, how can we tell whether the data (the two variables) are related linearly? One method for determining whether two variables are linearly related is to draw a picture of the data, called a *scatter diagram*, to learn whether the variables *might* be linearly related.

Scatter Diagrams

The graph consisting of the ordered pairs that make up the relation in the Cartesian plane is called a **scatter diagram.**

EXAMPLE 6 Drawing a Scatter Diagram

The on-base percentage for a baseball team is the percent of time that the team safely reaches base. Table 15 shows the number of runs scored and the on-base percentage for various teams in the 2015 season.

Table 15

Team	On-Base Percentage, x	Runs Scored, y	x, y
NY Yankees	32.3	764	(32.3, 764)
Los Angeles Angels	30.7	661	(30.7, 661)
Texas Rangers	32.5	751	(32.5, 751)
Toronto Blue Jays	34.0	891	(34.0, 891)
Minnesota Twins	30.5	696	(30.5, 696)
Oakland A's	31.2	694	(31.2, 694)
Kansas City Royals	32.2	724	(32.2, 724)
Baltimore Orioles	30.7	713	(30.7, 713)

Source: espn.com

(a) Draw a scatter diagram of the data.

(b) Describe what happens as the on-base percentage increases.

Solution

(a) To draw a scatter diagram, plot the ordered pairs listed in Table 15. See Figure 48.

(b) The scatter diagram reveals that as the on-base percentage increases, the number of runs scored also increases. While this relation is not perfectly linear (because the points don't all fall on a straight line), the *pattern* of the data is linear.

Figure 48

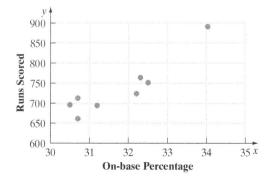

Quick ✓

12. The data listed below represent the total cholesterol (in mg/dL) and age of males.

Age, x	Total Cholesterol, y	Age, x	Total Cholesterol, y
25	180	38	239
25	195	48	204
28	186	51	243
32	180	62	228
32	197	65	269

(a) Draw a scatter diagram of the data.
(b) Describe the relation between age and total cholesterol.

Recognizing the Type of Relation Between Two Variables

Scatter diagrams help us see the type of relation between two variables. In this section, we will focus distinguishing linear relations from nonlinear relations. See Figure 49.

EXAMPLE 7   Distinguishing Between Linear and Nonlinear Relations

Determine whether each relation in Figure 50 is linear or nonlinear. If it is linear, state whether the slope is positive or negative.

Figure 49

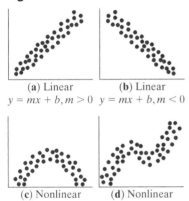

Figure 50

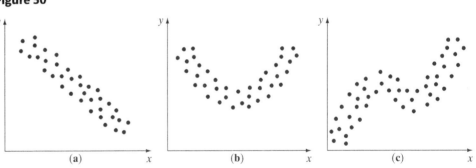

Solution

(a) Linear with negative slope   **(b)** Nonlinear   **(c)** Nonlinear

Quick ✓

13. Determine whether each relation is linear or nonlinear. If it is linear, state whether the slope is positive or negative.

(a)
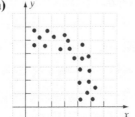

(b)
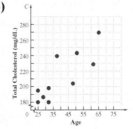

Fitting a Line to Data

Suppose a scatter diagram indicates a linear relationship, as in Figure 49(a) or (b). To find a linear equation for such data, draw a line through two points on the scatter diagram. Then determine the equation of that line by using the point-slope form of a line, $y - y_1 = m(x - x_1)$.

EXAMPLE 8   **Finding a Model for Linearly Related Data**

Using the data in Table 15 from Example 6,

(a) Select two points and find the equation for the line connecting the points.

(b) Graph the line on the scatter diagram from Example 6(a).

(c) Use the linear equation found in part (a) to predict the number of runs scored by a team whose on-base percentage is 33.5%.

(d) Interpret the slope. Does it make sense to interpret the y-intercept?

Solution

(a) Select two points—for example, $(30.7, 661)$ and $(32.5, 751)$. (You should select your own two points and work through the solution.) The slope of the line joining these points is

$$m = \frac{751 - 661}{32.5 - 30.7} = \frac{90}{1.8} = 50$$

Use the point-slope form to find the equation of the line that has slope 50 passing through $(30.7, 661)$:

Point-slope form: $y - y_1 = m(x - x_1)$

$m = 50;\ x_1 = 30.7,\ y_1 = 661$: $y - 661 = 50(x - 30.7)$

Distribute 50: $y - 661 = 50x - 1535$

Add 661 to both sides: $y = 50x - 874$

(b) Figure 51 shows the scatter diagram with the graph of the line found in part (a). We graphed the line through the two points selected in part (a).

Figure 51

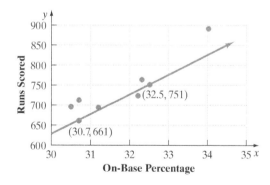

(c) Evaluate $y = 50x - 874$ at $x = 33.5$.

$$y = 50(33.5) - 874$$
$$= 801$$

We predict that a team whose on-base percentage is 33.5% will score 801 runs.

(d) The slope of the linear equation is 50, so if the on-base percentage increases by 1, then the number of runs scored will increase by about 50. The y-intercept, -874, represents the runs scored when the on-base percentage is 0. Since negative runs scored does not make sense and we have no observations near zero, interpreting the y-intercept does not make sense.

Section 3.IR5 Point-Slope Form of a Line   IR-163

Quick ✓

14. Use the data from Quick Check Problem 12 on page IR-161.

  (a) Select two points and find a linear model that describes the relation between the points.

  (b) Graph the line on the scatter diagram obtained in Quick Check Problem 12 (page IR-161).

  (c) Predict the total cholesterol of a 39-year-old male.

  (d) Interpret the slope. Does it make sense to interpret the y-intercept?

3.IR5 Exercises   MyLab Statistics

Underlined exercises have complete video solutions in MyLab.

Problems 1–14 are the Quick ✓ s that follow the EXAMPLES.

Building Skills

In Problems 15–30, find the equation of the line that contains the given point and has the given slope. Write the equation in slope-intercept form and graph the line. See Objective 1.

15. $(2, 5)$; slope $= 3$

16. $(4, 1)$; slope $= 6$

17. $(-1, 2)$; slope $= -2$

18. $(6, -3)$; slope $= -5$

19. $(8, -1)$; slope $= \dfrac{1}{4}$

20. $(-8, 2)$; slope $= -\dfrac{1}{2}$

21. $(0, 13)$; slope $= -6$

22. $(0, -4)$; slope $= 9$

23. $(5, -7)$; slope $= 0$

24. $(3, 12)$; undefined slope

25. $(-4, 5)$; undefined slope

26. $(-7, -1)$; slope $= 0$

27. $(-3, 0)$; slope $= 0.667$

28. $(-10, 0)$; slope $= -0.8$

29. $(-8, 6)$; slope $= -0.75$

30. $(-4, -6)$; slope $= 1.5$

In Problems 31–36, find the equation of the line that contains the given point and satisfies the given information. Write the equation in slope-intercept form, if possible. See Objectives 1 and 2.

31. Vertical line that contains $(-3, 10)$

32. Horizontal line that contains $(-6, -1)$

33. Horizontal line that contains $(-1, -5)$

34. Vertical line that contains $(4, -3)$

35. Horizontal line that contains $(0.2, -4.3)$

36. Vertical line that contains $(3.5, 2.4)$

In Problems 37–52, find the equation of the line that contains the given points. Write the equation in slope-intercept form, if possible. See Objective 2.

37. $(0, 4)$ and $(-2, 0)$

38. $(0, 3)$ and $(6, 0)$

39. $(1, 2)$ and $(0, 6)$

40. $(2, 4)$ and $(0, 8)$

41. $(-3, 2)$ and $(1, -4)$

42. $(-2, 4)$ and $(2, -2)$

43. $(-3, -11)$ and $(2, -1)$

44. $(4, 18)$ and $(-1, 3)$

45. $(4, -3)$ and $(-3, -3)$

46. $(-6, 5)$ and $(7, 5)$

47. $(2, -1)$ and $(2, -9)$

48. $(-3, 8)$ and $(-3, 1)$

49. $(0.1, 0.6)$ and $(0.5, 0.7)$

50. $(0.7, 0.8)$ and $(0.2, 0.4)$

51. $\left(\dfrac{1}{2}, -\dfrac{9}{4}\right)$ and $\left(\dfrac{5}{2}, -\dfrac{1}{4}\right)$

52. $\left(\dfrac{1}{3}, \dfrac{12}{5}\right)$ and $\left(\dfrac{4}{3}, \dfrac{2}{5}\right)$

Mixed Practice

In Problems 53–70, find the equation of the line described. Write the equation in slope-intercept form, if possible. Graph the line.

53. Contains $(4, -2)$ with slope $= 5$

54. Contains $(3, 2)$ with slope $= 4$

55. Horizontal line that contains $(-3, 5)$

56. Vertical line that contains $(-4, 2)$

57. Contains $(1, 3)$ and $(-4, -2)$

58. Contains $(-2, -8)$ and $(2, -6)$

59. Contains $(-2, 3)$ with slope $= \dfrac{1}{2}$

60. Contains $(-8, 3)$ with slope $= \dfrac{1}{4}$

61. Vertical line that contains $(5, 2)$

62. Horizontal line that contains $(-2, -6)$

63. Contains $(3, -19)$ and $(-1, 9)$

64. Contains $(-3, 13)$ and $(4, -22)$

65. Contains $(6, 3)$ with slope $= -\dfrac{2}{3}$

66. Contains $(-6, 3)$ with slope $= -\dfrac{2}{3}$

67. Contains $(-2, 3)$ and $(4, -6)$

68. Contains $(5, -3)$ and $(-3, 3)$

69. x-intercept: $(5, 0)$; y-intercept: $(0, -2)$

70. x-intercept: $(-6, 0)$; y-intercept: $(0, 4)$

Applying the Concepts

71. **Shipping Packages** The shipping department for a warehouse has noted that if 60 packages are shipped during a month, the total expenses for the department are $1635. If 120 packages are shipped during a month, the total expenses for the shipping department are $1770. Let x represent the number of packages and y represent the total expenses for the shipping department.
    (a) Interpret the meaning of the point $(60, 1635)$ in the context of this problem.
    (b) Plot the ordered pairs $(60, 1635)$ and $(120, 1770)$ in a rectangular coordinate system and graph the line through the points.
    (c) Find the linear equation, in slope-intercept form, that relates the total expenses for the shipping department, y, to the number of packages sent, x.
    (d) Use the equation found in part (c) to find the total expenses during a month when 200 packages were sent.
    (e) Interpret the slope.

72. **Retirement Plans** Based on the retirement plan available from his employer, Kei knows that if he retires after 20 years, his monthly retirement income will be $3150. If he retires after 30 years, his monthly income increases to $3600. Let x represent the number of years of service and y represent the monthly retirement income.
    (a) Interpret the meaning of the point $(30, 3600)$ in the context of this problem.
    (b) Plot the ordered pairs $(20, 3150)$ and $(30, 3600)$ in a rectangular coordinate system and graph the line through the points.
    (c) Find the linear equation, in slope-intercept form, that relates the monthly retirement income, y, to the number of years of service, x.
    (d) Use the equation found in part (c) to find the monthly income for 15 years of service.
    (e) Interpret the slope.

73. **Credit Scores** Your Fair Isaacs Corporation (FICO) credit score is used to determine your ability to get credit (such as a car loan or a credit card). FICO scores have a range of 300 to 850, with a higher score indicating a better credit history. Suppose a bank offers a person with a credit score of 600 a 15% interest rate on a 3-year car loan, while a person with a credit score of 750 is offered a 6% interest rate. Let x represent a person's credit score and y represent the interest rate.
    (a) Fill in the ordered pairs: (___, 15); (___, 6)
    (b) Plot the ordered pairs from part (a) in a rectangular coordinate system, and graph the line through the points.
    (c) Find the linear equation, in slope-intercept form, that relates the interest rate (in percent), y, to the credit score, x.
    (d) Use the equation found in part (c) to predict the interest rate for a person with a credit score of 700.
    (e) Interpret the slope.

74. **U.S. Traffic Fatalities** Nationwide, the statistics for traffic fatalities show a decline. In 1999, the United States had 41,717 fatal crashes, and in 2009 the number dropped to 33,808. Let y represent the number of traffic fatalities and x represent the number of years since 1999.
    (*Source: National Highway Traffic Safety Administration*)
    (a) Fill in the ordered pairs: (___, 41,717); (___, 33,808).
    (b) Plot the ordered pairs from part (a) in a rectangular coordinate system and graph the line through the points.
    (c) Find the linear equation, in slope-intercept form, that relates the number of traffic fatalities, y, to the number of years since 1999, x.
    (d) Use the equation found in part (c) to find the number of traffic fatalities in 2019.
    (e) Interpret the slope.

75. **Diamonds** The relation between the cost of a diamond and its weight is linear. In looking at two diamonds, we find that one of the diamonds weighs 0.7 carat and costs $3340, while the other diamond weighs 0.8 carat and costs $4065. *Source: diamonds.com*
    (a) Find a linear equation that relates the price of a diamond, y, to its weight, x.

(b) Predict the price of a diamond that weighs 0.77 carat.
(c) Interpret the slope.
(d) If a diamond costs $5300, what do you think it should weigh? Round your answer to the nearest thousandth.

76. Apartments In the North Chicago area, an 820-square-foot apartment rents for $1507 per month. A 970-square-foot apartment rents for $1660. Suppose that the relation between area and rent is linear. *Source: apartments.com*
(a) Find a linear equation that relates the rent of a North Chicago apartment, y, to its area, x.
(b) Predict the rent of a 900-square-foot apartment in North Chicago.
(c) Interpret the slope.
(d) If the rent of a North Chicago apartment is $1300 per month, how big would you expect it to be?

77. The Consumption Function A famous theory in economics developed by John Maynard Keynes states that personal consumption expenditures are a linear function of disposable income. An economist wishes to develop a model that relates income and consumption and obtains the following information from the United States Bureau of Economic Analysis. In 2010, personal disposable income was $11,380 billion and personal consumption expenditures were $10,349 billion. In 2015, personal disposable income was $15,605 billion and personal consumption expenditures were $13,618 billion.
(a) Find a linear equation that relates personal consumption expenditures, y, to disposable income, x.
(b) In 2013, personal disposable income was $14,221 billion. Use this information to find personal consumption expenditures in 2013.
(c) Interpret the slope. In economics, this slope is called the **marginal propensity to consume.**
(d) If personal consumption expenditures were $13,117 billion, what do you think that disposable income was?

78. Birth Weight According to the National Center for Health Statistics, the average birth weight of babies born to 22-year-old mothers is 3280 grams. The average birth weight of babies born to 32-year-old mothers is 3370 grams. Suppose that the relation between age of mother and birth weight is linear.
(a) Find a linear equation that relates age of mother x to birth weight y.

(b) Predict the birth weight of a baby born to a mother who is 30 years old.
(c) Interpret the slope.
(d) If a baby weighs 3310 grams, how old do you expect the mother to be?

79. Concrete As concrete cures, it gains strength. The data below represent the 7-day and 28-day strength (in pounds per square inch) of a certain type of concrete.

7-day Strength, x	28-day Strength, y
2300	4070
3390	5220
2430	4640
2890	4620
3330	4850
2480	4120
3380	5020
2660	4890
2620	4190
3340	4630

(a) Draw a scatter diagram of the data, treating 7-day strength as the x-variable.
(b) What type of relation appears to exist between 7-day strength and 28-day strength?
(c) Select two points and find an equation of the line containing the points.
(d) Graph the line on the scatter diagram drawn in part (a).
(e) Predict the 28-day strength of a slab of concrete if its 7-day strength is 3000 psi.
(f) Interpret the slope of the line found in part (c).

80. Candy The following data represent the weight (in grams) of various candy bars and the corresponding number of calories.

Candy Bar	Weight, x	Calories, y
Hershey's Milk Chocolate	44.28	230
Nestle Crunch	44.84	230
Butterfinger	61.30	270
Baby Ruth	66.45	280
Almond Joy	47.33	220
Twix (with Caramel)	58.00	280
Snickers	61.12	280
Heath	39.52	210

Source: Megan Pocius, student at Joliet Junior College

(a) Draw a scatter diagram of the data, treating weight as the *x*-variable.
(b) What type of relation appears to exist between the weight of a candy bar and the number of calories?
(c) Select two points and find an equation of the line containing the points.
(d) Graph the line on the scatter diagram drawn in part (a).
(e) Predict the number of calories in a candy bar that weighs 62.3 grams.
(f) Interpret the slope of the line found in part (c).

81. Raisins The following data represent the weight (in grams) of a box of raisins and the number of raisins in the box.

Weight, x	Number of Raisins, y
42.3	87
42.7	91
42.8	93
42.4	87
42.6	89
42.4	90
42.3	82
42.5	86
42.7	86
42.5	86

Source: Jennifer Maxwell, student at Joliet Junior College

(a) Draw a scatter diagram of the data, treating weight as the *x*-variable.
(b) Select two points and find the equation of the line containing the points.
(c) Graph the line on the scatter diagram drawn in part (b).
(d) Predict the number of raisins in a box that weighs 42.5 grams.
(e) Interpret the slope of the line found in part (c).

82. Height versus Head Circumference The following data represent the height (in inches) and head circumference (in inches) of 9 randomly selected children.

Height, x	Head Circumference, y
25.25	16.4
25.75	16.9
25	16.9
27.75	17.6
26.50	17.3
27.00	17.5
26.75	17.3
26.75	17.5
27.5	17.5

Source: Denise Slucki, student at Joliet Junior College

(a) Draw a scatter diagram of the data, treating height as the *x*-variable.
(b) Select two points and find the equation of the line containing the points.
(c) Graph the line on the scatter diagram drawn in part (b).
(d) Predict the head circumference of a child who is 26.5 inches tall.
(e) Interpret the slope of the line found in part (c).

83. A strain of *E. coli* Beu 397-recA441 is placed into a Petri dish at 30° Celsius and allowed to grow. The population is estimated by means of an optical device in which the amount of light that passes through the Petri dish is measured. The data below are collected. Do you think that a linear function could be used to describe the relation between the two variables? Why or why not?

Time, x	Population, y
0	0.09
2.5	0.18
3.5	0.26
4.5	0.35
6	0.50

Source: Dr. Polly Lavery, Joliet Junior College

CHAPTER 4

Integrated Review: Getting Ready for Probability

Outline

4.IR1 Scientific Notation
4.IR2 Interval Notation; Intersection and Union of sets
4.IR3 Linear Inequalities

4.IR1 Scientific Notation

Objectives

1. Convert Decimal Notation to Scientific Notation
2. Convert Scientific Notation to Decimal Notation
3. Use Scientific Notation to Multiply and Divide

Did you know that the mass of the Sun is 1,989,000,000,000,000,000,000,000,000,000 kg? Did you know that the mass of a dust particle is 0.000000000753 kg? These numbers are difficult to write and difficult to read, so we use exponents to rewrite them.

1 Convert Decimal Notation to Scientific Notation

Decimal notation is the notation we commonly see in newspapers or magazines. The numbers 1,989,000,000,000,000,000,000,000,000,000 kg and 0.000000000753 kg are written in decimal notation. Scientific notation expresses a number as the product of two factors: One factor is a number between 1 and 10, including 1 but not including 10, and the other factor is an integer power of 10.

> **Definition**
>
> A number written as the product of a number x, where $1 \leq x < 10$, and a power of 10 is said to be written in **scientific notation.** That is, a number is written in scientific notation when it is in the form
>
> $$x \times 10^N$$
>
> where
>
> $$1 \leq x < 10 \text{ and } N \text{ is an integer}$$

Notice in the definition, $x < 10$. That's because when $x = 10$, we have 10^1, a power of 10. For example, in scientific notation,

$$\text{Mass of Sun} = 1.989 \times 10^{30} \text{ kilograms}$$

$$\text{Mass of a dust particle} = 7.53 \times 10^{-10} \text{ kilograms}$$

> **Converting from Decimal Notation to Scientific Notation**
>
> To change a positive number to scientific notation:
>
> **Step 1:** Count the number N of decimal places that the decimal point must be moved in order to arrive at a number x, where $1 \leq x < 10$.
>
> **Step 2:** If the original number is greater than or equal to 1, the scientific notation is $x \times 10^N$. If the original number is between 0 and 1, the scientific notation is $x \times 10^{-N}$.

In Words
For a number greater than or equal to 1, use $10^{\text{positive exponent}}$.
For a number between 0 and 1, use $10^{\text{negative exponent}}$.

EXAMPLE 1 How to Convert from Decimal Notation to Scientific Notation

Write 5283 in scientific notation.

Step-by-Step Solution

For a number to be in scientific notation, the decimal must be moved so that there is a single nonzero digit to the left of the decimal point. All remaining digits must appear to the right of the decimal point.

Step 1: The "understood" decimal point in 5283 follows the 3. Therefore, move the decimal to the left by $N = 3$ places until it is between the 5 and the 2. Do you see why?

$$5\,2\,8\,3.$$
$$\quad\;\; 3\,2\,1$$

Step 2: The original number is greater than 1, so write 5283 in scientific notation as

$$5.283 \times 10^3$$

EXAMPLE 2 How to Convert from Decimal Notation to Scientific Notation

Write 0.054 in scientific notation.

Step-by-Step Solution

Step 1: Because 0.054 is less than 1, move the decimal point to the right $N = 2$ places until it is between the 5 and the 4.

$$0.054$$
$$\;\;\; 1\,2$$

Step 2: The original number is between 0 and 1, so write 0.054 in scientific notation as

$$5.4 \times 10^{-2}$$

Quick ✓

1. A number written as 3.2×10^{-6} is said to be written in _____ notation.

2. When 47,000,000 is written in scientific notation, the power of 10 will be _____ (positive or negative).

3. *True or False* When a number is expressed in scientific notation, it is expressed as the product of a number x, $0 \le x < 1$, and a power of 10.

In Problems 4–9, write each number in scientific notation.

4. 432
5. 10,302
6. 5,432,000
7. 0.093
8. 0.0000459
9. 0.00000008

② Convert Scientific Notation to Decimal Notation

Now we are going to convert a number from scientific notation to decimal notation. Study Table 1 to discover the pattern.

Table 1

Scientific Notation	Product	Decimal Notation	Location of Decimal Point
3.69×10^2	3.69×100	369	moved 2 places to the right
3.69×10^1	3.69×10	36.9	moved 1 place to the right
3.69×10^0	3.69×1	3.69	didn't move
3.69×10^{-1}	3.69×0.1	0.369	moved 1 place to the left
3.69×10^{-2}	3.69×0.01	0.0369	moved 2 places to the left

Section 4.IR1 Scientific Notation   IR-169

The pattern in Table 1 leads to the following steps for converting a number from scientific notation to decimal notation.

> **Converting a Number from Scientific Notation to Decimal Notation**
>
> **Step 1:** Determine the exponent, N, on the number 10.
>
> **Step 2:** If the exponent is positive, then move the decimal point N decimal places to the right. If the exponent is negative, then move the decimal point $|N|$ decimal places to the left. Add zeros, as needed.

EXAMPLE 3 **How to Convert from Scientific Notation to Decimal Notation**

Write 2.3×10^3 in decimal notation.

Step-by-Step Solution

Step 1: Determine the exponent on the number 10.   The exponent on the 10 is 3.

Step 2: Because the exponent is positive, move the decimal point three places to the right. Notice we add zeros to the right of 3, as needed. $2.3\,0\,0.$

So $2.3 \times 10^3 = 2300$.

EXAMPLE 4 **How to Convert from Scientific Notation to Decimal Notation**

Write 4.57×10^{-5} in decimal notation.

Step-by-Step Solution

Step 1: Determine the exponent on the number 10.   The exponent on the 10 is -5.

Step 2: Because the exponent is negative, move the decimal point five places to the left.   Add zeros to the left of the original decimal point. $0.0000\,4.57$

So $4.57 \times 10^{-5} = 0.0000457$.

Quick ✓

10. *True or False* To write 3.2×10^{-6} in decimal notation, move the decimal point in 3.2 six places to the left.

11. *True or False* To convert 2.4×10^3 to decimal notation, move the decimal point three places to the right.

In Problems 12–16, write each number in decimal notation.

12. 3.1×10^2 **13.** 9.01×10^{-1} **14.** 1.7×10^5

15. 7×10^0 **16.** 8.9×10^{-4}

▶ ❸ Use Scientific Notation to Multiply and Divide

To multiply and divide numbers written in scientific notation, we use two Laws of Exponents: the Product Rule, $a^m \cdot a^n = a^{m+n}$, and the Quotient Rule, $\dfrac{a^m}{a^n} = a^{m-n}$. Use these laws where the base is 10 as follows:

$$10^m \cdot 10^n = 10^{m+n} \quad \text{and} \quad \frac{10^m}{10^n} = 10^{m-n}$$

EXAMPLE 5 Multiplying Using Scientific Notation

Perform the indicated operation. Express the answer in scientific notation.

$$(3 \times 10^2) \cdot (2.5 \times 10^5)$$

Solution

$$(3 \times 10^2) \cdot (2.5 \times 10^5) = (3 \cdot 2.5) \times (10^2 \cdot 10^5)$$

$3 \cdot 2.5 = 7.5$; use $a^m \cdot a^n = a^{m+n}$: $= 7.5 \times 10^7$

EXAMPLE 6 Multiplying Using Scientific Notation

Perform the indicated operation. Express the answer in scientific notation.

(a) $(4 \times 10^{-2}) \cdot (6 \times 10^8)$ 　　　(b) $(3.2 \times 10^{-3}) \cdot (4.8 \times 10^{-4})$

Solution

(a)
$$(4 \times 10^{-2}) \cdot (6 \times 10^8) = (4 \cdot 6) \times (10^{-2} \cdot 10^8)$$

$4 \cdot 6 = 24$; use $a^m \cdot a^n = a^{m+n}$: $= 24 \times 10^6$

Convert 24 to scientific notation: $= (2.4 \times 10^1) \times 10^6$

Use $a^m \cdot a^n = a^{m+n}$: $= 2.4 \times 10^7$

(b)
$$(3.2 \times 10^{-3}) \cdot (4.8 \times 10^{-4}) = (3.2 \cdot 4.8) \times (10^{-3} \cdot 10^{-4})$$

$3.2 \cdot 4.8 = 15.36$; use $a^m \cdot a^n = a^{m+n}$: $= 15.36 \times 10^{-7}$

Convert 15.36 to scientific notation: $= (1.536 \times 10^1) \times 10^{-7}$

Use $a^m \cdot a^n = a^{m+n}$: $= 1.536 \times 10^{-6}$

Quick ✓

In Problems 17–20, perform the indicated operation. Express the answer in scientific notation.

17. $(3 \times 10^4) \cdot (2 \times 10^3)$ 　　　**18.** $(2 \times 10^{-2}) \cdot (4 \times 10^{-1})$

19. $(5 \times 10^{-4}) \cdot (3 \times 10^7)$ 　　　**20.** $(8 \times 10^{-4}) \cdot (3.5 \times 10^{-2})$

EXAMPLE 7 Dividing Using Scientific Notation

Perform the indicated operation. Express the answer in scientific notation.

(a) $\dfrac{6 \times 10^6}{2 \times 10^2}$ 　　　(b) $\dfrac{2.4 \times 10^4}{3 \times 10^{-2}}$

Solution

(a)
$$\frac{6 \times 10^6}{2 \times 10^2} = \frac{6}{2} \times \frac{10^6}{10^2}$$

$\dfrac{6}{2} = 3$; use $\dfrac{a^m}{a^n} = a^{m-n}$: $= 3 \times 10^4$

(b)
$$\frac{2.4 \times 10^4}{3 \times 10^{-2}} = \frac{2.4}{3} \times \frac{10^4}{10^{-2}}$$

$\dfrac{2.4}{3} = 0.8$; use $\dfrac{a^m}{a^n} = a^{m-n}$: $= 0.8 \times 10^{4-(-2)}$

Convert 0.8 to scientific notation: $= (8 \times 10^{-1}) \times 10^6$

Use $a^m \cdot a^n = a^{m+n}$: $= 8 \times 10^5$

Quick ✓

In Problems 21–24, perform the indicated operation. Express the answer in scientific notation.

21. $\dfrac{8 \times 10^6}{2 \times 10^1}$

22. $\dfrac{2.8 \times 10^{-7}}{1.4 \times 10^{-3}}$

23. $\dfrac{3.6 \times 10^3}{7.2 \times 10^{-1}}$

24. $\dfrac{5 \times 10^{-2}}{8 \times 10^2}$

EXAMPLE 8 **Visits to Facebook**

In 2015, Facebook had 1.04×10^9 visitors each day. How many visitors to Facebook were there in April 2015? Express the answer in scientific and decimal notation. (*Source: Facebook*)

Solution

To find the number of visitors to Facebook for April 2015, multiply the number of visitors each day by the number of days, 30.
  In scientific notation, $30 = 3 \times 10^1$. So,

$$(1.04 \times 10^9)(3 \times 10^1) = (1.04 \cdot 3) \times (10^9 \cdot 10^1)$$
$$= 3.12 \times 10^{10} \quad \text{scientific notation}$$
$$= 31{,}200{,}000{,}000 \text{ visitors} \quad \text{decimal notation}$$

In April 2015, there were 31,200,000,000 visitors to Facebook.

Quick ✓

25. The United States consumes 3.49×10^8 gallons of gasoline each day. How many gallons of gasoline does the United States consume in a 30-day month? Express the answer in scientific and decimal notation.

26. McDonald's has 3.6×10^6 customers each day. How many customers does McDonald's get in a 365-day year?

4.IR1 Exercises MyLab Statistics
Underlined exercises have complete video solutions in MyLab.

Problems **1–26** *are the Quick ✓s that follow the* **EXAMPLES**.

Building Skills

In Problems 27–42, write each number in scientific notation. See Objective 1.

27. 300,000

28. 421,000,000

29. 64,000,000

30. 8,000,000,000

31. 0.00051

32. 0.0000001

33. 0.000000001

34. 0.0000283

35. 8,007,000,000

36. 401,000,000

37. 0.0000309

38. 0.000201

39. 620

40. 8

41. 4

42. 120

In Problems 43–54, a number is given in decimal notation. Write the number in scientific notation. See Objective 1.

43. **World Population** According to the U.S. Census Bureau, the population of the world on February 4, 2016, was approximately 7,303,000,000 persons.

44. **United States Population** According to the U.S. Census Bureau, the population of the United States on February 4, 2016, was approximately 319,000,000 persons.

45. **Federal Debt** According to the United States Treasury, the federal debt of the United States as of February 2016 was $18,900,000,000,000.

46. **Interest Payments on Debt** In 2015, total interest payments on the federal debt (Problem 45) totaled $200,000,000,000.

47. **Distance to the Sun** The average distance from Earth to the Sun is 93,000,000 miles.

48. **Distance to the Moon** The average distance from Earth to the moon (of Earth) is 384,000 kilometers.

49. **Smallpox Virus** The diameter of a smallpox virus is 0.00003 mm.

50. **Human Blood Cell** The diameter of a human blood cell is 0.0075 mm.

51. **Dust Mites** Dust mites are microscopic bugs that are a major cause of allergies and asthma. They are approximately 0.00000025 m in length.

52. **DNA** The length of a DNA molecule can exceed 0.025 inch in some organisms.

53. **Mass of a Penny** The mass of the United States' Lincoln penny is approximately 0.00311 kg.

54. **Physics** The radius of a typical atom is approximately 0.0000000001 m.

In Problems 55–70, write each number in decimal notation. See Objective 2.

55. 4.2×10^5
56. 3.75×10^2
57. 1×10^8
58. 6×10^6
59. 3.9×10^{-3}
60. 6.1×10^{-6}
61. 4×10^{-1}
62. 5×10^{-4}
63. 3.76×10^3
64. 4.9×10^{-1}
65. 8.2×10^{-3}
66. 5.4×10^5
67. 6×10^{-5}
68. 5.123×10^{-3}
69. 7.05×10^6
70. 7×10^8

In Problems 71–76, a number is given in scientific notation. Write the number in decimal notation. See Objective 2.

71. **Time** A femtosecond is equal to 1×10^{-15} second.

72. **Dust Particle** The mass of a dust particle is 7.53×10^{-10} kg.

73. **Vitamin** A One-A-Day vitamin pill contains 2.25×10^{-3} gram of zinc.

74. **Water Molecule** The diameter of a water molecule is 3.85×10^{-7} m.

75. **Coffee Drinkers** In 2015, there were 1.83×10^6 coffee drinkers in the United States.

76. **Fatal Accidents** In 2014, there were 3.3×10^4 drivers in fatal auto accidents in the United States. (*Source: U.S. Department of Transportation*)

In Problems 77–88, perform the indicated operations. Express your answer in scientific notation. See Objective 3.

77. $(2 \times 10^6)(1.5 \times 10^3)$
78. $(3 \times 10^{-4})(8 \times 10^{-5})$
79. $(1.2 \times 10^0)(7 \times 10^{-3})$
80. $(4 \times 10^7)(2.5 \times 10^{-4})$
81. $\dfrac{9 \times 10^4}{3 \times 10^{-4}}$
82. $\dfrac{6 \times 10^3}{1.2 \times 10^5}$
83. $\dfrac{2 \times 10^{-3}}{8 \times 10^{-5}}$
84. $\dfrac{4.8 \times 10^7}{1.2 \times 10^2}$
85. $\dfrac{56{,}000}{0.00007}$
86. $\dfrac{0.000275}{2500}$
87. $\dfrac{300{,}000 \times 15{,}000{,}000}{0.0005}$
88. $\dfrac{24{,}000{,}000{,}000}{0.00006 \times 2000}$

Applying the Concepts

89. **Speed of Light** Light travels at the rate of 1.86×10^5 miles per second. How far does light travel in 1 minute $(6.0 \times 10^1 \text{ seconds})$?

90. **Speed of Sound** Sound travels at the rate of 1.127×10^3 feet per second. How far does sound travel in 1 minute $(6.0 \times 10^1 \text{ seconds})$?

91. **Hair Growth** Human hair grows at a rate of 1×10^{-8} mile per hour. How many miles will human hair grow after 100 hours? After one week? Express each answer in decimal notation.

92. **Disneyland** The average number of visitors to Disneyland each day is 4×10^4. How many visitors visit Disneyland in a 30-day month? Express your answer in decimal notation.

93. **Ice Cream Consumption** Total U.S. consumption of ice cream in 2014 amounted to about 4.1 billion pounds. (*Source: U.S. Department of Agriculture*)

    (a) Write 3.9 billion pounds in scientific notation.
    (b) The population of the United States in 2014 was approximately 319,000,000 persons. Express this number in scientific notation.
    (c) Use your answers to parts (a) and (b) to find, to the nearest tenth of a pound, the average number of pounds of ice cream that were consumed per person in the United States in 2014.

94. **M&M'S® Candy** Over 400 million M&M'S® candies are produced in the United States every day. (*Source: Mars.com*)

    (a) Write 400 million in scientific notation.
    (b) Assuming a 30-day month, how many M&M'S® candies are produced in a month in the United States? Write this answer in scientific notation.

(c) The population of the United States is approximately 319,000,000 persons. Express this number in scientific notation.

(d) Use your answers to parts (b) and (c) to find, to the nearest whole number, the average number of M&M'S candies eaten per person per month in the United States.

Extending the Concepts

Scientists often need to measure very small things, such as cells. They use the following units of measure:

$$millimeter\ (mm) = 1 \times 10^{-3}\ meter$$
$$micron\ (\mu m) = 1 \times 10^{-6}\ meter$$
$$nanometer\ (nm) = 1 \times 10^{-9}\ meter$$
$$picometer\ (pm) = 1 \times 10^{-12}\ meter$$

Write the following measurements in meters using scientific notation:

95. 250 μm
96. 60.4 nm
97. 800 pm
98. 40 mm
99. 71.5 nm
100. 200 μm

Assume a cell is in the shape of a sphere. Given that the volume of a sphere is $V = \frac{4}{3}\pi r^3$, find the volume of each cell whose radius is given. Express the answer as a multiple of π in cubic meters.

101. 21 μm
102. 0.75 nm
103. 6 nm
104. 108 μm

Explaining the Concepts

105. Explain how to convert a number written in decimal notation to scientific notation.

106. Explain how to convert a number written in scientific notation to decimal notation.

107. Dana thinks that the number 34.5×10^4 is correctly written in scientific notation. Is Dana correct? If so, explain why. If not, explain why the answer is wrong and write the correct answer.

108. Explain why scientific notation is used to perform calculations that involve multiplying and dividing but not adding and subtracting.

4.IR2 Interval Notation; Intersection and Union of Sets

Objectives

1. Represent Inequalities Using the Real Number Line and Interval Notation
2. Determine the Intersection or Union of Two Sets

Work Smart

Remember that the inequalities $<$ and $>$ are called strict inequalities, while $\leq$ and $\geq$ are called nonstrict inequalities.

1 Represent Inequalities Using the Real Number Line and Interval Notation

Suppose that a and b are two real numbers and $a < b$. The notation

$$a < x < b$$

means that x is a number *between* a and b. So the expression $a < x < b$ is equivalent to the two inequalities $a < x$ and $x < b$. Similarly, the expression $a \leq x \leq b$ is equivalent to the two inequalities $a \leq x$ and $x \leq b$. We define $a \leq x < b$ and $a < x \leq b$ similarly. Expressions such as $-2 < x < 5$ and $x \geq 5$ are in **inequality notation.**

Even though the expression $3 \geq x \geq 2$ is technically correct, we prefer that the numbers in an inequality go from smaller values to larger values. So we write $3 \geq x \geq 2$ as $2 \leq x \leq 3$.

A statement such as $3 \leq x \leq 1$ is false because there is no number x for which $3 \leq x$ and $x \leq 1$. Never mix the direction of inequalities, as in $2 \leq x \geq 3$.

Now let's see how to use **interval notation** to represent the solution set of an inequality.

Definitions

Let a and b represent two real numbers with $a < b$.

A **closed interval,** denoted by $[a, b]$, consists of all real numbers x for which $a \leq x \leq b$.

An **open interval,** denoted by (a, b), consists of all real numbers x for which $a < x < b$.

The **half-open,** or **half-closed, intervals** are $(a, b]$, consisting of all real numbers x for which $a < x \leq b$, and $[a, b)$, consisting of all real numbers x for which $a \leq x < b$.

In each of these definitions, a is the **left endpoint** and b is the **right endpoint** of the interval.

The symbol ∞ (which is read "infinity") is not a real number but a notational device that indicates unboundedness in the positive direction. In other words, the symbol ∞ means the inequality has no right endpoint. The symbol −∞ (which is read "minus infinity" or "negative infinity") is also not a number but a notational device that indicates unboundedness in the negative direction. The symbol −∞ means the inequality has no left endpoint. The symbols ∞ and −∞ allow us to define five other kinds of intervals.

Work Smart

The symbols ∞ and −∞ are never included as endpoints because they are not real numbers. Therefore, we use parentheses when −∞ or ∞ is an endpoint.

Interval Notation

$[a, \infty)$ consists of all real numbers x where $x \geq a$
(a, ∞) consists of all real numbers x where $x > a$
$(-\infty, a]$ consists of all real numbers x where $x \leq a$
$(-\infty, a)$ consists of all real numbers x where $x < a$
$(-\infty, \infty)$ consists of all real numbers x (that is, $-\infty < x < \infty$)

Figure 1
$x > 3$

Figure 2
$x \geq 3$

Another way to represent inequalities is to draw a graph on the real number line. The inequality $x > 3$ or the interval $(3, \infty)$ consists of all numbers x that lie to the right of 3 on the real number line. Graph these values by shading the real number line to the right of 3. Use a *parenthesis* on the endpoint to indicate that 3 is not included in the set. See Figure 1.

To graph the inequality $x \geq 3$ or the interval $[3, \infty)$, shade to the right of 3 but use a *bracket* on the endpoint to indicate that 3 is included in the set. See Figure 2.

Table 2 summarizes interval notation, inequality notation, and their graphs.

Table 2

Interval Notation	Inequality Notation	Graph
The open interval (a, b)	$\{x \mid a < x < b\}$	
The closed interval $[a, b]$	$\{x \mid a \leq x \leq b\}$	
The half-open interval $[a, b)$	$\{x \mid a \leq x < b\}$	
The half-open interval $(a, b]$	$\{x \mid a < x \leq b\}$	
The interval $[a, \infty)$	$\{x \mid x \geq a\}$	
The interval (a, ∞)	$\{x \mid x > a\}$	
The interval $(-\infty, a]$	$\{x \mid x \leq a\}$	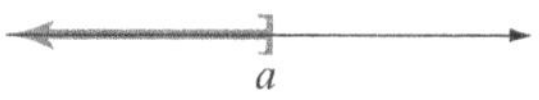
The interval $(-\infty, a)$	$\{x \mid x < a\}$	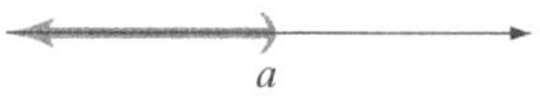
The interval $(-\infty, \infty)$	$\{x \mid x$ is a real number $\}$	

EXAMPLE 1 **Using Interval Notation and Graphing Inequalities**

Write each inequality using interval notation. Graph the inequality.

(a) $-2 \leq x \leq 4$ **(b)** $1 < x \leq 5$

Solution

(a) $-2 \le x \le 4$ describes all numbers x between -2 and 4, inclusive. In interval notation, write $[-2, 4]$. To graph $-2 \le x \le 4$, place brackets at -2 and 4 and shade in between. See Figure 3.

Figure 3

(b) $1 < x \le 5$ describes all numbers x greater than 1 and less than or equal to 5. In interval notation, write $(1, 5]$. To graph $1 < x \le 5$, place a parenthesis at 1 and a bracket at 5 and shade in between. See Figure 4.

Figure 4

EXAMPLE 2 Using Interval Notation and Graphing Inequalities

Write each inequality using interval notation. Graph the inequality.

(a) $x < 2$ (b) $x \ge -3$

Solution

(a) $x < 2$ describes all numbers x less than 2. In interval notation, write $(-\infty, 2)$. To graph $x < 2$, place a parenthesis at 2 and then shade to the left. See Figure 5.

Figure 5

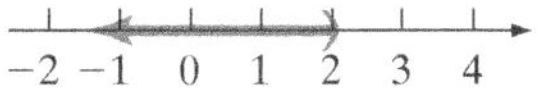

(b) $x \ge -3$ describes all numbers x greater than or equal to -3. In interval notation, write $[-3, \infty)$. To graph $x \ge -3$, place a bracket at -3 and then shade to the right. See Figure 6.

Figure 6

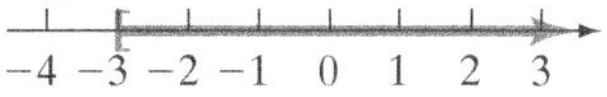

Quick ✓

1. A(n) _____ _____, denoted $[a, b]$, consists of all real numbers x for which $a \le x \le b$.

2. In the interval (a, b), a is called the _____ _____ and b is called the _____ _____ of the interval.

3. *True or False* The inequality $x \le -11$ is written $[-11, -\infty)$ in interval notation.

In Problems 4–7, write each inequality in interval notation. Graph the inequality.

4. $-3 \le x \le 2$
5. $3 \le x < 6$
6. $x \le 3$
7. $\dfrac{1}{2} < x < \dfrac{7}{2}$

EXAMPLE 3 Using Inequality Notation and Graphing Inequalities

Write each interval in inequality notation involving x. Graph the inequality.

(a) $[-2, 4)$ (b) $(1, 5)$

Solution

(a) The interval $[-2, 4)$ consists of all numbers x for which $-2 \leq x < 4$. See Figure 7 for the graph.

(b) The interval $(1, 5)$ consists of all numbers x for which $1 < x < 5$. See Figure 8 for the graph.

Figure 7

Figure 8

EXAMPLE 4 Using Inequality Notation and Graphing Inequalities

Write each interval in inequality notation involving x. Graph the inequality.

(a) $\left[\dfrac{3}{2}, \infty\right)$ 　　　(b) $(-\infty, 1)$

Solution

(a) The interval $\left[\dfrac{3}{2}, \infty\right)$ consists of all numbers x for which $x \geq \dfrac{3}{2}$. See Figure 9 for the graph.

(b) The interval $(-\infty, 1)$ consists of all numbers x for which $x < 1$. See Figure 10 for the graph.

Figure 9
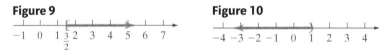

Figure 10

Quick ✓

In Problems 8–11, write each interval as an inequality. Graph the inequality.

8. $(0, 5]$ 　　　　　**9.** $(-6, 0)$

10. $(5, \infty)$ 　　　　**11.** $\left(-\infty, \dfrac{8}{3}\right]$

▶ ❷ Determine the Intersection or Union of Two Sets

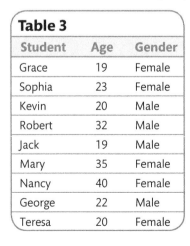

Table 3

Student	Age	Gender
Grace	19	Female
Sophia	23	Female
Kevin	20	Male
Robert	32	Male
Jack	19	Male
Mary	35	Female
Nancy	40	Female
George	22	Male
Teresa	20	Female

Table 3 contains information about students in an Intermediate Algebra course. We can classify these people in various sets. For example, we can define set A as the set of all students whose age is less than 25. Then

$$A = \{\text{Grace, Sophia, Kevin, Jack, George, Teresa}\}$$

If we define set B as the set of all students who are female, then

$$B = \{\text{Grace, Sophia, Mary, Nancy, Teresa}\}$$

Now list all the students who are in set A and set B, that is, students who are under 25 years of age and female.

$$A \text{ and } B = \{\text{Grace, Sophia, Teresa}\}$$

Now list all the students who are in set A or in set B or in both sets.

$$A \text{ or } B = \{\text{Grace, Sophia, Kevin, Jack, George, Teresa, Mary, Nancy}\}$$

Figure 11 shows a Venn diagram illustrating the relations among A, B, A and B, and A or B. Notice that Grace, Sophia, and Teresa are in both A and B, while Robert is in neither A nor B.

Figure 11

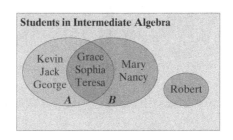

When we used the word "and," we listed elements common to both set A and set B. When we used the word "or," we listed elements in set A or in set B or in both sets. These results lead us to the following definitions.

Definitions
- The **intersection** of two sets A and B, denoted $A \cap B$, is the set of all elements in both set A and set B.
- The **union** of two sets A and B, denoted $A \cup B$, is the set of all elements in set A or in set B or in both set A and set B.
- The word **and** implies intersection, while the word **or** implies union.

EXAMPLE 5 **Finding the Intersection and Union of Sets**

Let $A = \{1, 3, 5, 7, 9\}$ and let $B = \{1, 2, 3, 4, 5\}$. Find

(a) $A \cap B$ (b) $A \cup B$

Solution

(a) $A \cap B$ is the set of all elements that are in both A and B. So,
$$A \cap B = \{1, 3, 5\}$$

(b) $A \cup B$ is the set of all elements that are in A or B, or both. So,
$$A \cup B = \{1, 2, 3, 4, 5, 7, 9\}$$

Work Smart

When finding the union of two sets, list each element only once, even if it occurs in both sets.

Quick ✓

12. The _____ of two sets A and B, denoted $A \cap B$, is the set of all elements that belong to both set A and set B.
13. The word _____ implies intersection. The word _____ implies union.
14. *True or False* The intersection of two sets can be the empty set.

In Problems 15–20, let $A = \{1, 2, 3, 4, 5, 6\}$, $B = \{1, 3, 5, 7\}$, and $C = \{2, 4, 6, 8\}$.

15. Find $A \cap B$.
16. Find $A \cap C$.
17. Find $A \cup B$.
18. Find $A \cup C$.
19. Find $B \cap C$.
20. Find $B \cup C$.

EXAMPLE 6 **Finding the Intersection and Union of Two Sets**

Suppose $A = \{x | x \leq 5\}$, $B = \{x | x \geq 1\}$, and $C = \{x | x < -2\}$.

(a) Determine $A \cap B$. Graph the set on a real number line and write it in set-builder notation and interval notation.

(b) Determine $B \cup C$. Graph the set on a real number line and write it in set-builder notation and interval notation.

Solution

(a) $A \cap B$ is the set of all real numbers less than or equal to 5 and greater than or equal to 1. We can identify this set by determining where the graphs of the inequalities overlap. See Figure 12.

Figure 12

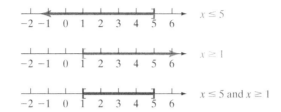

$A \cap B$ is $\{x | 1 \leq x \leq 5\}$ in set-builder notation, and $[1, 5]$ in interval notation.

Figure 13

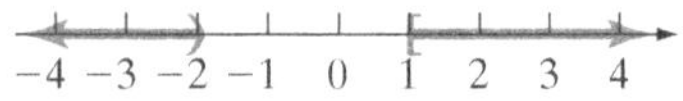

(b) $B \cup C$ is the set of all real numbers greater than or equal to 1 or less than -2. See Figure 13. Therefore, $B \cup C$ is $\{x | x < -2 \text{ or } x \geq 1\}$ in set-builder notation, and $(-\infty, -2) \cup [1, \infty)$ in interval notation.

Quick ✓

Let $A = \{x | x > 2\}$, $B = \{x | x < 7\}$, and $C = \{x | x \leq -3\}$.

21. Determine $A \cap B$. Graph the set on a real number line and write it in set-builder notation and interval notation.
22. Determine $A \cup C$. Graph the set on a real number line and write it in set-builder notation and interval notation.

4.IR2 Exercises — MyLab Statistics
Underlined exercises have complete video solutions in MyLab.

Problems 1–22 are the Quick ✓s that follow the EXAMPLES.

Building Skills

In Problems 23–30, write each inequality using interval notation. Graph the inequality. See Objective 1.

23. $2 \leq x \leq 10$
24. $1 < x < 7$
25. $-4 \leq x < 0$
26. $-8 < x \leq 1$
27. $x \geq 6$
28. $x < 0$
29. $x < \dfrac{3}{2}$
30. $x \geq -\dfrac{5}{2}$

In Problems 31–38, write each interval as an inequality involving x. Graph each inequality. See Objective 1.

31. $(1, 8)$
32. $[-2, 3]$
33. $(-5, 1]$
34. $[1, 4)$
35. $(-\infty, 5)$
36. $(2, \infty)$
37. $[3, \infty)$
38. $(-\infty, 8]$

In Problems 39–44 use $A = \{4, 5, 6, 7, 8, 9\}$, $B = \{1, 5, 7, 9\}$ and $C = \{2, 3, 4, 6\}$ to find each set. See Objective 1.

39. $A \cup B$
40. $A \cup C$
41. $A \cap B$
42. $A \cap C$
43. $B \cap C$
44. $B \cup C$

In Problems 45–48, use the graph of the inequality to find each set. See Objective 1.

45. $A = \{x | x \leq 5\}$; $B = \{x | x > -2\}$
Find (a) $A \cap B$ and (b) $A \cup B$.

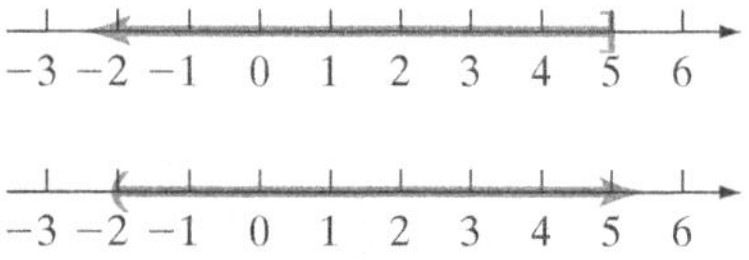

46. $A = \{x | x \geq 4\}$; $B = \{x | x < 1\}$
Find (a) $A \cap B$ and (b) $A \cup B$.

47. $E = \{x|x > 3\}; F = \{x|x < -1\}$
Find (a) $E \cap F$ and (b) $E \cup F$.

48. $E = \{x|x \leq 2\}; F = \{x|x \geq -2\}$
Find (a) $E \cap F$ and (b) $E \cup F$.

4.IR3 Linear Inequalities

Objectives
1. Solve Linear Inequalities Using Properties of Inequalities
2. Model Inequality Problems

Definition
A **linear inequality in one variable** is an inequality that can be written in the form

$$ax + b < c \text{ or } ax + b \leq c \text{ or } ax + b > c \text{ or } ax + b \geq c$$

where $a, b,$ and c are real numbers and $a \neq 0$.

Examples of linear inequalities in one variable, x, are

$$x + 2 > 7 \quad \frac{1}{2}x + 4 \leq 9 \quad 5x > 0 \quad 3x + 7 \geq 8x - 3$$

1 Solve Linear Inequalities Using Properties of Inequalities

To **solve an inequality** means to find all values of the variable for which the statement is true. These values are the **solutions** of the inequality. The set of all solutions is the **solution set**. As with equations, one method for solving a linear inequality is to replace it by a series of *equivalent inequalities* until an inequality with an obvious solution, such as $x > 2$, is obtained.

Two inequalities having the same solution set are called **equivalent inequalities.** We obtain equivalent inequalities by using some of the same operations used to find equivalent equations.

Consider the inequality $3 < 8$. If we add 2 to both sides of the inequality, the left side becomes 5 and the right side becomes 10. Because $5 < 10$, adding the same quantity to both sides of an inequality does not change the sense, or direction, of the inequality. This result is called the **Addition Property of Inequality**.

In Words
The Addition Property of Inequality states that the direction of the inequality does not change when the same quantity is added to each side of the inequality.

Addition Property of Inequality
For real numbers $a, b,$ and c

If $a < b$, then $a + c < b + c$
If $a > b$, then $a + c > b + c$

In Words
Subtracting a quantity from both sides of an inequality also does not change the direction of the inequality.

The Addition Property of Inequality also holds true for subtracting a real number from both sides of an inequality, because $a - b$ is equivalent to $a + (-b)$.

EXAMPLE 1 How to Solve an Inequality Using the Addition Property of Inequality

Solve the linear inequality $4y - 5 \leq 3y - 2$. Express the solution set using set-builder notation and interval notation. Graph the solution set.

Step-by-Step Solution

Step 1: Get the terms containing variables on the left side of the inequality.

Subtract 3y from both sides:

$$4y - 5 \leq 3y - 2$$
$$4y - 5 - 3y \leq 3y - 2 - 3y$$
$$y - 5 \leq -2$$

Step 2: Isolate the variable y on the left side.

Add 5 to both sides:

$$y - 5 + 5 \leq -2 + 5$$
$$y \leq 3$$

Figure 14

The solution set using set-builder notation is $\{y | y \leq 3\}$. The solution set using interval notation is $(-\infty, 3]$. The solution set is graphed in Figure 14.

Quick ✓

1. To _____ an inequality means to find the set of all values of the variable for which the statement is true.
2. Write the inequality that results from adding 7 to each side of $3 < 8$. What property of inequalities does this illustrate?

In Problems 3–6, find the solution of the linear inequality. Express the solution set in set-builder notation and interval notation. Graph the solution set.

3. $n - 2 > 1$
4. $-2x + 3 < 7 - 3x$
5. $5n + 8 \leq 4n + 4$
6. $3(4x - 8) + 12 > 11x - 13$

We've seen what happens when we add a real number to both sides of an inequality. Let's look at two examples from arithmetic to see if we can figure out what happens when we multiply or divide both sides of an inequality by a nonzero constant.

EXAMPLE 2 Multiplying or Dividing an Inequality by a Positive Number

(a) Write the inequality that results from multiplying both sides of the inequality $2 < 5$ by 3.

(b) Write the inequality that results from dividing both sides of the inequality $18 > 14$ by 2.

Solution

(a) Multiplying both sides of $2 < 5$ by 3 results in the numbers 6 and 15 on each side of the inequality, so we have $6 < 15$.

(b) Dividing both sides of $18 > 14$ by 2 results in the numbers 9 and 7 on each side of the inequality, so we have $9 > 7$.

From Example 2 it would seem that multiplying (or dividing) both sides of an inequality by a positive real number does not change the direction of the inequality.

EXAMPLE 3 Multiplying or Dividing an Inequality by a Negative Number

(a) Write the inequality that results from multiplying both sides of the inequality $2 < 5$ by -3.

(b) Write the inequality that results from dividing both sides of the inequality $18 > 14$ by -2.

Solution

(a) Multiplying both sides of $2 < 5$ by -3 results in the numbers -6 and -15 on each side of the inequality, so we have $-6 > -15$.

(b) Dividing both sides of $18 > 14$ by -2 results in the numbers -9 and -7 on each side of the inequality, so we have $-9 < -7$.

Example 3 suggests that multiplying (or dividing) both sides of an inequality by a *negative real number* results in an inequality that changes the direction of the original inequality. The results of Examples 2 and 3 lead us to the **Multiplication Properties of Inequality.**

> **In Words**
> The Multiplication Properties of Inequality state that if you multiply (or divide) both sides of an inequality by a positive number, the inequality symbol remains the same, but if you multiply (or divide) both sides by a negative number, the inequality symbol reverses.

Multiplication Properties of Inequality

Let a, b, and c be real numbers.

$$\text{If } a < b \text{ and } c \text{ is positive, then } ac < bc.$$
$$\text{If } a > b \text{ and } c \text{ is positive, then } ac > bc.$$
$$\text{If } a < b \text{ and } c \text{ is negative, then } ac > bc.$$
$$\text{If } a > b \text{ and } c \text{ is negative, then } ac < bc.$$

The Multiplication Property of Inequality also holds for dividing both sides of an inequality by a nonzero real number.

EXAMPLE 4 **Solving a Linear Inequality Using the Multiplication Properties of Inequality**

Solve each linear inequality. Express the solution set using set-builder notation and interval notation. Graph the solution set.

(a) $\frac{1}{3}x > -2$ (b) $-4x \geq 24$

Solution

(a)
$$\frac{1}{3}x > -2$$

Multiply both sides of the inequality by 3: $3 \cdot \frac{1}{3}x > 3 \cdot (-2)$

$$x > -6$$

The solution set using set-builder notation is $\{x \mid x > -6\}$. The solution set using interval notation is $(-6, \infty)$. The graph of the solution set is shown in Figure 15.

Figure 15

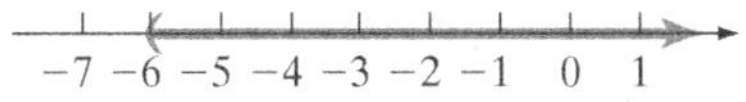

(b)
$$-4x \geq 24$$

Divide both sides of the inequality by -4. $\dfrac{-4x}{-4} \leq \dfrac{24}{-4}$
Remember to reverse the inequality symbol!

$$x \leq -6$$

The solution set using set-builder notation is $\{x \mid x \leq -6\}$. The solution set using interval notation is $(-\infty, -6]$. The graph of the solution set is shown in Figure 16.

Figure 16

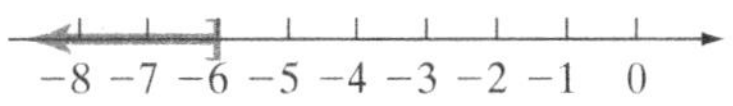

CHAPTER 4 Integrated Review: Getting Ready for Probability

Quick ✓

7. Write the inequality that results from multiplying both sides of the inequality $3 < 12$ by $\frac{1}{3}$. What property of inequalities does this illustrate?

8. Write the inequality that results from dividing both sides of $6 < 10$ by -2. What property of inequalities does this illustrate?

In Problems 9–12, find the solution of the linear inequality and express the solution set in set-builder notation and interval notation. Graph the solution set.

9. $\frac{1}{6}k < 2$

10. $2n \geq -6$

11. $-\frac{3}{2}k > 12$

12. $-\frac{4}{3}p \leq -\frac{4}{5}$

We can now solve inequalities using both the Addition and Multiplication Properties of Inequality. In each solution, isolate the variable on the left side of the inequality, so the inequality is easier to read. If the variable ends up on the right side, remember that

$a < x$ is equivalent to $x > a$ and
$a > x$ is equivalent to $x < a$

EXAMPLE 5 — How to Solve an Inequality Using Both the Addition and Multiplication Properties of Inequality

Solve the inequality $4(x + 1) - 2 < 8x - 26$. Express the solution set using set-builder notation and interval notation. Graph the solution set.

Step-by-Step Solution

Step 1: Remove parentheses.

$$4(x + 1) - 2 < 8x - 26$$

Use the Distributive Property:

$$4x + 4 - 2 < 8x - 26$$

Step 2: Combine like terms on each side of the inequality.

$$4x + 2 < 8x - 26$$

Step 3: Get the terms containing variables on the left side of the inequality and the constants on the right side.

Subtract 8x from both sides:

$$4x + 2 - 8x < 8x - 26 - 8x$$
$$-4x + 2 < -26$$

Subtract 2 from both sides:

$$-4x + 2 - 2 < -26 - 2$$
$$-4x < -28$$

Step 4: Get the coefficient of the variable term to be 1.

Divide both sides by -4:
Remember to reverse the inequality symbol!

$$\frac{-4x}{-4} > \frac{-28}{-4}$$
$$x > 7$$

Figure 17

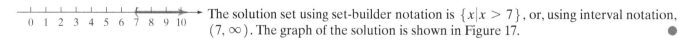

The solution set using set-builder notation is $\{x | x > 7\}$, or, using interval notation, $(7, \infty)$. The graph of the solution is shown in Figure 17.

After solving an inequality, we can substitute a value for the variable that is in the solution set into the original inequality and see whether we obtain a true statement. If we obtain a true statement, then we have some evidence that our solution is correct. However, this does *not* prove that our solution is correct. The check for Example 7 is shown below.

Check Because the solution of the inequality $4(x + 1) - 2 < 8x - 26$ is any real number greater than 7, let's replace x by 10.

$$4(x + 1) - 2 < 8x - 26$$

Replace x with 10: $\quad 4(10 + 1) - 2 \stackrel{?}{<} 8(10) - 26$

$$4(11) - 2 \stackrel{?}{<} 80 - 26$$

Perform the arithmetic: $\quad\quad\quad 42 < 54 \quad$ True

$42 < 54$ is a true statement, so $x = 10$ is in the solution set. Also, if we substitute 6 for x, which is less than 7, we obtain $26 < 22$, which is false. We have some evidence that our solution set, $(7, \infty)$, is correct.

Quick ✓

In Problems 13–16, find the solution of the linear inequality. Express the solution set using set-builder notation and interval notation. Graph the solution set.

13. $3x - 7 > 14$
14. $-4n - 3 < 9$
15. $2x - 6 < 3(x + 1) - 5$
16. $-4(x + 6) + 18 \geq -2x + 6$

EXAMPLE 6 **Solving a Linear Inequality Containing Fractions**

Solve the inequality $\frac{1}{2}(x - 4) \geq \frac{3}{4}(2x + 1)$. Express the solution using set-builder notation and interval notation. Graph the solution set.

Solution

Clear the fractions by multiplying both sides of the inequality by 4, the least common denominator of $\frac{1}{2}$ and $\frac{3}{4}$.

$$\frac{1}{2}(x - 4) \geq \frac{3}{4}(2x + 1)$$

$$4 \cdot \frac{1}{2}(x - 4) \geq 4 \cdot \frac{3}{4}(2x + 1)$$

$$2(x - 4) \geq 3(2x + 1)$$

Use the Distributive Property: $\quad 2x - 8 \geq 6x + 3$

Subtract 6x from both sides: $\quad 2x - 8 - 6x \geq 6x + 3 - 6x$

$$-4x - 8 \geq 3$$

Add 8 to both sides: $\quad -4x - 8 + 8 \geq 3 + 8$

$$-4x \geq 11$$

Divide both sides by -4 and remember to reverse the inequality symbol: $\quad \dfrac{-4x}{-4} \leq \dfrac{11}{-4}$

$$x \leq -\frac{11}{4}$$

The solution is $\left\{x \mid x \leq -\dfrac{11}{4}\right\}$, or, using interval notation, $\left(-\infty, -\dfrac{11}{4}\right]$. The graph of the solution is given in Figure 18.

Figure 18

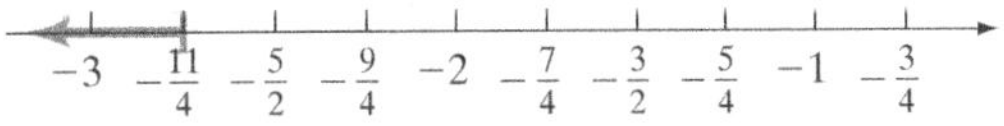

Quick ✓

In Problems 17 and 18, find the solution of the linear inequality and express the solution set using set-builder notation and interval notation. Graph the solution set.

17. $\dfrac{1}{2}(x + 2) > \dfrac{1}{5}(x + 17)$
18. $\dfrac{4}{3}x - \dfrac{2}{3} \leq \dfrac{4}{5}x + \dfrac{3}{5}$

Some inequalities are true for all values of the variable, and some are false for all values of the variable. We present these special cases now.

EXAMPLE 7 **Solving an Inequality Whose Solution Set Is All Real Numbers**

Solve the inequality $3(x + 4) - 5 > 7x - (4x + 2)$. Express the solution using set-builder notation and interval notation, and graph the solution set.

Solution

$$3(x + 4) - 5 > 7x - (4x + 2)$$

Use the Distributive Property: $$3x + 12 - 5 > 7x - 4x - 2$$

$$3x + 7 > 3x - 2$$

Subtract $3x$ from each side: $$3x + 7 - 3x > 3x - 2 - 3x$$

$$7 > -2$$

The statement $7 > -2$ is true, so the solution is all real numbers. The solution set is $\{x | x \text{ is any real number}\}$, or $(-\infty, \infty)$ in interval notation. Figure 19 shows the graph of the solution set.

Figure 19

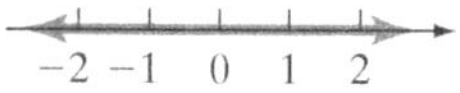

EXAMPLE 8 **Solving an Inequality Whose Solution Set Is the Empty Set**

Solve the inequality $8\left(\frac{1}{2}x - 1\right) + 2x \le 6x - 10$. Express the solution using set-builder notation and interval notation, if possible, and graph the solution set.

Solution

$$8\left(\frac{1}{2}x - 1\right) + 2x \le 6x - 10$$

Use the Distributive Property: $$4x - 8 + 2x \le 6x - 10$$

$$6x - 8 \le 6x - 10$$

Subtract $6x$ from each side: $$6x - 8 - 6x \le 6x - 10 - 6x$$

$$-8 \le -10$$

The statement $-8 \le -10$ is false, so this inequality has no solution. The solution set is the empty set, $\varnothing$ or $\{ \ \}$. Figure 20 shows the graph of the solution set on a real number line.

Figure 20

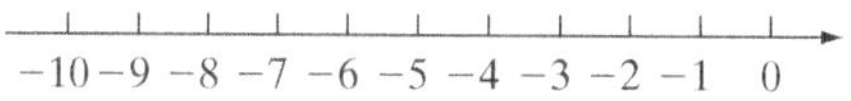

Quick ✓

19. (a) *True or False* When you are solving an inequality, if the variable is eliminated and the result is a true statement, the solution is all real numbers.

(b) *True or False* When you are solving an inequality, if the variable is eliminated and the result is a false statement, the solution is the empty set or $\varnothing$.

In Problems 20–23, find the solution of the linear inequality. Express the solution set using set-builder notation and interval notation, if possible. Graph the solution set.

20. $-2x + 7(x + 5) \le 6x + 32$

21. $-x + 7 - 8x \ge 2(8 - 5x) + x$

22. $\frac{3}{2}x + 5 - \frac{5}{2}x < 4x - 3(x + 1)$

23. $x + 3(x + 4) \ge 2x + 5 + 3x - x$

▶ ❷ Model Inequality Problems

When solving word problems modeled by inequalities, look for key words that indicate the type of inequality symbol to use. Some key words and phrases are listed in Table 4.

Table 4

Word or Phrase	Inequality Symbol	Word or Phrase	Inequality Symbol
at least	≥	at most	≤
no less than	≥	no more than	≤
more than	>	fewer than	<
greater than	>	less than	<

Work Smart

The context of the words "less than" is important! "3 **less than** x" is $x - 3$, but "3 is **less than** x" is $3 < x$. Do you see the difference?

To solve applications involving linear inequalities, use the same steps for setting up applied problems that we introduced in Section 2.IR5.

EXAMPLE 9   A Handyman's Fee

A handyman charges a flat fee of $60 plus $22 per hour for a job. How many hours does this handyman need to work to make at least $500?

Solution

Step 1: Identify We want to know the number of hours the handyman must work to make at least $500.

Step 2: Name Let h represent the number of hours that the handyman must work.

Step 3: Translate The sum of the flat fee that the handyman charges and the hourly charge must exceed $500. So

$$\underbrace{60}_{\text{flat fee}} \underbrace{+}_{\text{plus}} \underbrace{22}_{\text{hourly wage}} \underbrace{\cdot}_{\text{times}} \underbrace{h}_{\text{no. hours worked}} \underbrace{\geq}_{\text{at least}} \underbrace{500}_{500} \quad \text{The Model}$$

Step 4: Solve

$$60 + 22h \geq 500$$

Subtract 60 from each side: $60 - 60 + 22h \geq 500 - 60$

Simplify: $22h \geq 440$

Divide both sides by 22: $\dfrac{22h}{22} \geq \dfrac{440}{22}$

$$h \geq 20$$

Step 5: Check If the handyman works 23 hours (for example), will he earn more than $500? Since $60 + 22(23) = 566$ is greater than 500, we have evidence that our answer is correct.

Step 6: Answer The handyman must work no less than 20 hours to earn at least $500. ●

Quick ✓

24. A worker in a large apartment complex uses an elevator to move supplies. The elevator has a weight limit of 2000 pounds. The worker weighs 180 pounds, and each box of supplies weighs 91 pounds. Find the maximum number of boxes of supplies the worker can move on one trip in the elevator.

4.IR3 Exercises MyLab Statistics

Underlined exercises have complete video solutions in MyLab.

Problems 1–24 are the Quick ✓s that follow the EXAMPLES.

In Problems 25–32, fill in the blank with the correct symbol. State which property of inequality is being used. See Objective 3.

25. If $x - 7 < 11$, then x ____ 18

26. If $3x - 2 > 7$, then $3x$ ____ 9

27. If $\frac{1}{3}x > -2$, then x ____ -6

28. If $\frac{5}{3}x \geq -10$, then x ____ -6

29. If $3x + 2 \leq 13$, then $3x$ ____ 11

30. If $\frac{3}{4}x + 1 \leq 9$, then $\frac{3}{4}x$ ____ 8

31. If $-3x \geq 15$, then x ____ -5

32. If $-4x < 28$, then x ____ -7

In Problems 33–62, solve the inequality. Express the solution set in set-builder notation and interval notation. Graph the solution set on a real number line. See Objective 3.

33. $x + 1 < 5$

34. $x + 4 \leq 3$

35. $x - 6 \geq -4$

36. $x - 2 < 1$

37. $3x \leq 15$

38. $4x > 12$

39. $-5x < 35$

40. $-7x \geq 28$

41. $3x - 7 > 2$

42. $2x + 5 > 1$

43. $3x - 1 \geq 3 + x$

44. $2x - 2 \geq 3 + x$

45. $1 - 2x \leq 3$

46. $2 - 3x \leq 5$

47. $-2(x + 3) < 8$

48. $-3(1 - x) > x + 8$

49. $4 - 3(1 - x) \leq 1$

50. $8 - 4(2 - x) \leq -2x$

51. $\frac{1}{2}(x - 4) > x + 8$

52. $3x + 4 > \frac{1}{3}(x - 2)$

53. $4(x - 1) > 3(x - 1) + x$

54. $2y - 5 + y < 3(y - 2)$

55. $5(n + 2) - 2n \leq 3(n + 4)$

56. $3(p + 1) - p \geq 2(p + 1)$

57. $2n - 3(n - 2) < n - 4$

58. $4x - 5(x + 1) \leq x - 3$

59. $4(2w - 1) \geq 3(w + 2) + 5(w - 2)$

60. $3q - (q + 2) > 2(q - 1)$

61. $3y - (5y + 2) > 4(y + 1) - 2y$

62. $8x - 3(x - 2) \geq x + 4(x + 1)$

In Problems 63–72, write the given statement using inequality symbols. Let x represent the unknown quantity. See Objective 4.

63. Karen's salary this year will be at least $16,000.

64. Bob's salary this year will be at most $120,000.

65. There will be at most 20,000 fans at the Cleveland Indians game today.

66. The cost of a new lawnmower is at least $250.

67. The cost to remodel a kitchen is more than $12,000.

68. There are no more than 25 students in your math class on any given day.

69. x is a positive number

70. x is a nonnegative number

71. x is a nonpositive number

72. x is a negative number

Mixed Practice

In Problems 73–90, solve the inequality and express the solution set in set-builder notation and interval notation, if possible. Graph the solution set on a real number line.

73. $-1 < x - 5$

74. $6 \geq x + 15$

75. $-\frac{3}{4}x > -\frac{9}{16}$

76. $-\frac{5}{8}x > \frac{25}{48}$

77. $3(x + 1) > 2(x + 1) + x$

78. $5(x - 2) < 3(x + 1) + 2x$

79. $-4a + 1 > 9 + 3(2a + 1) + a$

80. $-5b + 2(b - 1) \leq 6 - (3b - 1) + 2b$

81. $n + 3(2n + 3) > 7n - 3$

82. $2k - (k - 4) \geq 3k + 10 - 2k$

83. $\frac{x}{2} \geq 1 - \frac{x}{4}$

84. $\frac{x}{3} \geq 2 + \frac{x}{6}$

85. $\dfrac{x+5}{2} + 4 > \dfrac{2x+1}{3} + 2$

86. $\dfrac{3z-1}{4} + 1 \le \dfrac{6z+5}{2} + 2$

87. $-5z - (3 + 2z) > 3 - 7z$

88. $2(4a - 3) \le 5a - (2 - 3a)$

89. $1.3x + 3.1 < 4.5x - 15.9$

90. $4.9 + 2.6x < 4.2x - 4.7$

Applying the Concepts

91. **Auto Rental** A car can be rented from Certified Auto Rental for $55 per week plus $0.18 per mile. How many miles can you drive if you have at most $280 to spend for weekly transportation?

92. **Truck Rental** A truck can be rented from Acme Truck Rental for $80 per week plus $0.28 per mile. How many miles can you drive if you have at most $100 to spend on truck rental?

93. **Final Grade** Yvette has scores of 72, 78, 66, and 81 on her algebra tests. What score must she make on the final exam to pass the course with at least 360 points? The final exam counts as two test grades.

94. **Final Grade** To earn an A in Mrs. Smith's elementary statistics class, Elizabeth must earn at least 540 points. Thus far, Elizabeth has earned scores of 85, 83, 90, and 96. The final exam counts as two test grades. How many points does Elizabeth have to score on the final exam to earn an A?

95. **Calling Plan** Imperial Telephone has a long-distance calling plan that has a monthly fee of $10 and a charge of $0.03 per minute. Mayflower Communications has a monthly fee of $6 and a fee of $0.04 per minute. For how many minutes is Imperial Telephone the cheaper plan?

96. **Commission** A recent college graduate had an offer of a sales position that pays $15,000 per year plus 1% of all sales.
    (a) Write an expression for the total annual salary based on sales of S dollars.
    (b) For what total sales amount will the college graduate earn in excess of $150,000 annually?

97. **Borrowing Money** The amount of money that a lending institution will allow you to borrow mainly depends on the interest rate and your annual income. The equation $L = 2.98I - 76.11$ describes the amount of money, L, that a bank will lend at an interest rate of 7.5% for 30 years, based upon annual income, I. For what annual income, I, will a bank lend at least $150,000? (*Source: Information Please Almanac*)

98. **Advertising** A marketing firm found that the equation $S = 2.1A + 224$ describes how the amount of sales S of a product depends on A, the amount spent on advertising the product. Both S and A are measured in thousands of dollars. For what amount, A, is the sales of a product at least 350 thousand dollars?

Extending the Concepts

99. **Grades** In your Economics 101 class, you have scores of 68, 82, 87, and 89 on the first four of five tests. To earn a grade of B or higher, the average of the first five test scores must be greater than or equal to 80. Find the minimum score that you can make on the last test and earn a B.

100. **Delivery Service** A messenger service charges $10 to make a delivery to an address. In addition, each letter delivered costs $3 and each package delivered costs $8. If there are 15 more letters than packages delivered to this address, what is the maximum number of items that can be delivered for $85?

Explaining the Concepts

101. In graphing an inequality, when is a left parenthesis used? When is a left bracket used?

102. Explain the circumstances in which the direction of the inequality symbol is reversed when one is solving a simple inequality.

103. Explain how you recognize when the solution of an inequality is all real numbers. Explain how you recognize when the solution of an inequality is the empty set.

104. Why is the interval notation $[-7, -\infty)$ for $x > -7$ incorrect? What is the correct notation?

CHAPTER 6

Integrated Review: Getting Ready for the Normal Probability Distribution

Outline

6.IR1 Perimeter and Area of Polygons and Circles

6.IR1 Perimeter and Area of Polygons and Circles

Objectives

1. Find the Perimeter and Area of a Rectangle and a Square
2. Find the Perimeter and Area of a Parallelogram and a Trapezoid
3. Find the Perimeter and Area of a Triangle
4. Find the Circumference and Area of a Circle

In Words
Vertices is the plural form of the word *vertex*. The word *polygon* comes from the Greek words *poly*, which means "many" and *gon*, which means "angle."

A **polygon** is a closed figure in a plane consisting of line segments that meet at the vertices. A **regular polygon** is one in which the sides have the same length and the angles have the same measure.

A polygon is named according to the number of side. See Table 1.

Table 1 Polygons

Number of Sides	Name of Polygon	Sketch
3	Triangle	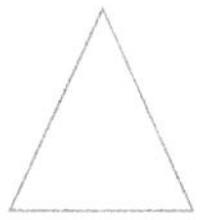
4	Quadrilateral	
5	Pentagon	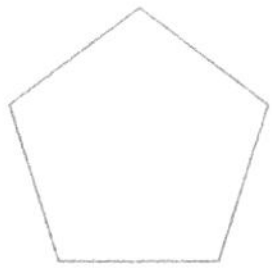
6	Hexagon	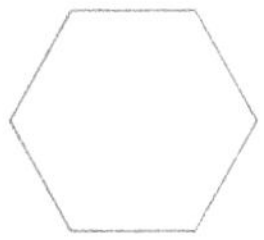

A **quadrilateral** is a polygon with four sides. A **parallelogram** is a quadrilateral in which both pairs of opposite sides are parallel. See Figure 1.

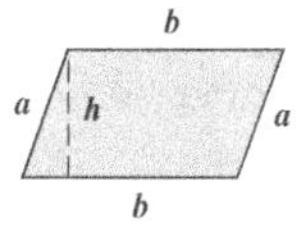

Figure 1
Parallelogram

A **rectangle** is a parallelogram that contains a right angle. See Figure 2.

Figure 2
Rectangle

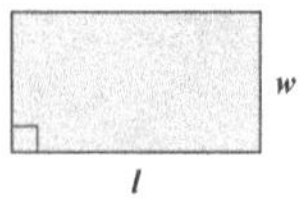

A **square** is a rectangle with all sides of equal length. See Figure 3.

Figure 3
Square

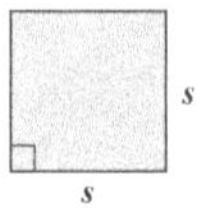

A **rhombus** is a parallelogram that has all sides equal in length. See Figure 4.

Work Smart
Any square is a rhombus. But a rhombus is not necessarily a square.

Figure 4
Rhombus

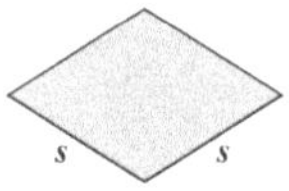

A **trapezoid** is a quadrilateral with exactly one pair of opposite sides that are parallel. See Figure 5.

Figure 5
Trapezoid

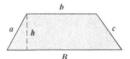

The **perimeter** of a polygon is the distance around the polygon. To put it another way, the perimeter of a polygon is the sum of the lengths of its sides.

① Find the Perimeter and Area of a Rectangle and a Square

A rectangle is a polygon, so the perimeter of a rectangle is the sum of the lengths of its sides.

EXAMPLE 1 | **Finding the Perimeter of a Rectangle**

Find the perimeter of the rectangle in Figure 6.

Figure 6

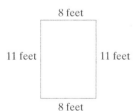

Solution

The perimeter of a rectangle is the sum of the lengths of its sides, so here,

$$\text{Perimeter} = 11 \text{ feet} + 8 \text{ feet} + 11 \text{ feet} + 8 \text{ feet}$$
$$= 38 \text{ feet}$$

Did you notice that the perimeter from Example 1 can also be written as follows?

$$\text{Perimeter} = 2 \cdot 11 \text{ feet} + 2 \cdot 8 \text{ feet}$$

In general, the perimeter, P, of a rectangle is written $P = 2l + 2w$, where l is the length and w is the width.

A different measure of a polygon is its *area*. The **area** of a polygon is the amount of surface of the polygon. Consider the rectangle shown in Figure 7. If we count the number of 1-unit-by-1-unit squares within the rectangle, we see that the area of the rectangle is 6 square units. The area can also be found by multiplying the number of units of length by the number of units of width. In other words, the area of a rectangle is the product of its length and width.

Figure 7

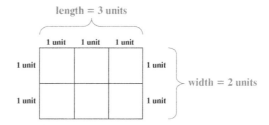

EXAMPLE 2 Finding the Area of a Rectangle

Figure 8

Find the area of the rectangle in Figure 8.

Solution

The area of a rectangle is the product of its length and its width, so here,

$$\text{Area} = 10 \text{ feet} \cdot 6 \text{ feet}$$
$$= 60 \text{ square feet}$$

We also use the notation 60 ft² for 60 square feet.

Table 2 gives a summary of the formulas for the perimeter and area of a rectangle.

Table 2 Perimeter and Area of a Rectangle

Figure	Sketch	Perimeter	Area
Rectangle	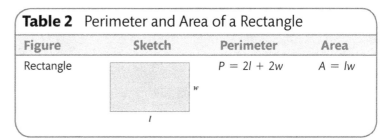	$P = 2l + 2w$	$A = lw$

EXAMPLE 3 Finding the Perimeter and Area of a Rectangle

Figure 9

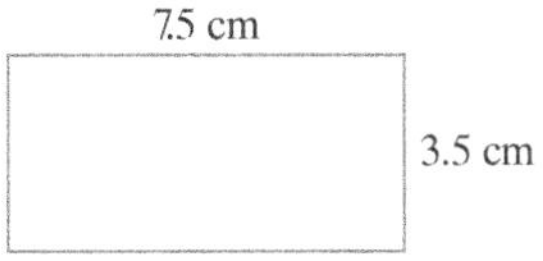

Find **(a)** the perimeter and **(b)** the area of the rectangle shown in Figure 9.

Solution

(a) From Figure 9, the length, l, is 7.5 cm and the width, w, is 3.5 cm. The perimeter of the rectangle is

$$P = 2l + 2w$$
$l = 7.5 \text{ cm}, w = 3.5 \text{ cm}: \quad = 2 \cdot 7.5 \text{ cm} + 2 \cdot 3.5 \text{ cm}$
$$= 15 \text{ cm} + 7 \text{ cm}$$
$$= 22 \text{ cm}$$

(b) The area of the rectangle is

$$A = lw$$
$l = 7.5 \text{ cm}, w = 3.5 \text{ cm}: \quad = 7.5 \text{ cm} \cdot 3.5 \text{ cm}$
$$= 26.25 \text{ square cm, or } 26.25 \text{ cm}^2$$

Quick ✓

1. The _____ of a polygon is the distance around the polygon.
2. The _____ of a polygon is the amount of surface the polygon covers.

In Problems 3 and 4, find the perimeter and area of each rectangle.

EXAMPLE 4 Fencing a Garden

To keep rabbits out of his vegetable garden, Alan wants to enclose the rectangular garden, which is 10 feet long and 6 feet wide, with a fence.

(a) How many feet of fencing should he buy to enclose the garden?

(b) Alan can buy a 20-foot roll of garden fencing for $12.97. How much will Alan have to spend to fence his yard? *Hint:* Alan can buy only full rolls of fencing.

Solution

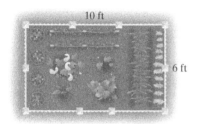

(a) To find the number of feet of fencing needed, find the perimeter of the rectangular garden.

$$P = 2l + 2w$$
$$l = 10 \text{ ft}, w = 6 \text{ ft}: \quad = 2(10 \text{ ft}) + 2(6 \text{ ft})$$
$$\text{Multiply:} \quad = 20 \text{ ft} + 12 \text{ ft}$$
$$= 32 \text{ ft}$$

The perimeter of Alan's garden is 32 feet, so he needs 32 feet of fencing.

(b) Because Alan's garden has a perimeter of 32 feet and the garden fencing is sold in 20-foot rolls, Alan needs to buy two rolls of fencing.

$$\text{Total cost} = 2 \text{ rolls of fencing} \cdot \$12.97 \text{ per roll}$$
$$= \$25.94$$

Alan will spend $25.94 to fence his garden.

EXAMPLE 5 Building a Patio

Kevin is building a rectangular patio that measures 10 feet by 15 feet. The patio pavers he will use to build the patio measure 9 inches by 12 inches.

(a) What is the area of the patio, in square feet?

(b) What is the area of a patio paver, in square inches?

(c) There are 144 square inches in a square foot. What is the size of the patio in square inches?

(d) How many patio pavers will Kevin need to buy?

Solution

(a) The area of the rectangular patio is

$$A = lw$$
$$l = 10 \text{ ft}, w = 15 \text{ ft}: \quad = 10 \text{ ft} \cdot 15 \text{ ft}$$
$$= 150 \text{ ft}^2$$

(b) The area of a patio paver is

$$A = lw$$
$$l = 9 \text{ in.}, w = 12 \text{ in.}: \quad = 9 \text{ in.} \cdot 12 \text{ in.}$$
$$= 108 \text{ in}^2$$

(c) To convert 150 sq. ft to square inches, use the unit fraction $1 = \dfrac{144 \text{ sq. in.}}{1 \text{ sq. ft}}$

$$150 \text{ sq. ft} = \frac{150 \text{ sq. ft}}{1} \cdot \frac{144 \text{ sq. in.}}{1 \text{ sq. ft}}$$
$$= 150 \cdot 144 \text{ sq. in.}$$
$$= 21{,}600 \text{ sq. in.}$$

(d) To find the number of pavers Kevin needs, divide the area of the patio, 21,600 sq. in., by the area of a paver, 108 sq. in.

$$\frac{21{,}600 \text{ sq. in.}}{108 \text{ sq. in.}} = 200$$

Kevin needs 200 pavers to construct his patio.

Quick ✓

5. Jennifer is buying a wallpaper border for her rectangular dining room. The room is 9 feet long and 6 feet wide. **(a)** Find the perimeter of the room. **(b)** If wallpaper costs $9.95 for each 10-foot roll, find the cost of the wallpaper border.

6. A rectangular room is 9 feet wide and 12 feet long. **(a)** Find the area of the room. **(b)** How much will it cost to carpet this room if carpeting costs $5.24 per square foot?

7. Steven is building a rectangular patio that measures 12 feet by 15 feet. The patio pavers he will use to build the patio measure 9 inches by 12 inches.

  (a) What is the area of the patio, in square feet?

  (b) There are 144 in² in a square foot. What is the size of the patio in square inches?

  (c) The area of a patio paver is 108 in². How many pavers does Steven need to build his patio?

The Perimeter and Area of a Square

A square is a rectangle that has four congruent sides, so the perimeter of a square is

$$\text{Perimeter} = \text{side} + \text{side} + \text{side} + \text{side}$$
$$= 4 \cdot \text{side}$$
$$= 4s$$

where s is the length of a side.

EXAMPLE 6   **Finding the Perimeter of a Square**

Find the perimeter of the square in Figure 10.

Figure 10

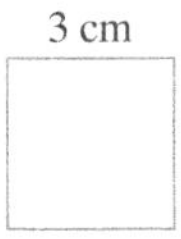

3 cm

Solution
From Figure 10, the length, s, of each side of the square is 3 cm. The perimeter of the square is

$$\text{Perimeter} = 4s$$
$$s = 3 \text{ cm:} \quad = 4 \cdot 3 \text{ cm}$$
$$= 12 \text{ cm}$$

The area of a rectangle is found by finding the product of the length and the width. In a square, the sides are congruent (have the same length), so

$$\text{Area} = \text{side} \cdot \text{side}$$
$$= \text{side}^2$$

EXAMPLE 7   **Finding the Area of a Square**

Find the area of the square in Figure 11.

Figure 11

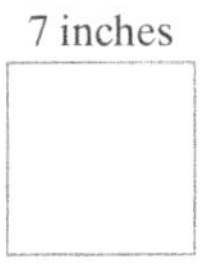

7 inches

Solution
From Figure 11, the length, s, of each side of the square is 7 inches. The area of the square is

$$\text{Area} = \text{side}^2$$
$$s = 7 \text{ inches:} \quad = (7 \text{ inches})^2$$
$$= 49 \text{ square inches, or } 49 \text{ in.}^2$$

Table 3 gives the formulas for the perimeter and area of a square.

Table 3 Perimeter and Area of a Square

Figure	Sketch	Perimeter	Area
Square		$P = 4s$	$A = s^2$

Quick ✓

8. *True or False* The area of a square is the sum of the lengths of its sides.

In Problems 9 and 10, find the perimeter and area of each square.

9. 4 cm

10. 1.5 yards

EXAMPLE 8 **Area of a Coffee Table**

Find the amount of glass that must be purchased to fit the top of a square coffee table that measures 30 inches on a side.

(a) Find the area of the glass table top in square feet.

(b) If glass costs $6.87 per square foot, what is the cost to fit the coffee table top?

Solution

(a) To find the area of the table top in square feet, change 30 inches to feet.

$$\frac{30 \text{ in.}}{1} \cdot \frac{1 \text{ ft}}{12 \text{ in.}} = \frac{30 \text{ in.}}{1} \cdot \frac{1 \text{ ft}}{12 \text{ in.}}$$

$$= \frac{30}{12} \text{ ft}$$

$$\frac{30}{12} = \frac{6 \cdot 5}{6 \cdot 2} = \frac{5}{2} = \frac{5}{2} \text{ ft}$$

$$= 2.5 \text{ ft}$$

Each side of the glass top is 2.5 feet. Now we can find the area.

$$A = s^2$$
$$= (2.5 \text{ ft})^2$$
$$= 6.25 \text{ ft}^2$$

(b) To find the cost of the tabletop, multiply the area of the square glass top, 6.25 square feet, by the cost per square foot, $6.87.

$$\text{Total cost} = \underbrace{6.25 \text{ square feet}}_{\text{area of table}} \cdot \underbrace{\$6.87}_{\text{cost per square foot}}$$

$$= \$42.9375$$

Round to the nearest penny: ≈ $42.94

It will cost $42.94 for glass to fit the top of the square coffee table.

Quick ✓

11. (a) Find the area, in square feet, of a square glass block window that measures 42 inches on a side.
    (b) Find the cost of replacing the window from part (a) if glass block costs $7.43 per square foot.

EXAMPLE 9 Finding the Perimeter of a Geometric Figure

Find **(a)** the perimeter and **(b)** the area of the region shown in Figure 12.

Figure 12

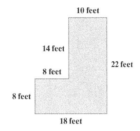

Solution

(a) The perimeter is the distance around the polygon, so

Perimeter = 8 feet + 8 feet + 14 feet + 10 feet + 22 feet + 18 feet
= 80 feet

Figure 13

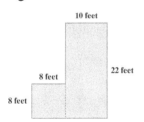

(b) The region can be divided into an 8-foot-by-8-foot square plus a 10-foot-by-22-foot rectangle. See Figure 13. The total area is the area of the square plus the area of the rectangle.

Area = area of square + area of rectangle
= $(8 \text{ feet})^2$ + (10 feet)(22 feet)
= 64 square feet + 220 square feet
= 284 square feet

Work Smart

Depending on the geometric figure, the space can often be divided in more than one way to get the area of the figure. For example, the region in Figure 13 could also be divided into a 10-foot-by-14-foot rectangle and an 8-foot-by-18-foot rectangle.

Quick ✓

12. Find the perimeter and area of the figure.

EXAMPLE 10 Painting a Room

You've decided to paint your rectangular bedroom. Two walls are 14 feet long and 7 feet high, and the other two walls are 10 feet long and 7 feet high.

(a) Ignoring the window and door openings in the bedroom, find the total surface area of the bedroom.

(b) The clerk at the paint store says that 1 gallon of paint will cover 300 square feet. How many 1-gallon cans of paint must you purchase to paint your bedroom?

Solution

(a) The bedroom has two walls that are 14 feet long and 7 feet high. The surface area of these two walls is

$$\text{Area} = 2 \cdot l \cdot w$$
$$l = 14 \text{ ft}; w = 7 \text{ ft:} \quad = 2 \cdot 14 \text{ feet} \cdot 7 \text{ feet}$$
$$= 196 \text{ feet}^2$$

The surface area of the other two walls is

$$\text{Area} = 2 \cdot l \cdot w$$
$$l = 10 \text{ ft}; w = 7 \text{ ft:} \quad = 2 \cdot 10 \text{ feet} \cdot 7 \text{ feet}$$
$$= 140 \text{ feet}^2$$

The total area to be painted is

$$\text{Area} = 196 \text{ square feet} + 140 \text{ square feet}$$
$$= 336 \text{ square feet}$$
$$= 336 \text{ ft}^2$$

(b) A 1-gallon can of paint covers 300 square feet. You have 336 square feet to paint, so you will need to purchase 2 gallons of paint.

Quick ✓

13. Anna decides to paint her rectangular family room. Two walls are 12 feet long and 7 feet high, and the other two walls are 15 feet long and 7 feet high.

   (a) Ignoring any window and door openings in the bedroom, find the total surface area of the bedroom.

   (b) How many 1-gallon cans of paint must she purchase if a 1-gallon can of paint will cover 300 ft²?

▶ ❷ Find the Perimeter and Area of a Parallelogram and a Trapezoid

Recall that a parallelogram is a quadrilateral with parallel opposite sides. A trapezoid is a quadrilateral with exactly one pair of opposite sides that are parallel. Table 4 gives the formulas for the perimeter and area of a parallelogram and a trapezoid.

Table 4 Perimeter and Area of a Parallelogram and a Trapezoid

Figure	Sketch	Perimeter	Area
Parallelogram		$P = 2a + 2b$	$A = bh$
Trapezoid		$P = a + b + c + B$	$A = \dfrac{1}{2}h(b + B)$

EXAMPLE 11   Finding the Perimeter and Area of a Parallelogram

Find **(a)** the perimeter and **(b)** the area of the parallelogram shown in Figure 14.

Figure 14

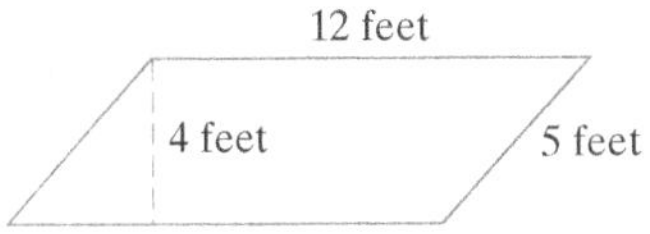

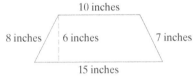

Solution

(a) From Figure 14, the length of the parallelogram is 12 feet and the width is 5 feet. The perimeter of the parallelogram is

$$P = 2a + 2b$$
$$= 2 \cdot 5 \text{ feet} + 2 \cdot 12 \text{ feet}$$
$$= 10 \text{ feet} + 24 \text{ feet}$$
$$= 34 \text{ feet}$$

(b) From Figure 14, the area of the parallelogram is

$$A = bh$$
$$b = 12 \text{ feet}; h = 4 \text{ feet}: \quad = 12 \text{ feet} \cdot 4 \text{ feet}$$
$$= 48 \text{ square feet}$$

EXAMPLE 12 **Finding the Perimeter and Area of a Trapezoid**

Figure 15

Find **(a)** the perimeter and **(b)** the area of the trapezoid shown in Figure 15.

Solution

(a) The perimeter of the trapezoid is

$$\text{Perimeter} = 8 \text{ inches} + 10 \text{ inches} + 7 \text{ inches} + 15 \text{ inches}$$
$$= 40 \text{ inches}$$

(b) The area of the trapezoid is

$$A = \frac{1}{2}h(b + B)$$
$$= \frac{1}{2} \cdot 6 \text{ inches} \cdot (10 \text{ inches} + 15 \text{ inches})$$

Add inside parentheses: $\quad = \frac{1}{2} \cdot 6 \text{ inches} \cdot 25 \text{ inches}$

$$= 75 \text{ square inches}$$

Quick ✓

14. To find the area of a trapezoid, we use the formula $A =$ _____, where _____ is the height of the trapezoid and the bases have lengths _____ and _____.

In Problems 15 and 16, find the perimeter and area of each figure.

15.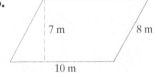

16.

17. Leon needs to purchase a sheet of Plexiglas in the shape of a parallelogram, as shown in the figure. Find the cost of the Plexiglas if the cost is $7.50 per square foot.

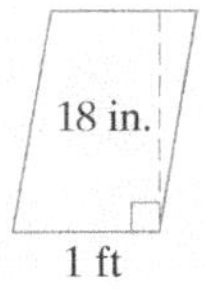

③ Find the Perimeter and Area of a Triangle

Recall that a triangle is a polygon with three sides. Table 5 gives the formulas for the perimeter and area of a triangle.

Table 5 Perimeter and Area of a Triangle

Figure	Sketch	Perimeter	Area
Triangle	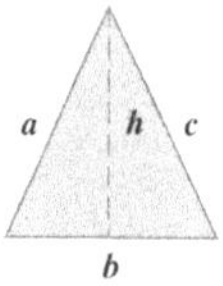 (triangle with sides a, c, base b, and height h)	$P = a + b + c$	$A = \dfrac{1}{2}bh$

EXAMPLE 13 Finding the Perimeter and Area of a Triangle

Figure 16

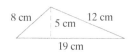

Find **(a)** the perimeter and **(b)** the area of the triangle shown in Figure 16.

(a) To find the perimeter of the triangle, add the lengths of the three sides of the triangle.

$$\begin{aligned}\text{Perimeter} &= a + b + c \\ &= 8\text{ cm} + 12\text{ cm} + 19\text{ cm} \\ &= 39\text{ cm}\end{aligned}$$

(b) To find the area, use $A = \dfrac{1}{2}bh$ with base $= b = 19$ cm and height $= h = 5$ cm.

$$\begin{aligned}A &= \dfrac{1}{2}bh \\ &= \dfrac{1}{2} \cdot 19\text{ cm} \cdot 5\text{ cm} \\ &= 47.5 \text{ square cm}\end{aligned}$$

Quick ✓

18. *True or False* The area of a triangle is given by the formula $A = \dfrac{1}{2}bh$, where b is the base and h is the height.

In Problems 19 and 20, find the perimeter and area of each triangle.

19. (triangle with sides 6 mm, 5 mm, base 8 mm, height 3 mm)

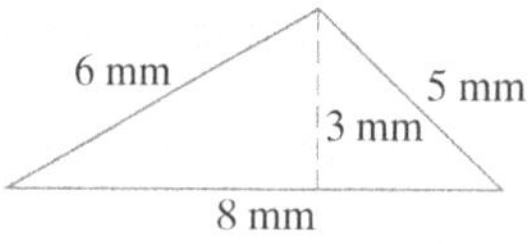

20. (right triangle with legs 5 feet and 12 feet, hypotenuse 13 feet)

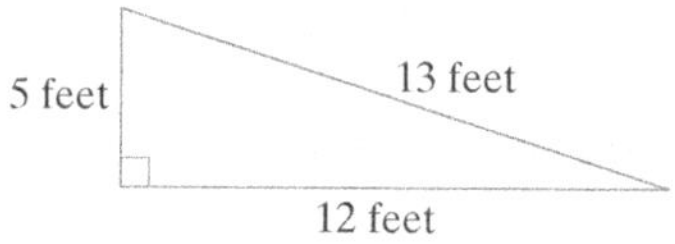

21. Find the area of a triangular tent flap that is 4 feet long and $6\dfrac{1}{2}$ feet high.

④ Find the Circumference and Area of a Circle

Figure 17

A **circle** is a figure made up of all points in the plane that are a fixed distance from a point called the **center**. The **radius** of the circle is any line segment drawn from the center of the circle to any point on the circle. The **diameter** of the circle is any line segment that has endpoints on the circle and passes through the center of the circle. See Figure 17. Note that the length of the diameter, d, of a circle is twice the length of the radius, r. That is, $d = 2r$.

EXAMPLE 14   Finding the Length of the Diameter of a Circle

Find the length of the diameter of a circle with radius 4 cm.

Solution
The length of the diameter is twice the length of the radius.

$$d = 2 \cdot r$$
$$d = 2 \cdot 4 \text{ cm}$$
$$d = 8 \text{ cm}$$

The diameter of the circle is 8 cm.

EXAMPLE 15   Finding the Length of the Radius of a Circle

Find the length of the radius of a circle with diameter 18 yards.

Solution
The length of the radius is one-half the length of the diameter.

$$r = \frac{1}{2} \cdot d$$
$$r = \frac{1}{2} \cdot 18 \text{ yards}$$
$$r = 9 \text{ yards}$$

The radius of the circle is 9 yards.

Quick ✓

22. The _____ of a circle is any line segment drawn from the center of the circle to any point on the circle.

23. *True or False* The length of the diameter of a circle is exactly twice the length of the radius.

In Problems 24–27, find the length of the radius or diameter of each circle.

24. $d = 24$ feet, find r.

25. $d = 15$ in., find r.

26. $r = 9$ cm, find d.

27. $r = 3.6$ yards, find d.

The **circumference** of a circle is the distance around a circle. We use the diameter or the radius of the circle to find the circumference of a circle according to the formulas given in Table 6. Table 6 also gives the formula for the area of a circle.

The ratio of the circumference of a circle to its diameter is the special number represented by the Greek letter π. The number π (pronounced *pie*) is a nonrepeating, nonterminating decimal that is approximately equal to 3.141592654. In this section we use $\pi \approx 3.14$ in calculations.

Table 6 Circumference and Area of a Circle

Figure	Sketch	Perimeter	Area
Circle		$C = \pi d$ where d is the length of the diameter $C = 2\pi r$ where r is the length of the radius	$A = \pi r^2$

Figure 18

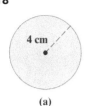
(a)

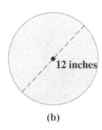
(b)

EXAMPLE 16 Finding the Circumference of a Circle

Find the circumference of each circle in Figure 18. Give the exact circumference and then an approximate circumference. Use $\pi \approx 3.14$.

Solution

(a) The radius is 4 cm, so use the formula $C = 2\pi r$.

$$C = 2\pi r$$
$r = 4\text{ cm}:\quad = 2 \cdot \pi \cdot 4 \text{ cm}$
$\quad\quad\quad\quad = 8\pi \text{ cm} \quad \text{Exact}$
$\quad\quad\quad\quad \approx 8 \cdot 3.14 \text{ cm}$
$\quad\quad\quad\quad \approx 25.12 \text{ cm} \quad \text{Approximate}$

The circumference of the circle is exactly 8π cm, or approximately 25.12 cm.

(b) The diameter is 12 inches, so use the formula $C = \pi d$.

$$C = \pi d$$
$d = 12 \text{ in.}:\quad = \pi \cdot 12 \text{ in.}$
$\quad\quad\quad\quad = 12\pi \text{ in.} \quad \text{Exact}$
$\quad\quad\quad\quad \approx 12 \cdot 3.14 \text{ in.}$
$\quad\quad\quad\quad \approx 37.68 \text{ in.} \quad \text{Approximate}$

The circumference of the circle is exactly 12π in., or approximately 37.68 in.

EXAMPLE 17 Finding the Circumference of a Circular Table

Tamara needs to buy a circular tablecloth for her patio table. The radius of the table is 12 inches, and she wants the tablecloth to extend 6 inches past the edge of the table. To the nearest inch, what should be the circumference of the tablecloth?

Solution

The radius of the table is 12 inches, and we add an additional 6 inches so that the tablecloth will fall below the edge of the table. The total radius, therefore, is 18 inches. Because we know the radius of the tablecloth, use $C = 2\pi r$ to find the circumference of the circular tablecloth to the nearest inch.

$$C = 2\pi r$$
$r = 18 \text{ in.}:\quad = 2 \cdot \pi \cdot 18 \text{ in.}$
$\pi \approx 3.14:\quad = 2 \cdot 3.14 \cdot 18 \text{ in.}$
$\quad\quad\quad\quad \approx 113.04 \text{ in.}$

The circumference of the tablecloth is approximately 113 inches.

EXAMPLE 18 Finding the Area of a Circle

Figure 19

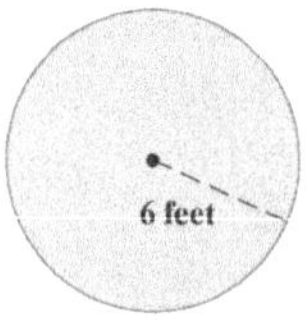

Find the area of the circle in Figure 19. Give the exact area and an approximate area rounded to the nearest hundredth. Use $\pi \approx 3.14$.

Solution

The circle in Figure 19 has radius 6 feet, so substitute $r = 6$ feet in the equation $A = \pi r^2$.

$$A = \pi r^2$$
$r = 6 \text{ ft}:\quad = \pi \cdot (6 \text{ feet})^2$
$\quad\quad\quad\quad = \pi \cdot 36 \text{ square feet}$
$\quad\quad\quad\quad = 36\pi \text{ square feet} \quad \text{Exact}$
$\pi \approx 3.14:\quad \approx 36 \cdot 3.14 \text{ square feet}$
$\quad\quad\quad\quad \approx 113.04 \text{ square feet} \quad \text{Approximate}$

The area of the circle is exactly 36π square feet, or approximately 113.04 square feet.

EXAMPLE 19 Finding the Area Enclosed by an Interstate

Figure 20

Interstate 540 forms a circle around Raleigh, North Carolina, and its suburbs; it is known as the Raleigh Outer Loop. See Figure 20. To the nearest square mile, find the area that Interstate 540 encloses, given that the radius of the circle with Raleigh at its center is 22.6 miles. (*Source:* North Carolina Department of Transportation)

Solution
The circle has radius 22.6 miles, so we have

$$A = \pi r^2$$
$$= \pi (22.6 \text{ mi})^2$$
$$\pi \approx 3.14: \approx 1603.79 \text{ mi}^2$$

The Raleigh Outer Loop encloses approximately 1604 square miles.

Quick ✓

28. The _____ of a circle is the distance around the circle.

29. *True or False* The area of a circle is given by the formula $A = \pi d^2$.

In Problems 30 and 31, find the circumference and area of each circle. Give an exact and approximate answer rounded to the nearest hundredth. Use $\pi \approx 3.14$.

30.

31.

32. In August 2012, Hurricane Isaac's tropical-storm-force winds extended outward in a circle 240 miles from its center. Find the approximate area that felt the tropical-storm-force winds to the nearest square mile. (*Source:* NASA)

6.IR1 Exercises — MyLab Statistics

Underlined exercises have complete video solutions in MyLab.

Problems 1–32 are the Quick ✓s that follow the EXAMPLES.

In Problems 33–36, find the perimeter and area of each rectangle. See Objective 1.

33.

34.

35.

36.

In Problems 37–40, find the perimeter and area of each figure. See Objective 1.

37.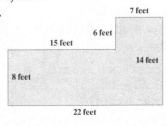

38.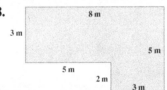

39.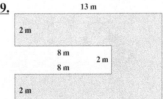

40.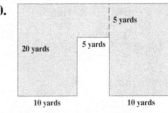

In Problems 41–48, find the perimeter and area of each quadrilateral. See Objective 2.

41.
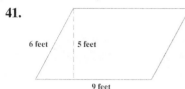

42.
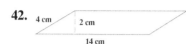

43. **44.**
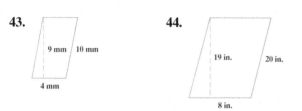

45. **46.**
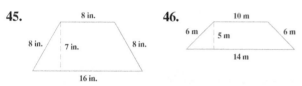

47. **48.**
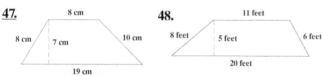

In Problems 49–52, find the perimeter and area of each triangle. See Objective 3.

49. **50.**

51. **52.**
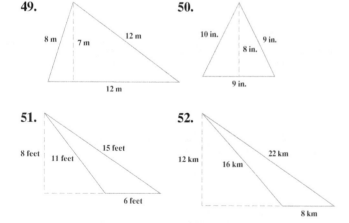

In Problems 53–56, find the length of the diameter of the circle. See Objective 4.

53. $r = 5$ in.   **54.** $r = 16$ feet

55. $r = 2.5$ cm   **56.** $r = 5.9$ in.

In Problems 57–60, find the length of the radius of the circle. See Objective 4.

57. $d = 14$ cm   **58.** $d = 58$ inches

59. $d = 11$ yards   **60.** $d = 27$ feet

In Problems 61–64, find (a) the circumference and (b) the area of each circle. For both the circumference and the area, provide exact answers and approximate answers rounded to the nearest hundredth. See Objective 4.

61. **62.**

63. **64.**
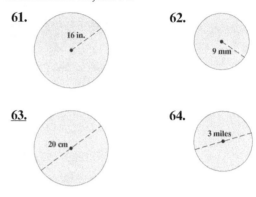

Applying the Concepts

65. Home Repairs Find the cost to replace a square glass block window that measures 2 feet on a side, if glass block costs $4.97 for an 8-inch-square block.

66. Home Repairs Find the cost to replace a rectangular glass block window that is 40 inches long and 2 feet wide, if glass block costs $5.83 for an 8-inch-square block.

67. Scrapbooking Ruth is making a rectangular scrapbook that is 12 inches long and 8 inches wide. She will attach decorative trim to the perimeter of the front and back of the scrapbook. Trim costs $4.19 for 2 yards. How much will the trim for the scrapbook cost? Trim must be purchased in increments of 2 yards.

68. Making Curtains Dorothy is making curtains for her bedroom. She plans to sew decorative braid around the edge of four curtain panels. Each panel is $5\frac{1}{2}$ feet long and 4 feet wide, and braid costs $0.99 per yard. How much will Dorothy spend for the braid? *Hint:* Dorothy cannot buy a partial yard of braid.

69. Ceiling Fan Ceiling fans are commonly measured in terms of their diameter. A 52-in. ceiling fan has a diameter of 52 inches. What is the radius of a 52-in. ceiling fan?

70. Ceiling Fan See Problem 69. What is the radius of a 29-in. ceiling fan?

71. Traffic Lights The diameter of a traffic light in the United States is 12 inches. What is the radius of a U.S. traffic light?

72. Traffic Lights In the United States, the previous standard for traffic light size was 8 inches. What is the radius of that traffic light?

73. Radius of Earth The radius of Earth is approximately 3959 miles. What is the diameter of Earth in miles?

74. Radius of Earth The radius of Earth is approximately 6371 km. What is the diameter of Earth in kilometers?

75. Diameter of the Sun The diameter of the sun is approximately 1,392,684 km. What is the radius of the sun?

76. Radius of the Moon The radius of the moon is approximately 1737.4 km. What is the diameter of the moon?

77. Diameter of a Pipe The outside diameter of a pipe is 1 inch, and the wall of the pipe is $\frac{1}{8}$ inch thick. What is the inside radius of the pipe?

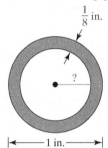

78. Diameter of a Pipe The outside diameter of a pipe is 2.5 inches, and the wall of the pipe is 0.75 inch thick. What is the inside radius of the pipe?

79. Earthquake The August 23, 2011, earthquake centered in Mineral, Virginia, was felt up to 300 miles away in all directions from the epicenter. See the figure. Approximately how many square miles were affected by the earthquake?

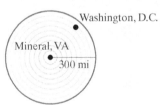

80. Earthquake The September 5, 2012, earthquake in Costa Rica was felt 100 miles away in all directions from the epicenter. Approximately how many square miles were affected by the earthquake?

81. Big Ben The minute hand on each of Big Ben's clock faces in London measures 15 feet. Assuming the length of the minute hand is also the radius of the clock, find **(a)** the circumference and **(b)** the area of Big Ben's clock face. Round answers to the nearest tenth.

82. Big Ben Big Ben's clock face is on an iron frame that is 23 ft. in diameter. To the nearest tenth, find **(a)** the circumference and **(b)** the area of the iron frame that holds Big Ben.

83. Irrigation System Farmers use a pivot irrigation system to irrigate large fields of crops. See the figure. To the nearest foot, how many square feet does a pivot system water if it has a radius of 1250 feet?

84. Garden Sprinkler Your garden sprinkler sprays water in a circular pattern. If the radius of the area covered is 1.8 meters, find the number of square meters the sprinkler waters, to the nearest square meter.

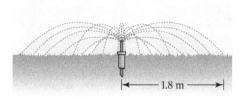

85. Washington, D.C., Beltway Interstate 495 encircles Washington, D.C., for 64 miles and is known as the "Beltway." Assuming the beltway is a circle, to the nearest hundredth, find **(a)** the diameter of the circle that surrounds Washington, D.C., and **(b)** the area it encloses.

86. Tree Diameter An arborist measures the circumference of a tree using a technique called "diameter at breast height," or DBH, where the tree is measured at a height of $4\frac{1}{2}$ feet from the ground. The circumference of a particular tree at this height is 57 inches. To the nearest inch, what is the DBH of the tree?

87. Remodel a Bathroom You plan to remodel your bathroom, and you've chosen 1-foot-by-1-foot ceramic tiles for the floor. The bathroom is 7 feet 6 inches long and 8 feet 2 inches wide.
  (a) How many tiles do you need to cover the floor of your bathroom?
  (b) Each tile costs $6. How much will it cost to tile your floor?
  (c) The store from which you purchase the tile offers a discount of 10% on orders over $350. Does your order qualify for the discount?

88. Painting a Room? A gallon of paint can cover about 500 square feet. Find the number of gallons of paint that must be purchased to paint two coats on each wall of a rectangular room measuring 8 feet by 12 feet, with a 10-foot ceiling. *Note:* You cannot purchase a partial can of paint!

89. Landscaping a Back Yard A circular swimming pool with a diameter of 24 feet is to be installed in a rectangular yard that measures 60 feet by 90 feet. Once the pool is installed, sod is to be installed on the remaining land.

(a) Determine the area of land that is to receive sod. Round your answer to the nearest square foot. Use $\pi \approx 3.14$.

(b) If sod costs $0.25 per square foot installed, what will be the cost of the lawn?

90. Landscaping a Back Yard A rectangular swimming pool whose dimensions are 12 feet by 24 feet is to be installed in a rectangular yard that measures 80 feet by 40 feet. Once the pool is installed, sod is to be installed on the remaining land.

(a) Determine the area of land that is to receive sod. Round your answer to the nearest square foot.

(b) A pallet of sod covers 500 square feet. How many pallets of sod are required?

(c) Each pallet of sod costs $96. What is the cost of the sod?

Extending the Concepts

In Problems 91 and 92, find the exact area of the shaded region.

91.

92.

93. Approximately how many feet will a wheel with a diameter of 20 inches have traveled after five revolutions? Round your answer to the nearest hundredth.

94. Approximately how many feet will a wheel with a diameter of 18 inches have traveled after three revolutions? Round your answer to the nearest hundredth.

95. Window Find the area of the window, given that the upper portion is a semicircle. Round your answer to the nearest hundredth.

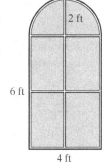

96. Area of a Region Find the area of the following figure.

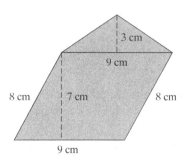

97. Area of a Region Suppose the area under the entire curve shown below is 1 and the area of the blue shaded region is 0.1587. Find the area of the nonshaded region.

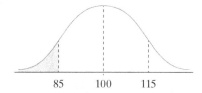

98. Area of a Patio Dan and Kathy plan to use patio pavers to build a patio in the shape shown in the accompanying figure.

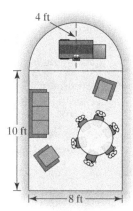

(a) Find the total area of the patio, to the nearest square foot.

(b) Pavers cost $14.75 per square foot. Find the cost to build the patio using pavers.

Explaining the Concepts

99. Find and explain the error Ted made when finding the area of a triangle that is 12 feet long and 4 feet high. Then correctly solve the problem.

$$A = \frac{1}{2}bh$$
$$= \frac{1}{2}(12 \text{ ft})(4 \text{ ft})$$
$$= \frac{1}{2} \cdot 12 \text{ ft} \cdot \frac{1}{2} \cdot 4 \text{ ft}$$
$$= 6 \text{ ft} \cdot 2 \text{ ft}$$
$$= 12 \text{ ft}^2$$

100. Which figure has the greater area: a circle with diameter 6 inches or a square with side length 6 inches? Explain your answer.

CHAPTER

8 Integrated Review: Getting Ready for Confidence Intervals

Outline

8.IR1 Compound Inequalities
8.IR2 Absolute Value Equations and Inequalities

8.IR1 Compound Inequalities

Objectives

① Solve Compound Inequalities Involving "and"
② Solve Compound Inequalities Involving "or"
③ Solve Problems Using Compound Inequalities

① Solve Compound Inequalities Involving "and"

A **compound inequality** is formed by joining two inequalities with the word "and" or "or." For example,

$$3x + 1 > 4 \quad \text{and} \quad 2x - 3 < 7$$
$$5x - 2 \leq 13 \quad \text{or} \quad 2x - 5 > 3$$

are compound inequalities. To **solve a compound inequality** means to find all possible values of the variable such that the compound inequality results in a true statement. For example, the compound inequality

$$3x + 1 > 4 \quad \text{and} \quad 2x - 3 < 7$$

is true for $x = 2$ but false for $x = 0$.

Let's look at an example that illustrates how to solve compound inequalities involving the word "and."

EXAMPLE 1 How to Solve a Compound Inequality Involving "and"

Solve $3x + 2 > -7$ and $4x + 1 \leq 9$. Express your solution using set-builder and interval notation. Graph the solution set.

Step-by-Step Solution

Step 1: Solve each inequality separately.

$$3x + 2 > -7 \qquad\qquad 4x + 1 \leq 9$$

Subtract 2 from both sides: $3x > -9$ Subtract 1 from both sides: $4x \leq 8$

Divide both sides by 3: $x > -3$ Divide both sides by 4: $x \leq 2$

Step 2: Find the intersection of the solution sets, which will represent the solution set to the compound inequality.

To find the intersection of the two solution sets, graph each inequality separately. See Figures 1(a) and (b).

Figure 1

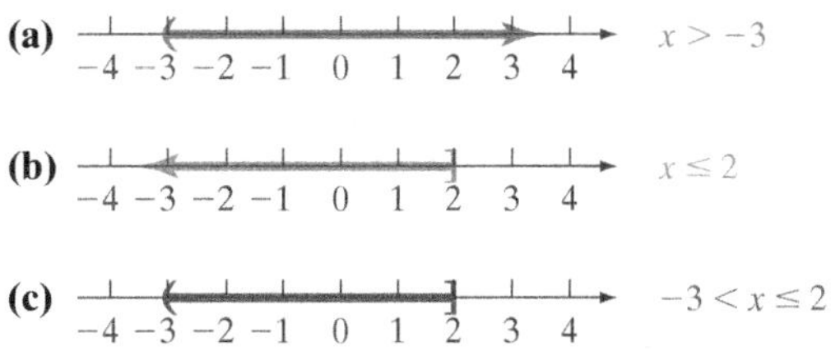

(a) $x > -3$

(b) $x \leq 2$

(c) $-3 < x \leq 2$

IR-205

The intersection of $x > -3$ and $x \leq 2$ is $-3 < x \leq 2$, so the solution set is $\{x | -3 < x \leq 2\}$ or, using interval notation, $(-3, 2]$. The graph of the solution set is shown in Figure 1(c) on the previous page.

The steps below summarize the procedure for solving compound inequalities involving "and."

Work Smart

The words "and" and "intersection" suggest overlap. When solving these types of problems, look for the overlap of the graphs.

> **Solving Compound Inequalities Involving "and"**
>
> **Step 1:** Solve each inequality separately.
>
> **Step 2:** Find the INTERSECTION of the solution sets of each inequality in Step 1.

EXAMPLE 2 **Solving a Compound Inequality with "and"**

Solve $-2x + 5 > -1$ and $5x + 6 \leq -4$. Express the solution using set-builder notation and interval notation. Graph the solution set.

Solution

Solve each inequality separately:

$$-2x + 5 > -1 \qquad\qquad 5x + 6 \leq -4$$

Subtract 5 from both sides: $\quad -2x > -6 \quad$ Subtract 6 from both sides: $\quad 5x \leq -10$

Divide both sides by -2 and reverse the direction of the inequality! $\quad x < 3 \qquad$ Divide both sides by 5: $\quad x \leq -2$

The intersection of the solution sets is the solution set to the compound inequality. See Figures 2(a) and (b).

Figure 2

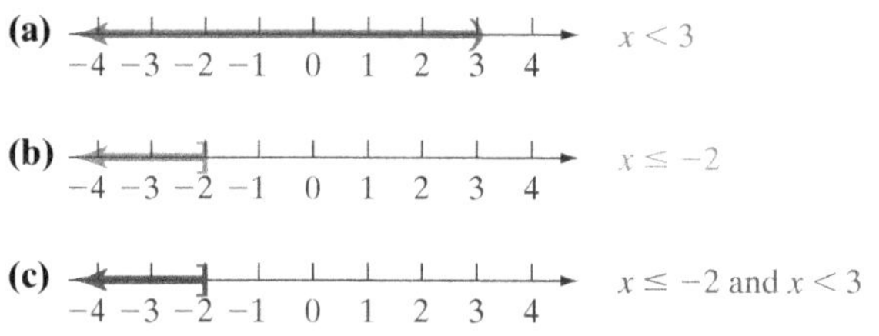

(a) $x < 3$

(b) $x \leq -2$

(c) $x \leq -2$ and $x < 3$

The intersection of $x < 3$ and $x \leq -2$ is $x \leq -2$. The solution set is $\{x | x \leq -2\}$ or, using interval notation, $(-\infty, -2]$. The graph of the solution set is shown in Figure 2(c).

Quick ✓

In Problems 1–3, solve each compound inequality. Express the solution using set-builder notation and interval notation. Graph the solution set.

1. $2x + 1 \geq 5$ and $-3x + 2 < 5$
2. $4x - 5 < 7$ and $3x - 1 > -10$
3. $-8x + 3 < -5$ and $\dfrac{2}{3}x + 1 < 3$

EXAMPLE 3 **Solving a Compound Inequality with "and"**

Solve $x - 5 > -1$ and $2x - 3 \leq -5$. Express the solution using set-builder and interval notation. Graph the solution set.

Figure 3

Solution

Solve each inequality separately:

	$x - 5 > -1$		$2x - 3 \leq -5$
Add 5 to both sides:	$x > 4$	Add 3 to both sides:	$2x \leq -2$
		Divide both sides by 2:	$x \leq -1$

The intersection of the solution sets is the solution set to the compound inequality. See Figure 3. The intersection is the empty set, so the solution set is $\{\ \}$ or $\emptyset$.

Quick ✓

In Problems 4 and 5, solve each compound inequality. Express the solution using set-builder notation and interval notation. Graph the solution set.

4. $3x - 5 < -8$ and $2x + 1 > 5$
5. $5x + 1 \leq 6$ and $3x + 2 \geq 5$

Sometimes, we can combine "and" inequalities into a more streamlined notation.

> **Writing Inequalities Involving "and" Compactly**
>
> If $a < b$, then we can write
>
> $$a < x \quad \text{and} \quad x < b$$
>
> more compactly as
>
> $$a < x < b$$

For example, we can write

$$-3 < -4x + 1 \quad \text{and} \quad -4x + 1 < 13$$

as

$$-3 < -4x + 1 < 13$$

We solve such compound inequalities by isolating the variable "in the middle" with a coefficient of 1.

EXAMPLE 4   Solving a Compound Inequality

Solve $-3 < -4x + 1 < 13$. Express the solution using set-builder notation and interval notation. Graph the solution set.

Solution

Isolate the variable "in the middle" with a coefficient of 1:

$$-3 < -4x + 1 < 13$$

Subtract 1 from all three parts (Addition Property): $\quad -3 - 1 < -4x + 1 - 1 < 13 - 1$

$$-4 < -4x < 12$$

Divide all three parts by -4 and reverse the inequalities' direction.
$$\frac{-4}{-4} > \frac{-4x}{-4} > \frac{12}{-4}$$

$$1 > x > -3$$

If $b > x > a$, then $a < x < b$: $\quad -3 < x < 1$

The solution set is $\{x \mid -3 < x < 1\}$ or, in interval notation, $(-3, 1)$. Figure 4 shows the graph of the solution set.

Figure 4

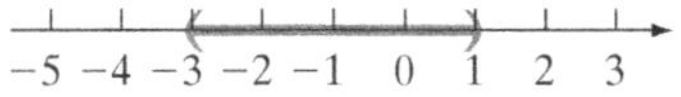

Figure 5

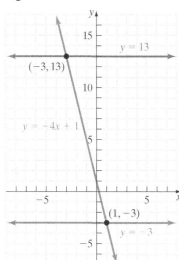

To visualize the results of Example 3, look at Figure 5, which shows the graph of $y = -3$, $y = -4x + 1$, and $y = 13$. Note that the graph of $y = -4x + 1$ is between the graphs of $y = -3$ and $y = 13$ for $-3 < x < 1$. Thus the solution set of $-3 < -4x + 1 < 13$ is $\{x | -3 < x < 1\}$.

Quick ✓

In Problems 6–8, solve each compound inequality. Express the solution using set-builder notation and interval notation. Graph the solution set.

6. $-2 < 3x + 1 < 10$
7. $0 < 4x - 5 \leq 3$
8. $3 \leq -2x - 1 \leq 11$

② Solve Compound Inequalities Involving "or"

The solution to compound inequalities involving the word "or" is the union of the solutions to each inequality.

EXAMPLE 5 How to Solve a Compound Inequality Involving "or"

Solve $3x - 5 < -2$ or $4 - 5x \leq -16$. Express the solution using set-builder notation and interval notation. Graph the solution set.

Step-by-Step Solution

Step 1: Solve each inequality separately.

$$3x - 5 < -2$$
Add 5 to each side: $\quad 3x < 3$
Divide both sides by 3: $\quad x < 1$

$$4 - 5x \leq -16$$
Subtract 4 from both sides: $\quad -5x \leq -20$
Divide both sides by -5; Reverse the direction of the inequality: $\quad x \geq 4$

Step 2: Find the union of the solution sets, which will represent the solution set to the compound inequality.

The union of the two solution sets is $x < 1$ or $x \geq 4$. The solution set using set-builder notation is $\{x | x < 1 \text{ or } x \geq 4\}$. The solution set using interval notation is $(-\infty, 1) \cup [4, \infty)$. Figure 6 shows the graph of the solution set.

Figure 6

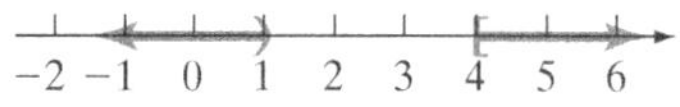

```
  -2 -1  0  1  2  3  4  5  6
```

Steps for Solving Compound Inequalities Involving "or"

Step 1: Solve each inequality separately.

Step 2: Find the UNION of the solution sets of each inequality from Step 1.

Work Smart

A common error is to write the solution $x < 1$ or $x > 4$ as $1 > x > 4$, which is incorrect. There are no real numbers that are less than 1 *and* greater than 4. Another common error is to "mix" symbols, as in $1 < x > 4$. This notation makes no sense!

Quick ✓

In Problems 9–12, solve each compound inequality. Express the solution using set-builder notation and interval notation. Graph the solution set.

9. $x + 3 < 1$ or $x - 2 > 3$
10. $3x + 1 \leq 7$ or $2x - 3 > 9$
11. $2x - 3 \geq 1$ or $6x - 5 \geq 1$
12. $\dfrac{3}{4}(x + 4) < 6$ or $\dfrac{3}{2}(x + 1) > 15$

EXAMPLE 6 Solving Compound Inequalities Involving "or"

Solve $\frac{1}{2}x - 1 < 1$ or $\frac{2x - 1}{3} \geq -1$. Express the solution using set-builder notation and interval notation. Graph the solution set.

Solution

Solve each inequality separately:

	$\frac{1}{2}x - 1 < 1$		$\frac{2x - 1}{3} \geq -1$
Add 1 to each side:	$\frac{1}{2}x < 2$	Multiply both sides by 3:	$2x - 1 \geq -3$
Multiply both sides by 2:	$x < 4$	Add 1 to both sides:	$2x \geq -2$
		Divide both sides by 2:	$x \geq -1$

Find the union of these solution sets. The graphs in Figure 7 show that the union of the two solution sets is the set of all real numbers.

Figure 7

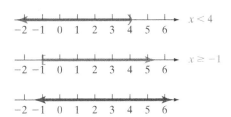

The solution set is $\{x \mid x \text{ is any real number}\}$ in set-builder notation, and $(-\infty, \infty)$ in interval notation.

Quick ✓

In Problems 13 and 14, solve each compound inequality. Express the solution using set-builder notation and interval notation. Graph the solution set.

13. $3x - 2 > -5$ or $2x - 5 \leq 1$ **14.** $-5x - 2 \leq 3$ or $7x - 9 > 5$

❸ Solve Problems Using Compound Inequalities

We now look at an application involving compound inequalities.

EXAMPLE 7 Federal Income Taxes

In 2015 married couples filing a joint federal tax return who were in the 25% tax bracket paid between $10,313 and $29,388 in federal income taxes. These couples' federal income taxes equal $10,313 plus 25% of their taxable income over $74,900. Find the range of taxable incomes into which a married couple must fall to have been in the 25% tax bracket. *Source: Internal Revenue Service*

Solution

Step 1: Identify We want to find the range of taxable incomes for married couples in the 25% tax bracket. This direct translation problem involves an inequality.

Step 2: Name Let t represent the taxable income.

Step 3: Translate The federal tax bill equals $10,313 plus 25% of the taxable income over $74,900. If the couple had taxable income equal to $75,900, their tax bill was $10,313 plus 25% of $1000 ($1000 is the amount over $74,900). In general, if the couple has taxable income t, then their tax bill will be

$\underbrace{\$10,313}_{\$10,313}$ plus $\underbrace{25\%}_{0.25}$ $\underbrace{\text{of the amount over } \$74,900}_{(t - \$74,900)}$

Work Smart

The word "range" tells us that an inequality is to be solved.

Because the tax bill was between $10,313 and $29,388, we have

$$10{,}313 \leq 10{,}313 + 0.25(t - 74{,}900) \leq 29{,}388 \quad \text{The Model}$$

Step 4: Solve

$$10{,}313 \leq 10{,}313 + 0.25(t - 74{,}900) \leq 29{,}388$$

Distribute 0.25: $\quad 10{,}313 \leq 10{,}313 + 0.25t - 18{,}725 \leq 29{,}388$

Combine like terms: $\quad 10{,}313 \leq -8412 + 0.25t \leq 29{,}388$

Add 7940 to all three parts: $\quad 18{,}725 \leq 0.25t \leq 37{,}800$

Divide all three parts by 0.25: $\quad 74{,}900 \leq t \leq 151{,}200$

Step 5: Check If a married couple had taxable income of $74,900, then their tax bill was $10,313 + 0.25($74,900 − $74,900) = $10,313. If a married couple had taxable income of $151,200, then their tax bill was $10,313 + 0.25($151,200 − $74,900) = $29,388.

Step 6: Answer the Question A married couple who filed a federal joint income tax return with a tax bill between $10,313 and $29,388 had taxable income between $74,900 and $151,200.

Quick ✓

15. In 2015 an individual filing a federal tax return whose income placed him or her in the 25% tax bracket paid federal income taxes between $5156.25 and $18,481.25. The individual had to pay federal income taxes equal to $5156.25 plus 25% of the amount over $37,450. Find the range of taxable income in order for an individual to have been in the 25% tax bracket. (*Source: Internal Revenue Service*)

16. AT&T offers a long-distance phone plan that charges $2.00 per month plus $0.10 per minute. During the course of a year, Sophia's long-distance phone bill ranges from $6.50 to $26.50. What was the range of monthly minutes?

8.IR1 Exercises MyLab Statistics

Underlined exercises have complete video solutions in MyLab.

Problems 1–16 are the Quick ✓ s that follow the EXAMPLES.

Building Skills

In Problems 17–20, use the graph to solve the compound inequality. Graph the solution set. See Objectives 2 and 3.

17. (a) $-5 \leq 2x - 1 \leq 3$

(b) $2x - 1 < -5$ or $2x - 1 > 3$

18. (a) $-1 \leq \dfrac{1}{2}x + 1 \leq 3$

(b) $\dfrac{1}{2}x + 1 < -1$ or $\dfrac{1}{2}x + 1 > 3$

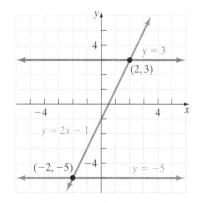

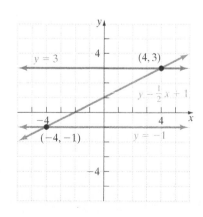

19. (a) $-4 < -\dfrac{5}{3}x + 1 < 6$

(b) $-\dfrac{5}{3}x + 1 \le -4$ or $-\dfrac{5}{3}x + 1 \ge 6$

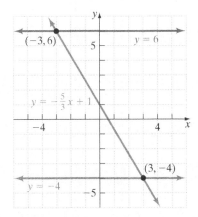

20. (a) $-3 < \dfrac{5}{4}x + 2 < 7$

(b) $\dfrac{5}{4}x + 2 \le -3$ or $\dfrac{5}{4}x + 2 \ge 7$

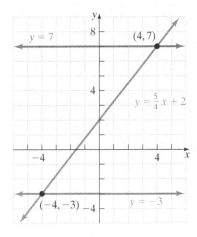

In Problems 21–44, solve each compound inequality. Graph the solution set. See Objective 1.

21. $x < 3$ and $x \ge -2$
22. $x \le 5$ and $x > 0$
23. $4x - 4 < 0$ and $-5x + 1 \le -9$
24. $6x - 2 \le 10$ and $10x > -20$
25. $4x - 3 < 5$ and $-5x + 3 > 13$
26. $x - 3 \le 2$ and $6x + 5 \ge -1$
27. $7x + 2 \ge 9$ and $4x + 3 \le 7$
28. $-4x - 1 < 3$ and $-x - 2 > 3$
29. $-3 \le 5x + 2 < 17$
30. $-10 < 6x + 8 \le -4$
31. $-3 \le 6x + 1 \le 10$
32. $-12 < 7x + 2 \le 6$
33. $3 \le -5x + 7 < 12$
34. $-6 < -3x + 6 \le 4$
35. $-1 \le \dfrac{1}{2}x - 1 \le 3$
36. $0 < \dfrac{3}{2}x - 3 \le 3$
37. $3 \le -2x - 1 \le 11$
38. $-3 < -4x + 1 < 17$
39. $\dfrac{2}{3}x + \dfrac{1}{2} < \dfrac{5}{6}$ and $-\dfrac{1}{5}x + 1 < \dfrac{3}{10}$
40. $x - \dfrac{3}{2} \le \dfrac{5}{4}$ and $-\dfrac{2}{3}x - \dfrac{2}{9} < \dfrac{8}{9}$
41. $-2 < \dfrac{3x + 1}{2} \le 8$
42. $-4 \le \dfrac{4x - 3}{3} < 3$
43. $-8 \le -2(x + 1) < 6$
44. $-6 < -3(x - 2) < 15$

In Problems 45–58, solve each compound inequality. Graph the solution set. See Objective 2.

45. $x < -2$ or $x > 3$
46. $x < 0$ or $x \ge 6$
47. $x - 2 < -4$ or $x + 3 > 8$
48. $x + 3 \le 5$ or $x - 2 \ge 3$
49. $6(x - 2) < 12$ or $4(x + 3) > 12$
50. $4x + 3 > -5$ or $8x - 5 < 3$
51. $-8x + 6x - 2 > 0$ or $5x > 3x + 8$
52. $3x \ge 7x + 8$ or $x < 4x - 9$
53. $2x + 5 \le -1$ or $\dfrac{4}{3}x - 3 > 5$
54. $-\dfrac{4}{5}x - 5 > 3$ or $7x - 3 > 4$
55. $\dfrac{1}{2}x < 3$ or $\dfrac{3x - 1}{2} > 4$
56. $\dfrac{2}{3}x + 2 \le 4$ or $\dfrac{5x - 3}{3} \ge 4$
57. $3(x - 1) + 5 < 2$ or $-2(x - 3) < 1$
58. $2(x + 1) - 5 \le 4$ or $-(x + 3) \le -2$

Mixed Practice

In Problems 59–72, solve each compound inequality. Graph the solution set.

59. $3a + 5 < 5$ and $-2a + 1 \le 7$
60. $5x - 1 < 9$ and $5x > -20$
61. $5(x + 2) < 20$ or $4(x - 4) > -20$

62. $3(x + 7) < 24$ or $6(x - 4) > -30$
63. $-4 \leq 3x + 2 \leq 10$
64. $-8 \leq 5x - 3 \leq 4$
65. $2x + 7 < -13$ or $5x - 3 > 7$
66. $3x - 8 < -14$ or $4x - 5 > 7$
67. $5 < 3x - 1 < 14$
68. $-5 < 2x + 7 \leq 5$
69. $\dfrac{x}{3} \leq -1$ or $\dfrac{4x - 1}{2} > 7$
70. $\dfrac{x}{2} \leq -4$ or $\dfrac{2x - 1}{3} \geq 2$
71. $-3 \leq -2(x + 1) < 8$
72. $-15 < -3(x + 2) \leq 1$

Applying the Concepts

In Problems 73–78, use the Addition Property and/or Multiplication Properties to find a and b.

73. If $-3 < x < 4$, then $a < x + 4 < b$.
74. If $-2 < x < 3$, then $a < x - 3 < b$.
75. If $4 < x < 10$, then $a < 3x < b$.
76. If $2 < x < 12$, then $a < \dfrac{1}{2}x < b$.
77. If $-2 < x < 6$, then $a < 3x + 5 < b$.
78. If $-4 < x < 3$, then $a < 2x - 7 < b$.

79. **Systolic Blood Pressure** Blood pressure is measured using two numbers. One of the numbers measures systolic blood pressure. The systolic blood pressure represents the pressure while the heart is beating. In a healthy person, the systolic blood pressure should be greater than 90 and less than 140. If we let the variable x represent a person's systolic blood pressure, express the systolic blood pressure of a healthy person using a compound inequality.

80. **Diastolic Blood Pressure** Blood pressure is measured using two numbers. One of the numbers measures diastolic blood pressure. The diastolic blood pressure represents the pressure while the heart is resting between beats. In a healthy person, the diastolic blood pressure should be greater than 60 and less than 90. If we let the variable x represent a person's diastolic blood pressure, express the diastolic blood pressure of a healthy person using a compound inequality.

81. **Computing Grades** Joanna desperately wants to earn a *B* in her history class. Her current test scores are 74, 86, 77, and 89. Her final exam is worth two test scores. In order to earn a *B*, Joanna's average must lie between 80 and 89, inclusive. What range of scores can Joanna receive on the final and earn a *B* in the course?

82. **Computing Grades** Jack needs to earn a *C* in his sociology class. His current test scores are 67, 72, 81, and 75. His final exam is worth three test scores. In order to earn a *C*, Jack's average must lie between 70 and 79, inclusive. What range of scores can Jack receive on the final exam and earn a *C* in the course?

83. **Federal Tax Withholding** The percentage method of withholding for federal income tax (2014) states that a single person whose weekly wages, after subtracting withholding allowances, are over $436 but not over $1506, shall have $34.90 plus 25% of the excess over $436 withheld. Over what range does the amount withheld vary if the weekly wages vary from $800 to $900, inclusive? *Source: Internal Revenue Service*

84. **Federal Tax Withholding** Rework Problem 83 if the weekly wages vary from $1000 to $1100, inclusive.

85. **Commission** Gerard had an offer for a medical equipment sales position that pays $2500 per month plus 1% of all sales. What total sales is required to earn between $3000 and $5000 per month?

86. **Commission** Juanita had a job offer to be an automobile sales position that pays $1500 per month plus 2.5% of all sales. What total sales is required to earn between $4000 and $6000 per month?

87. **Electric Bill** In North Carolina, Duke Energy charges $42.41 plus $0.092897 for each additional kilowatt hour (kwh) used during the months from July through October for usage in excess of 350 kwh. Suppose one homeowner's electric bill ranged from a low of $88.86 to a high of $137.16 during this time period. Over what range (in kwh) did the usage vary?

88. **Electric Bill** In North Carolina, Duke Energy charges $42.41 plus $0.084192 for each additional kilowatt hour (kwh) used during the months from November through June for usage in excess of 350 kwh. Suppose one homeowner's electric bill ranged from a low of $55.04 to a high of $89.56 during this time period. Over what range (in kwh) did the usage vary?

89. **The Arithmetic Mean** If $a < b$, show that $a < \dfrac{a + b}{2} < b$. We call $\dfrac{a + b}{2}$ the **arithmetic mean** of a and b.

90. Identifying Triangles A triangle has the property that the length of the longest side is greater than the difference of the other sides, and the length of the longest side is less than the sum of the other sides. That is, if a, b, and c are sides such that $a \leq b \leq c$, then $b - a < c < b + a$. Determine which of the following could be lengths of the sides of a triangle.

(a) 3, 4, 5 (b) 4, 7, 12
(c) 3, 3, 5 (d) 1, 9, 10

Extending the Concepts

91. Solve $2x + 1 \leq 5x + 7 \leq x - 5$.

92. Solve $x - 3 \leq 3x + 1 \leq x + 11$.

93. Solve $4x + 1 > 2(2x + 1)$. Provide an explanation that generalizes the result.

94. Solve $4x - 2 \geq 2(2x - 1)$. Provide an explanation that generalizes the result.

95. Consider the following analysis, assuming that $x < 2$.

$$5 > 2$$
$$5(x - 2) > 2(x - 2)$$
$$5x - 10 > 2x - 4$$
$$3x > 6$$
$$x > 2$$

How can it be that the final line in the analysis states that $x > 2$, when the original assumption stated that $x < 2$?

8.IR2 Absolute Value Equations and Inequalities

Objectives

1. Solve Absolute Value Equations
2. Solve Absolute Value Inequalities Involving $<$ or $\leq$
3. Solve Absolute Value Inequalities Involving $>$ or $\geq$
4. Solve Applied Problems Involving Absolute Value Inequalities

Recall that the absolute value of a number is its distance from the origin on the real number line. For example, $|-5| = 5$ because the distance on the real number line from 0 to -5 is 5 units. See Figure 8.

Figure 8
$|-5| = 5$

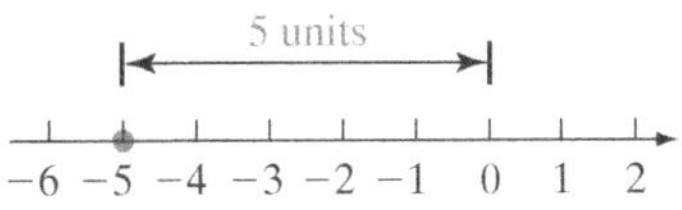

1 Solve Absolute Value Equations

EXAMPLE 1 **Solving an Absolute Value Equation**

Solve the equation $|x| = 4$.

Solution

We will present two geometric solutions and one algebraic solution.

Geometric Solution 1: The equation $|x| = 4$ asks, "Which real numbers x are 4 units from 0 on the number line?" Figure 9 shows the two numbers, -4 and 4. The solution set is $\{-4, 4\}$.

Figure 9

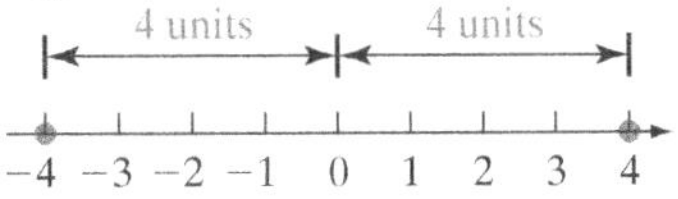

Geometric Solution 2: We can solve $|x| = 4$ by graphing $y = |x|$ and $y = 4$ on the same xy-plane, as in Figure 10. The x-coordinates of the points of intersection are -4 and 4, which are the solutions to the equation $|x| = 4$. Again, the solution set is $\{-4, 4\}$.

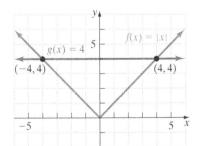

Figure 10

Algebraic Solution: We know that $|a| = a$ if $a \geq 0$ and $|a| = -a$ if $a < 0$. Because we do not know whether x is positive or negative in the equation $|x| = 4$, we solve the problem for $x \geq 0$ or $x < 0$.

If $x \geq 0$	If $x < 0$
$\|x\| = 4$	$\|x\| = 4$
$\|x\| = x$ since $x \geq 0$: $x = 4$	$\|x\| = -x$ since $x < 0$: $-x = 4$
	Multiply both sides by -1: $x = -4$

The solution set is $\{-4, 4\}$.

Quick ✓

In Problems 1 and 2, solve the equation geometrically and algebraically.

1. $|x| = 7$ **2.** $|z| = 1$

The results of Example 1 and Quick Checks 1 and 2 lead to the following result.

Equations Involving Absolute Value

If a is a positive real number and u is any algebraic expression, then

$$|u| = a \quad \text{is equivalent to} \quad u = a \quad \text{or} \quad u = -a$$

Note: If $a = 0$, the equation $|u| = 0$ is equivalent to $u = 0$. If $a < 0$, the equation $|u| = a$ has no solution, as explained below.

Figure 11

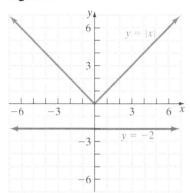

In the equation $|u| = a$, a must be nonnegative (greater than or equal to 0). If a is negative, the equation has no real solution. To see why, consider the equation $|x| = -2$. Figure 11 shows the graph of $y = |x|$ and $y = -2$. Notice that the graphs do not intersect, which shows that the equation $|x| = -2$ has no real solution. The solution set is the empty set, $\varnothing$ or $\{\ \}$.

EXAMPLE 2 **How to Solve an Equation Involving Absolute Value**

Solve the equation: $|2x - 1| + 3 = 12$

Step-by-Step Solution

Step 1: Isolate the expression containing the absolute value.

$$|2x - 1| + 3 = 12$$
Subtract 3 from both sides: $\quad |2x - 1| = 9$

Step 2: Rewrite the equation $|u| = a$ as $u = a$ or $u = -a$, where u is the algebraic expression in the absolute value symbol.

In the equation $|2x - 1| = 9$, $u = 2x - 1$ and $a = 9$.

$$2x - 1 = 9 \quad \text{or} \quad 2x - 1 = -9$$

Step 3: Solve each equation.

Add 1 to each side: $\quad 2x = 10 \qquad$ Add 1 to each side: $\quad 2x = -8$
Divide both sides by 2: $\quad x = 5 \qquad$ Divide both sides by 2: $\quad x = -4$

Step 4: Check: Verify each solution.

$$|2x - 1| + 3 = 12 \qquad\qquad |2x - 1| + 3 = 12$$
Let $x = 5$: $\quad |2(5) - 1| + 3 \stackrel{?}{=} 12 \qquad$ Let $x = -4$: $\quad |2(-4) - 1| + 3 \stackrel{?}{=} 12$
$$|10 - 1| + 3 \stackrel{?}{=} 12 \qquad\qquad |-8 - 1| + 3 \stackrel{?}{=} 12$$
$$9 + 3 \stackrel{?}{=} 12 \qquad\qquad 9 + 3 \stackrel{?}{=} 12$$
$$12 = 12 \quad \text{True} \qquad\qquad 12 = 12 \quad \text{True}$$

Both solutions check, so the solution set is $\{-4, 5\}$.

> **Solving Absolute Value Equations with One Absolute Value**
>
> **Step 1:** Isolate the expression containing the absolute value.
> **Step 2:** Rewrite the absolute value equation as two equations: $u = a$ and $u = -a$, where u is the expression in the absolute value symbol.
> **Step 3:** Solve each equation.
> **Step 4:** Verify your solution.

Quick ✓

3. $|u| = a$ is equivalent to $u =$ ___ or $u =$ ___.
4. $|2x + 3| = 5$ is equivalent to $2x + 3 = 5$ or _____.

In Problems 5–8, solve each equation.

5. $|2x - 3| = 7$
6. $|3x - 2| + 3 = 10$
7. $|-5x + 2| - 2 = 5$
8. $3|x + 2| - 4 = 5$

EXAMPLE 3   **Solving an Equation Involving Absolute Value with No Solution**

Solve the equation: $|x + 5| + 7 = 5$

Solution

$$|x + 5| + 7 = 5$$

Subtract 7 from both sides: $|x + 5| = -2$

Work Smart

The equation $|u| = a$, where a is a negative real number, has no real solution. Why? See Figure 11.

Since the absolute value of any real number is always nonnegative (greater than or equal to zero), the equation has no real solution. The solution set is { } or ∅.

Quick ✓

9. *True or False* $|x| = -4$ has no real solution.

In Problems 10–12, solve each equation.

10. $|5x + 3| = -2$
11. $|2x + 5| + 7 = 3$
12. $|x + 1| + 3 = 3$

What if an absolute value equation has two absolute values, as in $|3x - 1| = |x + 5|$? The signs of the algebraic expressions in the absolute value symbol have four possibilities.

1. both algebraic expressions are positive,
2. both are negative,
3. the left is positive and the right is negative, or
4. the left is negative and the right is positive.

To see how the solution works, use the fact that $|a| = a$, if $a \geq 0$ and $|a| = -a$, if $a < 0$.

Thus, if $3x - 1 \geq 0$, then $|3x - 1| = 3x - 1$. However, if $3x - 1 < 0$, then $|3x - 1| = -(3x - 1)$. This leads us to a method for solving absolute value equations with two absolute values.

Case 1: Both Expressions Are Positive	Case 2: Both Expressions Are Negative	Case 3: The Expression on the Left Is Positive and That on the Right Is Negative	Case 4: The Expression on the Left Is Negative and That on the Right Is Positive
$\|3x - 1\| = \|x + 5\|$ $3x - 1 = x + 5$	$\|3x - 1\| = \|x + 5\|$ $-(3x - 1) = -(x + 5)$ $3x - 1 = x + 5$	$\|3x - 1\| = \|x + 5\|$ $3x - 1 = -(x + 5)$	$\|3x - 1\| = \|x + 5\|$ $-(3x - 1) = x + 5$

When both algebraic expressions are positive, or both negative, we end up with equivalent equations. Therefore, Case 1 and Case 2 result in equivalent equations. Also, if one side is positive and the other is negative, we end up with equivalent equations. Therefore, Case 3 and Case 4 result in equivalent equations. The four possibilities reduce to two possibilities.

Equations Involving Two Absolute Values

If u and v are any algebraic expression, then

$$|u| = |v| \quad \text{is equivalent to} \quad u = v \quad \text{or} \quad u = -v$$

EXAMPLE 4 Solving an Absolute Value Equation Involving Two Absolute Values

Solve the equation: $|2x - 3| = |x + 6|$

Solution

The equation is in the form $|u| = |v|$, where $u = 2x - 3$ and $v = x + 6$. Rewrite the equation in the form $u = v$ or $u = -v$ and then solve each equation.

$u = v$: $2x - 3 = x + 6$ $u = -v$: $2x - 3 = -(x + 6)$
Add 3 to both sides: $2x = x + 9$ Distribute the -1: $2x - 3 = -x - 6$
Subtract x from both sides: $x = 9$ Add 3 to both sides: $2x = -x - 3$
    Add x to both sides: $3x = -3$
    Divide both sides by 3: $x = -1$

Check

$x = 9$: $|2(9) - 3| \stackrel{?}{=} |9 + 6|$ $x = -1$: $|2(-1) - 3| \stackrel{?}{=} |-1 + 6|$
     $|18 - 3| \stackrel{?}{=} |15|$ $|-2 - 3| \stackrel{?}{=} |5|$
     $|15| \stackrel{?}{=} 15$ $|-5| \stackrel{?}{=} 5$
     $15 = 15$ True $5 = 5$ True

Both solutions check, so the solution set is $\{-1, 9\}$.

Quick ✓

13. $|u| = |v|$ is equivalent to ___ = ___ or ___ = ___.

In Problems 14–17, solve each equation.

14. $|x - 3| = |2x + 5|$ **15.** $|8z + 11| = |6z + 17|$

16. $|3 - 2y| = |4y + 3|$ **17.** $|2x - 3| = |5 - 2x|$

❷ Solve Absolute Value Inequalities Involving $<$ or $\leq$

The method for solving absolute value equations relies on the geometric interpretation of absolute value. That is, the absolute value of a real number x is its distance from the origin on the real number line. We also use the geometric interpretation of absolute value to solve absolute value inequalities.

EXAMPLE 5 **Solving an Absolute Value Inequality**

Solve the inequality $|x| < 4$. Graph the solution set.

Solution

The inequality $|x| < 4$ asks for all real numbers x that are less than 4 units from the origin on the real number line. See Figure 12. We can see from the figure that any number between -4 and 4 satisfies the inequality. The solution set is $\{x | -4 < x < 4\}$ or, using interval notation, $(-4, 4)$.

Figure 12

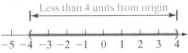

We could also visualize these results by graphing $y = |x|$ and $y = 4$. See Figure 13. To solve $|x| < 4$, look for all x-coordinates such that the graph of $y = |x|$ is below the graph of $y = 4$. We can see that this is true for all x between -4 and 4, as found above.

Figure 13

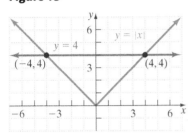

The results of Example 5 lead to the following.

Inequalities of the Form $<$ or $\leq$ Involving Absolute Value

If a is a positive real number and if u is an algebraic expression, then

$$|u| < a \quad \text{is equivalent to} \quad -a < u < a$$
$$|u| \leq a \quad \text{is equivalent to} \quad -a \leq u \leq a$$

Note: If $a = 0$, $|u| < 0$ has no solution; $|u| \leq 0$ is equivalent to $u = 0$. If $a < 0$, the inequality has no solution.

Quick ✓

18. If $a > 0$, then $|u| < a$ is equivalent to _____.
19. To solve $|3x + 4| < 10$, solve _____ $< 3x + 10 <$ _____.

In Problems 20 and 21, solve each inequality. Graph the solution set.

20. $|x| \leq 5$ 21. $|x| < \dfrac{3}{2}$

EXAMPLE 6 **How to Solve an Absolute Value Inequality Involving $\leq$**

Solve the inequality $|2x + 3| \leq 5$. Express the solution set in set-builder notation and interval notation. Graph the solution set.

Step-by-Step Solution

Step 1: The inequality is in the form $|u| \leq a$, where $u = 2x + 3$ and $a = 5$. Rewrite the inequality as a compound inequality that does not involve absolute value.

Use the fact that $|u| \leq a$ means $-a \leq u \leq a$:

$$|2x + 3| \leq 5$$
$$-5 \leq 2x + 3 \leq 5$$

Step 2: Solve the resulting compound inequality.

Subtract 3 from all three parts: $-5 - 3 \leq 2x + 3 - 3 \leq 5 - 3$
$$-8 \leq 2x \leq 2$$

Divide all three parts by 2: $\dfrac{-8}{2} \leq \dfrac{2x}{2} \leq \dfrac{2}{2}$
$$-4 \leq x \leq 1$$

The solution is $\{x | -4 \leq x \leq 1\}$ or, in interval notation, $[-4, 1]$. Figure 14 shows the graph of the solution set.

Figure 14

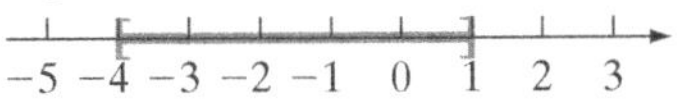

Work Smart

As a partial check of the solution of Example 6, we can check a number in the interval. Let's try $x = -3$.

$$|2(-3) + 3| \stackrel{?}{\leq} 5$$
$$|-6 + 3| \stackrel{?}{\leq} 5$$
$$|-3| \stackrel{?}{\leq} 5$$
$$3 \leq 5 \quad \text{True}$$

Quick ✓

In Problems 22–24, solve each inequality. Express the solution set in set-builder notation and interval notation. Graph the solution set.

22. $|x + 3| < 5$

23. $|2x - 3| \leq 7$

24. $|7x + 2| < -3$

EXAMPLE 7 Solving an Absolute Value Inequality Involving <

Solve the inequality $|-3x + 2| + 4 < 14$. Express the solution set in set-builder notation and interval notation. Graph the solution set.

Solution

First, isolate the absolute value.

$$|-3x + 2| + 4 < 14$$

Subtract 4 from both sides: $\quad |-3x + 2| < 10$

$|u| < a$ means $-a < u < a$: $\quad -10 < -3x + 2 < 10$

Subtract 2 from all three parts: $-10 - 2 < -3x + 2 - 2 < 10 - 2$

$$-12 < -3x < 8$$

Divide all three parts by -3.
Reverse the direction of the inequalities.
$$\frac{-12}{-3} > \frac{-3x}{-3} > \frac{8}{-3}$$

$$4 > x > -\frac{8}{3}$$

$b > x > a$ is equivalent to $a < x < b$:
$$-\frac{8}{3} < x < 4$$

Figure 15

The solution is $\left\{ x \mid -\frac{8}{3} < x < 4 \right\}$ using set-builder notation and $\left(-\frac{8}{3}, 4 \right)$ using interval notation. Figure 15 shows the graph of the solution set.

Quick ✓

25. True or False $|x + 1| + 2 < 7$ is equivalent to $-7 < x + 1 + 2 < 7$.

In Problems 26–29, solve each inequality. Express the solution set in set-builder notation and interval notation. Graph the solution set.

26. $|x| + 4 < 6$

27. $|x - 3| + 4 \leq 8$

28. $3|2x + 1| \leq 9$

29. $|-3x + 1| - 5 < 3$

▶ ❸ **Solve Absolute Value Inequalities Involving > or ≥**

EXAMPLE 8 Solving an Absolute Value Inequality Involving >

Solve the inequality $|x| > 3$. Graph the solution set.

Solution

The inequality $|x| > 3$ asks for all real numbers x that are more than 3 units from the origin on the real number line. See Figure 16.

Figure 16

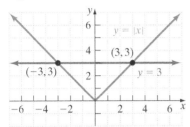

Figure 17

Any number less than -3 or greater than 3 satisfies the inequality. The solution set is $\{x | x < -3 \text{ or } x > 3\}$ or, using interval notation, $(-\infty, -3) \cup (3, \infty)$.

We could also visualize these results by graphing $y = |x|$ and $y = 3$. See Figure 17. To solve $|x| > 3$, we look for all x-coordinates such that the graph of $y = |x|$ is above the graph of $y = 3$. This is true for all x less than -3 or all x greater than 3, as we found above.

Based on Example 8, we have the following results.

Inequalities of the Form $>$ or $\geq$ Involving Absolute Value
If a is a positive real number and u is an algebraic expression, then

$|u| > a$ is equivalent to $u < -a$ or $u > a$
$|u| \geq a$ is equivalent to $u \leq -a$ or $u \geq a$

Quick ✓

30. $|u| > a$ is equivalent to _____ or _____.

31. $|5x - 2| \geq 7$ is equivalent to $5x - 2 \leq$ _____ or $5x - 2 \geq$ _____.

In Problems 32 and 33, solve each inequality. Graph the solution set.

32. $|x| \geq 6$ 33. $|x| > \dfrac{5}{2}$

EXAMPLE 9   **How to Solve an Inequality Involving $>$**

Solve the inequality $|2x - 5| > 3$. Express the solution set in set-builder notation and interval notation. Graph the solution set.

Step-by-Step Solution

Step 1: The inequality is in the form $|u| > a$, where $u = 2x - 5$ and $a = 3$. Rewrite the inequality as a compound inequality that does not involve absolute value.

$|u| > a$ means $u < -a$ or $u > a$: $|2x - 5| > 3$
         $2x - 5 < -3$ or $2x - 5 > 3$

Step 2: Solve each inequality separately.

$2x - 5 < -3$ $2x - 5 > 3$
Add 5 to both sides: $2x < 2$ Add 5 to both sides: $2x > 8$
Divide both sides by 2: $x < 1$ Divide both sides by 2: $x > 4$

Step 3: Find the union of the solution sets of each inequality.

The solution set is $\{x | x < 1 \text{ or } x > 4\}$ or, using interval notation, $(-\infty, 1) \cup (4, \infty)$. See Figure 18 for the graph of the solution set.

Figure 18

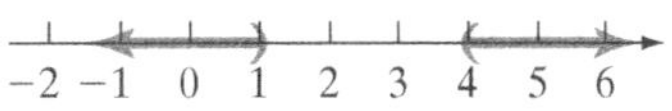

Work Smart

$|u| > a$
CANNOT be written as
$-a > u > a$

Quick ✓

34. $|x - 9| > 6$ is equivalent to $x - 9 > 6$ or $x - 9 \_\_ -6$.

In Problems 35–40, solve each inequality. Express the solution set in set-builder notation and interval notation. Graph the solution set.

35. $|x + 3| > 4$

36. $|4x - 3| \geq 5$

37. $|-3x + 2| > 7$

38. $|2x + 5| - 2 > -2$

39. $|6x - 5| \geq 0$

40. $|2x + 1| > -3$

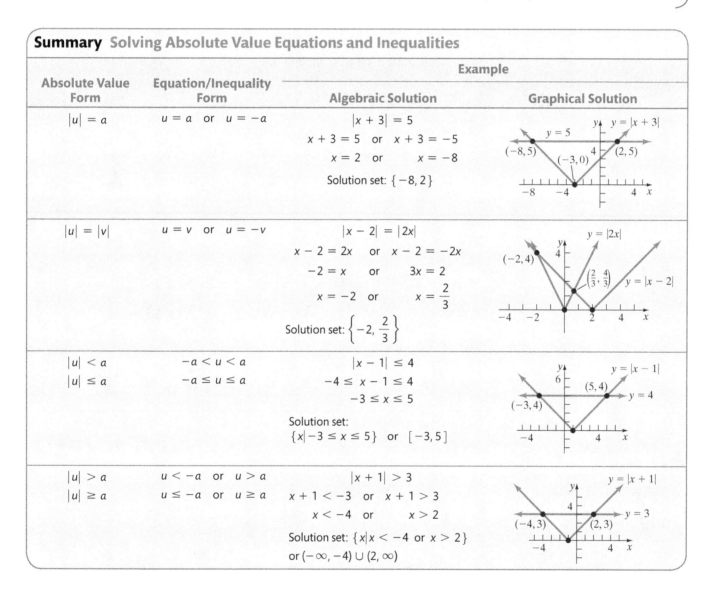

Summary Solving Absolute Value Equations and Inequalities

④ Solve Applied Problems Involving Absolute Value Inequalities

You may have read phrases such as "margin of error" and "tolerance" in the newspaper. For example, according to a recent Gallup poll, 67% of those polled said that recent increases in gas prices have caused financial hardship for their household. The poll had a margin of error of 4%. The 67% is an estimate of the true percentage of Americans who have experienced financial hardship as a consequence of increases in gas prices. If we let p represent the true percentage of Americans who have experienced financial hardship due to the increase in gas prices, then we can represent the poll's margin of error mathematically as

$$|p - 67| \leq 4$$

As another example, the tolerance of a belt used in a pulley system whose width is 6 inches is $\frac{1}{16}$ inch. If x represents the actual width of the belt, then we can represent the acceptable belt widths as

$$|x - 6| \leq \frac{1}{16}$$

EXAMPLE 10 **Analyzing the Margin of Error in a Poll**

The inequality

$$|p - 67| \leq 4$$

represents the percentage of Americans who said they experienced financial hardship due to increase in gas prices. Solve the inequality and interpret the results.

Solution

$$|p - 67| \leq 4$$

$|u| \leq a$ means $-a \leq u \leq a$: $\quad -4 \leq p - 67 \leq 4$

Add 67 to all three parts: $\quad 63 \leq p \leq 71$

The percentage of Americans who have experienced financial hardship due to increases in gas prices is between 63% and 71%, inclusive.

Quick ✓

41. The inequality $|x - 4| \leq \frac{1}{32}$ represents the acceptable belt width x (in inches) for a belt that is manufactured for a pulley system. Determine the acceptable belt width.

42. In a poll conducted by ABC News, 9% of Americans stated that they have been shot at. The margin of error in the poll was 1.7%. If we let p represent the true percentage of people who have been shot at, we can represent the margin of error as

$$|p - 9| \leq 1.7$$

Solve the inequality and interpret the results.

8.IR2 Exercises MyLab Statistics

Underlined exercises have complete video solutions in MyLab.

Problems 1–42 are the Quick ✓s that follow the EXAMPLES.

Building Skills

In Problems 43–64, solve each absolute value equation. See Objective 1.

43. $|x| = 10$
44. $|z| = 9$
45. $|y - 3| = 4$
46. $|x + 3| = 5$
47. $|-3x + 5| = 8$
48. $|-4y + 3| = 9$
49. $|y| - 7 = -2$
50. $|x| + 3 = 5$
51. $|2x + 3| - 5 = 3$
52. $|3y + 1| - 5 = -3$
53. $-2|x - 3| + 10 = -4$
54. $3|y - 4| + 4 = 16$
55. $|-3x| - 5 = -5$
56. $|-2x| + 9 = 9$
57. $\left|\frac{3x - 1}{4}\right| = 2$
58. $\left|\frac{2x - 3}{5}\right| = 2$
59. $|3x + 2| = |2x - 5|$
60. $|5y - 2| = |4y + 7|$
61. $|8 - 3x| = |2x - 7|$
62. $|5x + 3| = |12 - 4x|$
63. $|4y - 7| = |9 - 4y|$
64. $|5x - 1| = |9 - 5x|$

In Problems 65–78, solve each absolute value inequality. Graph the solution set on a real number line. See Objective 2.

65. $|x| < 9$
66. $|x| \leq \frac{5}{4}$
67. $|x - 4| \leq 7$
68. $|y + 4| < 6$

69. $|3x + 1| < 8$

70. $|4x - 3| \le 9$

71. $|6x + 5| < -1$

72. $|4x + 3| \le 0$

73. $2|x - 3| + 3 < 9$

74. $3|y + 2| - 2 < 7$

75. $|2 - 5x| + 3 < 10$

76. $|-3x + 2| - 7 \le -2$

77. $|(2x - 3) - 1| < 0.01$

78. $|(3x + 2) - 8| < 0.01$

In Problems 79–90, solve each absolute value inequality. Graph the solution set on the real number line. See Objective 3.

79. $|y - 5| > 2$

80. $|x + 4| \ge 7$

81. $|-4x - 3| \ge 5$

82. $|-5y + 3| > 7$

83. $2|y| + 3 > 1$

84. $3|z| + 8 > 2$

85. $|-5x - 3| - 7 > 0$

86. $|-9x + 2| - 11 \ge 0$

87. $4|-2x + 1| > 4$

88. $3|8x + 3| \ge 9$

89. $|1 - 2x| \ge |-5|$

90. $|3 - 5x| > |-7|$

Mixed Practice

In Problems 91–94, use the graphs to solve each problem.

91. (a) $|x| = 5$
    (b) $|x| \le 5$
    (c) $|x| > 5$

92. (a) $|x| = 6$
    (b) $|x| \le 6$
    (c) $|x| > 6$

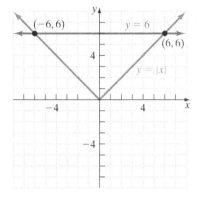

93. (a) $|x + 2| = 3$
    (b) $|x + 2| < 3$
    (c) $|x + 2| \ge 3$

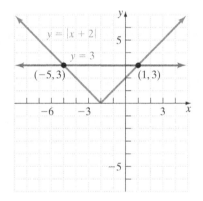

94. (a) $|2x| = 10$
    (b) $|2x| < 10$
    (c) $|2x| \ge 10$

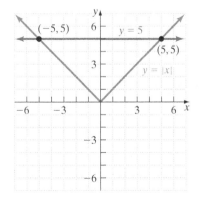

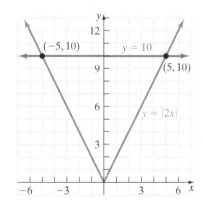

In Problems 95–114, solve each absolute value equation or inequality. For absolute value inequalities, graph the solution set on a real number line.

95. $|x| > 5$

96. $|x| \geq \dfrac{8}{3}$

97. $|2x + 5| = 3$

98. $|4x + 3| = 1$

99. $7|x| = 35$

100. $8|y| = 32$

101. $|5x + 2| \leq 8$

102. $|7y - 3| < 11$

103. $|-2x + 3| = -4$

104. $|3x - 4| = -9$

105. $|3x + 2| \geq 5$

106. $|5y + 3| > 2$

107. $|3x - 2| + 7 > 9$

108. $|4y + 3| - 8 \geq -3$

109. $|5x + 3| = |3x + 5|$

110. $|3z - 2| = |z + 6|$

111. $|4x + 7| + 6 < 5$

112. $|4x + 1| > 0$

113. $\left|\dfrac{x-2}{4}\right| = \left|\dfrac{2x+1}{6}\right|$

114. $\left|\dfrac{1}{2}x - 3\right| = \left|\dfrac{2}{3}x + 1\right|$

Applying the Concepts

115. Express the fact that x differs from 5 by less than 3 as an inequality involving absolute value. Solve for x.

116. Express the fact that x differs from -4 by less than 2 as an inequality involving absolute value. Solve for x.

117. Express the fact that twice x differs from -6 by more than 3 as an inequality involving absolute value. Solve for x.

118. Express the fact that twice x differs from 7 by more than 3 as an inequality involving absolute value. Solve for x.

119. Tolerance A certain rod in an internal combustion engine is supposed to be 5.7 inches long. The tolerance on the rod is 0.0005 inch. If x represents the length of a rod, the acceptable lengths of a rod can be expressed as $|x - 5.7| \leq 0.0005$. Determine the acceptable lengths of the rod.
Source: WiseCo Piston

120. Tolerance A certain rod in an internal combustion engine is supposed to be 6.125 inches long. The tolerance on the rod is 0.0005 inch. If x represents the length of a rod, the acceptable lengths of a rod can be expressed as $|x - 6.125| \leq 0.0005$. Determine the acceptable lengths of the rod.

121. IQ Scores According to the Stanford-Binet IQ test, a normal IQ score is 100. It can be shown that anyone with an IQ x that satisfies the inequality
$$\left|\dfrac{x - 100}{15}\right| > 1.96$$
has an unusual IQ score. Determine the IQ scores that would be considered unusual.

122. Gestation Period The length of human pregnancy is about 266 days. It can be shown that a mother whose gestation period x satisfies the inequality
$$\left|\dfrac{x - 266}{16}\right| > 1.96$$
has an unusual length of pregnancy. Determine the length of pregnancy that would be considered unusual.

Extending the Concepts

In Problems 123–130, solve each equation.

123. $|x| - x = 5$

124. $|y| + y = 3$

125. $z + |-z| = 4$

126. $y - |-y| = 12$

127. $|4x + 1| = x - 2$

128. $|2x + 1| = x - 3$

129. $|x + 5| = -(x + 5)$

130. $|y - 4| = y - 4$

Explaining the Concepts

131. Explain why $|2x - 3| + 1 = 0$ has no solution.

132. Explain why the solution set of $|5x - 3| > -5$ is the set of all real numbers.

133. Explain why $|4x + 3| + 3 < 0$ has the empty set as the solution set.

134. Solve $|x - 5| = |5 - x|$. Explain why the result is reasonable. What do we call this type of equation?

APPENDIX A

Functions, Exponential Functions, Logarithmic Functions

While the topics covered in this appendix may not directly apply to an Elementary Statistics course, it is content that students who are college educated should be exposed to. In addition, the skills learned by covering this material will apply to other courses that you take. For example, exponential and logarithmic functions are required in the fields of economics, biology, and chemistry. Therefore, exposure to this content will provide you with the mathematical skills you will need in order to be a success in those (and other) courses.

Outline

A.IR1 Relations
A.IR2 An Introduction to Functions
A.IR3 Functions and Their Graphs
A.IR4 Linear Functions and Models

A.IR5 Exponential Functions
A.IR6 Logarithmic Functions
A.IR7 Properties of Logarithms
A.IR8 Exponential Equations

A.IR1 Relations

Objectives

1. Understand Relations
2. Find the Domain and the Range of a Relation
3. Graph a Relation Defined by an Equation

Are You Prepared for This Section?

Before getting started, complete the following problems. If you get a problem wrong, go back to the section cited and review the material.

P1. Write the inequality $-4 \leq x \leq 4$ in interval notation. [Section 4.IR1, pp. IR-173–IR-176]

P2. Write the interval $[2, \infty)$ using an inequality. [Section 4.IR1, pp. IR-173–IR-176]

P3. Plot the ordered pairs $(-2, 4)$, $(3, -1)$, $(0, 5)$, and $(4, 0)$ in the rectangular coordinate system. [Section 3.IR1, pp. IR-111–IR-115]

P4. Graph the equation: $2x + 5y = 10$ [Section 3.IR2, pp. IR-129–IR-132]

P5. Graph the equation $y = x^2 - 3$ by plotting points. [Section 3.IR2, pp. IR-124–IR-125]

Prepared?...Answers **P1.** $[-4, 4]$
P2. $x \geq 2$
P3. **P4.**

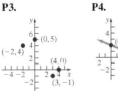

P5.

1 Understand Relations

When the value of one variable is related to the value of a second variable, it is called a *relation*. For example, an individual's level of education is related to annual income. Engine size is related to gas mileage.

Definition

A **relation** exists when the elements in one set are associated with elements in a second set. If x and y are two distinct elements in these sets, and if a relation exists between x and y, then x **corresponds** to y, or y **depends on** x, which is written as $x \to y$. A relation where y depends on x may also be written as an ordered pair (x, y).

EXAMPLE 1 Illustrating a Relation

The data presented in Figure 1 show a correspondence between states and senators in 2018 for randomly selected senators. A possible name for the relation might be "is represented in the U.S. Senate by." Thus the relation reveals that "Indiana is represented by Dan Coats," for example. In Figure 1, **mapping** is used to represent the relation by drawing an arrow from an element in the set "state" to an element in the set "senator."

Figure 1

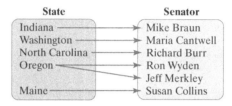

This relation could also be represented by using ordered pairs in the form (state, senator) as follows:

{ (Indiana, Mike Braun), (Washington, Maria Cantwell), (North Carolina, Richard Burr), (Oregon, Ron Wyden), (Oregon, Jeff Merkley), (Maine, Susan Collins) }

Quick ✓

1. If a relation exists between x and y, then say that x _____ to y or that y _____ on x, and write $x \to y$.

2. Use the map to represent the relation as a set of ordered pairs.

   Friend → Birthday
   Max, Alesia, Trent, Yolanda, Wanda, Elvis → January 20, March 3, July 6, November 8, January 8

3. Use the set of ordered pairs to represent the relation as a map.
   { (1, 3), (5, 4), (8, 4), (10, 13) }

❷ Find the Domain and the Range of a Relation

In a relation we say that y depends on x and can write the relation as a set of ordered pairs (x, y). The set of all x can be thought of as the **inputs** of the relation. The set of all y can be thought of as the **outputs** of the relation. This interpretation of a relation can be used to define *domain* and *range*.

Definition

The **domain** of a relation is the set of all inputs of the relation. The **range** is the set of all outputs of the relation.

EXAMPLE 2　Finding the Domain and the Range of a Relation

Find the domain and the range of the relation presented in Figure 1 from Example 1.

Solution
The domain of the relation is the set of all inputs, therefore, the domain of the relation is

{ Indiana, Washington, North Carolina, Oregon, Maine }

The range of the relation is the set of all outputs. Therefore, the range of the relation is

{ Mike Braun, Maria Cantwell, Richard Burr, Ron Wyden, Jeff Merkley, Susan Collins }

Work Smart
Never list elements in the domain or range more than once.

Notice that Oregon was not listed twice in the domain. The domain and the range are sets, and elements in a set should not be listed more than once. Also, the order in which the elements in the domain or range are listed does not matter.

Quick ✓

4. The _____ of a relation is the set of all inputs of the relation. The _____ is the set of all outputs of the relation.

5. State the domain and the range of the relation.

Friend	Birthday
Max	January 20
Alesia	March 3
Trent	July 6
Yolanda	November 8
Wanda	January 8
Elvis	

6. State the domain and the range of the relation.

{ (1, 3), (5, 4), (8, 4), (10, 13) }

Relations can also be represented by plotting a set of ordered pairs. The set of all x-coordinates represents the domain of the relation, and the set of all y-coordinates represents the range of the relation.

EXAMPLE 3　Finding the Domain and the Range of a Relation

Figure 2 shows the graph of a relation. Identify the domain and the range of the relation.

Figure 2

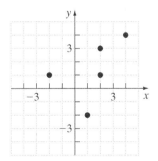

Work Smart
Write the points as ordered pairs to assist in finding the domain and range.

Solution
The ordered pairs in the graph are $(-2, 1)$, $(1, -2)$, $(2, 1)$, $(2, 3)$, and $(4, 4)$, so the domain is the set of all x-coordinates: $\{-2, 1, 2, 4\}$. The range is the set of all y-coordinates: $\{-2, 1, 3, 4\}$.

Quick ✓

7. Identify the domain and the range of the relation shown in the figure.

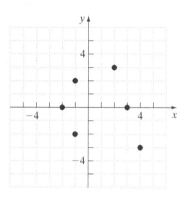

A third way to define a relation is by a graph. Remember that the graph of an equation is the set of all ordered pairs (x, y) that make the equation a true statement. If a graph exists for some ordered pair (x, y), then the x-coordinate is in the domain and the y-coordinate is in the range.

EXAMPLE 4 Identifying the Domain and the Range of a Relation from Its Graph

Figure 3 shows the graph of a relation. Determine the domain and the range of the relation.

Figure 3

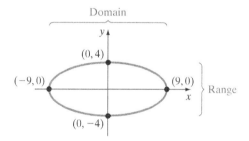

Work Smart

You can think of the domain as the shadow created by the graph on the x-axis by vertical beams of light. The range can be thought of as the shadow created by the graph on the y-axis by horizontal beams of light.

Solution

The domain of the relation consists of all x-coordinates at which the graph exists. The graph exists at all x-values between -9 and 9, inclusive. Therefore, the domain is $\{x \mid -9 \leq x \leq 9\}$ or, using interval notation, $[-9, 9]$.

The range of the relation consists of all y-coordinates at which the graph exists. The graph exists at all y-values between -4 and 4, inclusive. Therefore, the range is $\{y \mid -4 \leq y \leq 4\}$ or, using interval notation, $[-4, 4]$.

Quick ✓

8. *True or False* If the graph of a relation does not exist at $x = 3$, then 3 is not in the domain of the relation.

9. *True or False* The range of a relation is always the set of all real numbers.

In Problems 10 and 11, identify the domain and range of the relation from its graph.

10.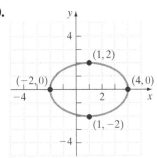

11.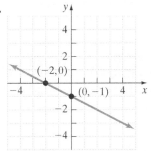

Work Smart

Relations can be defined by
1. Mapping
2. Sets of ordered pairs
3. Graphs
4. Equations

❸ Graph a Relation Defined by an Equation

Another way to define a relation (instead of a map, a set of ordered pairs, or a graph) is to use equations such as $x + y = 4$ or $x = y^2$. When relations are defined by equations, graph the relation in order to visualize how the variables are related. As seen in Example 4, the graph of the relation also helps identify its domain and range.

EXAMPLE 5 **Relations Defined by Equations**

Graph the relation $y = -x^2 + 4$. Find its domain and range using the graph.

Solution

The relation says to take the input x, square it, multiply this result by -1, and then add 4 to get the output y. Use the point-plotting method to graph the relation. Table 1 shows some points on the graph. Figure 4 shows a graph of the relation.

Table 1

x	$y = -x^2 + 4$	(x, y)
-3	$-(-3)^2 + 4 = -5$	$(-3, -5)$
-2	$-(-2)^2 + 4 = 0$	$(-2, 0)$
-1	3	$(-1, 3)$
0	4	$(0, 4)$
1	3	$(1, 3)$
2	0	$(2, 0)$
3	-5	$(3, -5)$

Figure 4

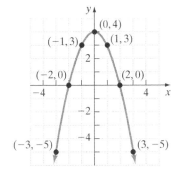

The graph extends indefinitely to the left and to the right (that is, the graph exists for all x-values). Therefore, the domain of the relation is the set of all real numbers, or $\{x \mid x \text{ is any real number}\}$, or, using interval notation, $(-\infty, \infty)$. Notice that there are no y-values greater than 4, but the graph exists everywhere for y-values less than or equal to 4. The range of the relation is $\{y \mid y \leq 4\}$ or, using interval notation, $(-\infty, 4]$.

Quick ✓

In Problems 12–14, graph each relation. Use the graph to identify the domain and range.

12. $y = 3x - 8$
13. $y = x^2 - 8$
14. $x = y^2 + 1$

A.IR1 Exercises — MyLab Statistics

Underlined exercises have complete video solutions in MyLab.

Problems 1–14 are the Quick ✓s that follow the EXAMPLES.

Building Skills

In Problems 15–18, write each relation as a set of ordered pairs. Then identify the domain and the range of the relation. See Objectives 1 and 2.

15.

Newspaper	Daily Circulation (in millions)
USA Today	2.28
Wall Street Journal	2.06
New York Times	1.1
Los Angeles Times	0.82
Washington Post	0.70

SOURCE: *Information Please Almanac*

16.

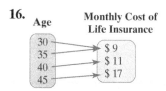

Age	Monthly Cost of Life Insurance
30	$9
35	$11
40	$17
45	

SOURCE: *wholesaleinsurance.net*

17.

Level of Education	Average Annual Income, 2015
Less than 9th Grade	$20,791
9th–12th Grade — No Diploma	$23,234
High School Graduate	$32,456
Associate's Degree	$42,508
Bachelor's Degree	$62,240

SOURCE: *United States Census Bureau*

18.

Region of the Country	Average Annual Income, 2014
Northeast	$59,210
Midwest	$54,267
South	$49,655
West	$57,688

SOURCE: *United States Census Bureau*

In Problems 19–24, write each relation as a map. Then identify the domain and the range of the relation. See Objectives 1 and 2.

19. $\{(-3, 4), (-2, 6), (-1, 8), (0, 10), (1, 12)\}$

20. $\{(-2, 6), (-1, 3), (0, 0), (1, -3), (2, 6)\}$

21. $\{(-2, 4), (-1, 2), (0, 0), (1, 2), (2, 4)\}$

22. $\{(-2, -8), (-1, -1), (0, 0), (1, 1), (2, 8)\}$

23. $\{(0, -4), (-1, -1), (-2, 0), (-1, 1), (0, 4)\}$

24. $\{(-3, 0), (0, 3), (3, 0), (0, -3)\}$

In Problems 25–32, identify the domain and the range of the relation from the graph. See Objective 2.

25.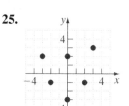

26.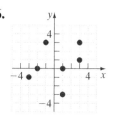

27.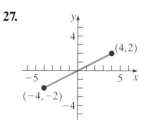

28.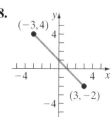

29.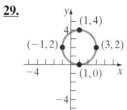

30.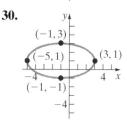

31.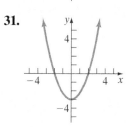

32.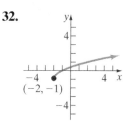

In Problems 33–54, graph the relation to identify the domain and the range of the relation. See Objective 3.

33. $y = -3x + 1$

34. $y = -4x + 2$

35. $y = \frac{1}{2}x - 4$

36. $y = -\frac{1}{2}x + 2$

37. $2x + y = 7$

38. $3x + y = 9$

39. $y = -x^2$

40. $y = x^2 - 2$

41. $y = 2x^2 - 8$

42. $y = -2x^2 + 8$

43. $y = |x|$

44. $y = |x| - 2$

45. $y = |x - 1|$

46. $y = -|x|$

47. $y = x^3$

48. $y = -x^3$

49. $y = x^3 + 1$

50. $y = x^3 - 2$

51. $x^2 - y = 4$ **52.** $x^2 + y = 5$
53. $x = y^2 - 1$ **54.** $x = y^2 + 2$

Applying the Concepts

55. Area of a Window Chip Gaines wishes to put a new window in his home. He wants the perimeter of the window to be 100 feet. The graph shows the relation between the width, x, of the opening (in feet) and the area of the opening.

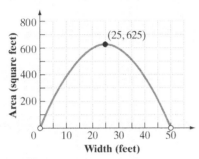

(a) Determine the domain and the range of the relation.

(b) Explain why the domain obtained in part (a) makes sense.

56. Projectile Motion The graph below shows the height, in feet, of a ball thrown straight up with an initial speed of 80 feet per second from an initial height of 96 feet after t seconds. Determine the domain and the range of the relation.

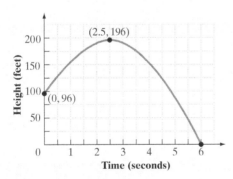

57. Cell Phones The graph below shows the relation between the monthly cost, C, of a cell phone and the number of anytime minutes used, m.

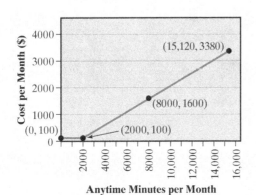

(a) Determine the domain and the range of the relation.

(b) If anytime minutes are from 7:00 A.M. to 7:00 P.M. Monday through Friday, explain why the domain obtained in part (a) makes sense, assuming there are 21 nonweekend days per month.

58. Wind Chill It is 10° Celsius outside. The wind is calm but then gusts up to 20 meters per second. You feel the chill go right through your bones. The graph below shows the relation between the wind chill temperature (in degrees Celsius) and wind speed, v (in meters per second). Determine the domain and the range of the relation.

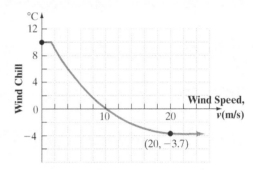

Extending the Concepts

59. Draw the graph of a relation whose domain is all real numbers, but whose range is a single real number. Compare your graph with those of your classmates. How are they similar?

60. Draw the graph of a relation whose domain is a single real number, but whose range is all real numbers. Compare your graph with those of your classmates. How are they similar?

Explaining the Concepts

61. Explain what a relation is. Be sure to include an explanation of domain and range.

62. State the four methods for describing a relation that are presented in this section. When is using ordered pairs most appropriate? When is using a graph most appropriate? Support your opinion.

A.IR2 An Introduction to Functions

Objectives

1. Determine Whether a Relation Expressed as a Map or Ordered Pairs Represents a Function
2. Determine Whether a Relation Expressed as an Equation Represents a Function
3. Determine Whether a Relation Expressed as a Graph Represents a Function
4. Find the Value of a Function
5. Find the Domain of a Function
6. Work with Applications of Functions

Are You Prepared for This Section?

Before getting started, complete the following problems. If you get a problem wrong, go back to the section cited and review the material.

P1. Evaluate the expression $2x^2 - 5x$ for

(a) $x = 1$ (b) $x = 4$ (c) $x = -3$ [Section 2.IR3, pp. IR-83–IR-84]

P2. Evaluate $\dfrac{3}{2x + 1}$ for $x = -\dfrac{1}{2}$. [Section 2.IR3, pp. IR-83–IR-84]

P3. Express the inequality $x \leq 5$ using interval notation. [Section 4.IR2, pp. IR-173–IR-176]

P4. Express the interval $(2, \infty)$ using set-builder notation. [Section 4.IR2, pp. IR-173–IR-176]

1 Determine Whether a Relation Expressed as a Map or Ordered Pairs Represents a Function

One of the most important concepts in algebra is now presented—the *function*. A function is a special type of relation. To understand functions, let's revisit the relation in Example 1 from Section A.IR1, shown again in Figure 5. In this correspondence between states and their senators, we named the relation "is represented in the U.S. Senate by."

Figure 5

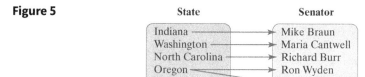

In this relation, if you were asked to name the senator who represents Oregon, you could respond, "Ron Wyden" or "Jeff Merkley." In other words, the input "state" does not correspond to a single output "senator."

Let's consider the relation in Figure 6, a correspondence between states and their populations. If asked for the population that corresponds to North Carolina, you could only respond, "10,384 thousand." In other words, each input "state" corresponds to exactly one output "population."

Figure 7 is a relation that shows a correspondence between animals and life expectancies. If asked to determine the life expectancy of a dog you could only respond, "11 years." You would have the same answer about the life expectancy of a cat.

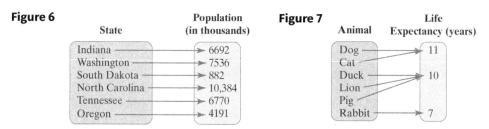

What do the relations in Figures 6 and 7 have in common that is missing from the relation in Figure 5? In the relations in Figure 6 and 7, each input corresponds to only one output. In Figure 5, however, the input Oregon corresponds to two outputs—Ron Wyden and Jeff Merkley. This leads to the definition of a *function*.

Prepared?...Answers **P1. (a)** -3
(b) 12 **(c)** 33 **P2.** undefined
P3. $(-\infty, 5]$ **P4.** $\{x | x > 2\}$

Section A.IR2 An Introduction to Functions   IR-233

> **In Other Words**
> For a relation to be a function, each input may have only one output.

Definition

A **function** is a relation in which each input, or element in the domain of the relation, corresponds to exactly one output, or element in the range of the relation.

The idea behind functions is predictability. If an input is known, a function can be used to determine the output with 100% certainty, as seen in Figures 6 and 7. Nonfunctions do not have this predictability (Figure 5).

EXAMPLE 1   Determining Whether a Relation Represents a Function

Determine whether the relation is a function. If so, state its domain and range.

(a) See Figure 8(a). The domain represents the length (mm) of the right humerus, and the range represents the length (mm) of the right tibia for each of five rats that had been sent to space.

(b) See Figure 8(b). The domain represents the age of six people, and the range represents their height.

(c) See Figure 8(c). The domain represents the age of five males, and the range represents their HDL (or "good") cholesterol (mg/dL).

Figure 8

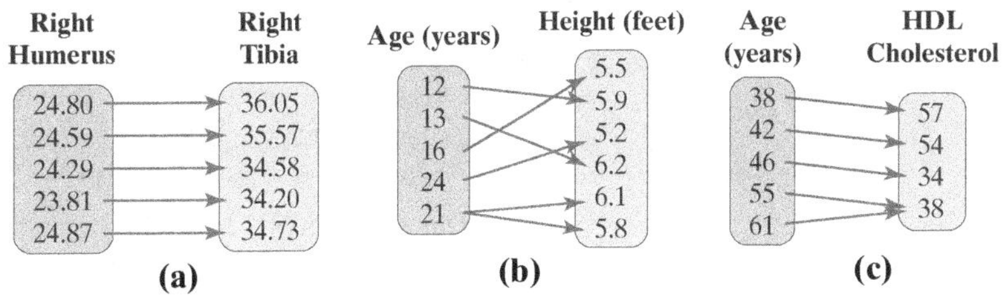

Right Humerus	Right Tibia
24.80	36.05
24.59	35.57
24.29	34.58
23.81	34.20
24.87	34.73

(a)
SOURCE: NASA Life Sciences

Age (years)	Height (feet)
12	5.5
13	5.9
16	5.2
24	6.2
21	6.1
	5.8

(b)
SOURCE: diamonds.com

Age (years)	HDL Cholesterol
38	57
42	54
46	34
55	38
61	

(c)

Solution

(a) The relation in Figure 8(a) is a function because each element in the domain corresponds to exactly one element in the range. The domain is { 24.80, 24.59, 24.29, 23.81, 24.87 }. The range is { 36.05, 35.57, 34.58, 34.20, 34.73 }.

(b) The relation in Figure 8(b) is not a function because an element in the domain, 21, corresponds to two elements in the range. A single height cannot be determined for the 21-year-olds.

(c) The relation in Figure 8(c) is a function because each element in the domain corresponds to exactly one element in the range. Notice that it is okay for more than one element in the function's domain to correspond to the same element in the range. The domain of the function is { 38, 42, 46, 55, 61 }. The range of the function is { 57, 54, 34, 38 }.

Quick ✓

1. A _____ is a relation in which each element in the domain of the relation corresponds to exactly one element in the range of the relation.

2. *True or False*   Every relation is a function.

Work Smart
All functions are relations, but not all relations are functions!

In Problems 3 and 4, determine whether the relation is a function. If so, state its domain and range.

3.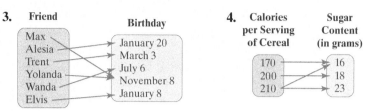

4.

A function can also be thought of as a set of ordered pairs (x, y) in which no ordered pairs have the same x-coordinate and different y-coordinates.

EXAMPLE 2 **Determining Whether a Relation Represents a Function**

Determine whether the relation is a function. If so, state its domain and range.

(a) $\{(1, 3), (-1, 4), (0, 6), (2, 8)\}$
(b) $\{(-2, 6), (-1, 3), (0, 2), (1, 3), (2, 6)\}$
(c) $\{(0, 3), (1, 4), (4, 5), (9, 5), (4, 1)\}$

Solution

(a) This relation is a function because no ordered pairs have the same x-coordinate and different y-coordinates. The domain of the function is the set of all x-coordinates, $\{-1, 0, 1, 2\}$. The range of the function is the set of all y-coordinates, $\{3, 4, 6, 8\}$.

(b) This relation is a function because no ordered pairs have the same x-coordinate and different y-coordinates. The domain is, $\{-2, -1, 0, 1, 2\}$. The range is, $\{2, 3, 6\}$.

(c) This relation is not a function because two ordered pairs, $(4, 5)$ and $(4, 1)$, have the same x-coordinate and different y-coordinates.

Work Smart
In a function, two different inputs can correspond to the same output, but two different outputs cannot be the result of a single input.

In Example 2(b), notice that inputs -2 and 2 each correspond to output 6. This does not violate the definition of a function – two different x-coordinates can have the same y-coordinate. A violation of the definition occurs when two ordered pairs have the same x-coordinate and different y-coordinates, as in Example 2(c).

Quick ✓

In Problems 5 and 6, determine whether each relation is a function. If so, state its domain and range.

5. $\{(-3, 3), (-2, 2), (-1, 1), (0, 0), (1, 1)\}$
6. $\{(-3, 2), (-2, 5), (-1, 8), (-3, 6)\}$

❷ Determine Whether a Relation Expressed as an Equation Represents a Function

We now know how to identify when a relation defined by a map or ordered pairs is a function. In Section A.IR1, relations were also expressed as equations. We will now address the circumstances under which equations are functions.

To determine whether an equation, where y depends on x, is a function, it is often easiest to solve the equation for y. If each value of x corresponds to exactly one value of y, the equation is a function; otherwise, it is not.

EXAMPLE 3 Determining Whether an Equation Is a Function

Determine whether the equation $y = 3x + 5$ shows y as a function of x.

Solution
The rule for getting from x to y is to multiply x by 3 and then add 5. Since only one output y can result by performing these operations on any input x, the equation is a function.

EXAMPLE 4 Determining Whether an Equation Is a Function

Determine whether the equation $y = \pm x^2$ shows y as a function of x.

Solution
Notice that for any single value of x (other than 0), two values of y result. For example, if $x = 2$, then $y = \pm 4$ (-4 or $+4$). Since a single x corresponds to more than one y, the equation is not a function.

In Other Words
The symbol $\pm$ is a shorthand device and is read "plus or minus." For example, ± 4 means "negative four or positive four."

Quick ✓
In Problems 7–9, determine whether each equation shows y as a function of x.

7. $y = -2x + 5$ **8.** $y = \pm 3x$ **9.** $y = x^2 + 5x$

❸ Determine Whether a Relation Expressed as a Graph Represents a Function

Remember that the graph of an equation is the set of all ordered pairs (x, y) that satisfy the equation. For a relation to be a function, each number x in the domain can correspond to only one number y in the range. This means that a graph is *not* a function if two points with the same x-coordinate have different y-coordinates. This leads to the following test.

Vertical Line Test
A set of points in the xy-plane is the graph of a function if and only if every vertical line intersects the graph in at most one point.

EXAMPLE 5 Using the Vertical Line Test to Identify Graphs of Functions

Which of the graphs in Figure 9 are graphs of functions?

Figure 9

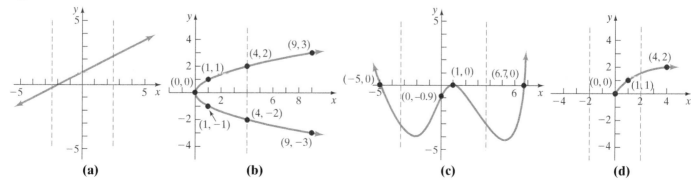

(a)   (b)   (c)   (d)

Solution

The graphs in Figures 9(a), (c), and (d) are functions because every vertical line intersects these graphs in at most one point. The graph in Figure 9(b) is not a function, because a vertical line intersects the graph in more than one point. •

Does Example 5 show you why the vertical line test works? If a vertical line intersects the graph of an equation in two or more points, then the same x-coordinate corresponds to two or more different y-coordinates, and we have violated the definition of a function.

Quick ✓

10. *True or False* For a graph to be a function, any vertical line can intersect the graph in at most one point.

In Problems 11 and 12, use the vertical line test to determine whether the graph is a function.

11.
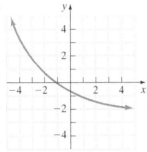

12.
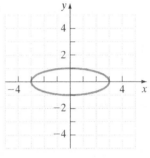

❹ Find the Value of a Function

Functions are often denoted by letters such as f, F, g, G, and so on. If f is a function, use special notation, called *function notation*, to represent the function. The following table shows some equations where y is a function of x. In each case, denote the function by the letter f.

Work Smart

Think of $f(x)$ as a different way of writing y.

Equation in Two Variables	Function Notation
$y = \dfrac{1}{2}x + 5$	$f(x) = \dfrac{1}{2}x + 5$
$y = -3x + 1$	$f(x) = -3x + 1$
$y = 2x^2 - 3x$	$f(x) = 2x^2 - 3x$

Work Smart

Be careful with function notation. In the expression $y = f(x)$, y is the dependent variable, x is the independent variable, and f is the name given to a rule that relates the input x to the output y.

Notice that $f(x)$ replaces the variable y. If f is a function, then for each number x in its domain, the corresponding value in the range is denoted by $f(x)$, which is read as "f of x" or as "f at x." Understand that $f(x)$ does not mean "f times x."

We call $f(x)$ the **value of the function f at the number x**; $f(x)$ is the number that results when the function is applied to x. Finding the value of a function is just like finding the value of y in an equation in two variables when y is a function of x. For example, "Given the equation $y = \dfrac{1}{2}x + 5$, find the value of y when $x = 4$" means the same as "Given the function $f(x) = \dfrac{1}{2}x + 5$, evaluate $f(4)$." In both cases, replace the value of x with 4 and evaluate.

EXAMPLE 6 Finding Values of a Function

If $f(x) = \frac{3}{2}x - 6$ and $g(x) = x^2 + 6x$, evaluate:

(a) $f(10)$ **(b)** $f(-8)$ **(c)** $g(3)$ **(d)** $g(-2)$

Solution

(a) Wherever an x is in the equation defining the function f, substitute 10.

$$f(10) = \frac{3}{2}(10) - 6$$
$$= 15 - 6$$
$$= 9$$

So $f(10) = 9$.

(b) Substitute -8 for x in the expression $\frac{3}{2}x - 6$ to get

$$f(-8) = \frac{3}{2}(-8) - 6$$
$$= -12 - 6$$
$$= -18$$

So $f(-8) = -18$.

(c) Substitute 3 for x in the expression $x^2 + 6x$ to get.

$$g(3) = (3)^2 + 6(3)$$
$$= 9 + 18$$
$$= 27$$

(d) Substitute -2 for x in the expression $x^2 + 6x$ to get

$$g(-2) = (-2)^2 + 6(-2)$$
$$= 4 + (-12)$$
$$= -8$$

Work Smart
Notice the use of parentheses when -2 is squared in Example 6(d) Remember, $-2^2 \neq (-2)^2$.

Quick ✓

In Problems 13–16, let $f(x) = 3x + 2$ and $g(x) = -2x^2 + x + 3$. Evaluate each function.

13. $f(4)$ **14.** $f(-2)$ **15.** $g(-3)$ **16.** $g(1)$

For a function $y = f(x)$, the variable x is called the **independent variable**, because it can be assigned any of the numbers in the domain. The variable y is called the **dependent variable**, because its value depends on x.

The independent variable is also called the **argument** of the function. For example, if f is the function defined by $f(x) = x^2$, then f tells us to square the argument. Thus, $f(2)$ means to square 2, $f(a)$ means to square a, and $f(x + h)$ means to square the quantity $x + h$.

The notation $f(x)$ plays a dual role—it represents the rule for getting from the input to the output and it represents the output y of the function. In Example 6(c), the rule for getting from the input to the output is given by $g(x) = x^2 + 6x$. In other words, the function says to "take some input x, square it, and add the result to six times the input x." If the input is 3, then $g(3)$ represents the output, 27.

EXAMPLE 7 Evaluate a Function

For the function $F(z) = 4z + 7$, evaluate:

(a) $F(z + 3)$
(b) $F(z) + F(3)$

Solution

(a) Wherever a z is in the equation defining F, substitute $z + 3$ to get

$$F(z + 3) = 4(z + 3) + 7$$
$$= 4z + 12 + 7$$
$$= 4z + 19$$

(b) $F(z) + F(3) = \underbrace{4z + 7}_{F(z)} + \underbrace{4 \cdot 3 + 7}_{F(3)}$
$$= 4z + 7 + 12 + 7$$
$$= 4z + 26$$

Quick ✓

17. In the function $H(q) = 2q^2 - 5q + 1$, H is called the _____ variable, and q is called the _____ variable or _____.

In Problems 18 and 19, let $f(x) = 2x - 5$. Evaluate each function.

18. $f(x - 2)$
19. $f(x) - f(2)$

Summary Important Facts About Functions

1. Each x in the domain has exactly one corresponding y in the range.
2. Letters such as f are used to denote the function. It represents the rule used to get from an x in the domain to $f(x)$ in the range.
3. If $y = f(x)$, then x is the independent variable or argument of f, and y is the dependent variable or the value of f at x.

⑤ Find the Domain of a Function

In working with functions, it is important to know which inputs are possible and produce an output.

In Other Words
The domain of a function is the set of all inputs for which the function gives an output that is a real number.

Definition
When only the equation of a function is given, the **domain of f** is the set of real numbers x for which $f(x)$ is a real number.

When identifying the domain of a function, do not forget that division by zero is undefined. Exclude values of the variable that result in division by zero.

EXAMPLE 8 Finding the Domain of a Function

Find the domain of each of the following functions:

(a) $G(x) = x^2 + 1$
(b) $g(z) = \dfrac{z - 3}{z + 1}$

Solution

(a) The function G squares a number x and then adds 1 to the result. These operations can be performed on any real number, so the domain of G is the set of all real numbers, which we can express as $\{x | x \text{ is a real number}\}$ or, using interval notation, $(-\infty, \infty)$.

(b) The function g involves division. Since division by 0 is not defined, the denominator $z + 1$ cannot be 0. Therefore, z cannot equal -1. The domain of g is $\{z | z \neq -1\}$.

Work Smart
In Example 8(b), it is understood that $\{z | z \neq -1\}$ includes all real numbers except -1.

Quick ✓

20. When only the equation of a function f is given, the _____ of f is the set of real numbers x for which $f(x)$ is a real number.

In Problems 21 and 22, find the domain of each function.

21. $f(x) = 3x^2 + 2$

22. $h(x) = \dfrac{x + 1}{x - 3}$

In an application, a function's domain may be restricted by physical or geometric considerations, in addition to mathematical restrictions. For example, the domain of $f(x) = x^2$ is the set of all real numbers. However, if f is used to find the area of a square given side x, then restrict the domain of f to the positive real numbers, since the length of a side cannot be 0 or negative.

EXAMPLE 9 Finding the Domain of a Function

The number N of computers produced at a Dell Computers' manufacturing facility in one day after t hours is given by the function, $N(t) = 336t - 7t^2$. What is the domain of this function?

Solution
The independent variable is t, the number of hours in the day. Therefore the function's domain is $\{t | 0 \leq t \leq 24\}$, or the interval $[0, 24]$.

Quick ✓

23. The function $A(r) = \pi r^2$ gives the area of a circle A as a function of the radius r. What is the domain of the function?

▶ 6 Work with Applications of Functions

In practice, the symbols used for the independent and dependent variables should remind us of what they represent. For example, in economics, C is used for cost and q for quantity, so that $C(q)$ represents the cost of manufacturing q units of a good. Here C is the dependent variable, q is the independent variable, and $C(q)$ is the rule that tells us how to get the output C from the input q.

EXAMPLE 10 Life-Cycle Hypothesis

The Life-Cycle Hypothesis from economics was presented by Franco Modigliani in 1954. It states that income is a function of age. The function

$$I(a) = -55a^2 + 5119a - 54{,}448$$

represents the relation between average annual income I and age a.

(a) Identify the dependent and independent variables.

(b) Evaluate $I(20)$ and explain what $I(20)$ represents.

Solution

(a) Because income depends on age, the dependent variable is income, I, and the independent variable is age, a.

(b) Let $a = 20$ in the function.

$$I(20) = -55(20)^2 + 5119(20) - 54{,}448$$
$$= 25{,}932$$

The average annual income of a 20-year-old is $25,932.

Quick ✓

24. In 2010, the *Deepwater Horizon* oil explosion spilled millions of gallons of oil in the Gulf of Mexico. The oil slick takes the shape of a circle. Suppose that the area A (in square miles) of the circle contaminated with oil can be determined using the function $A(t) = 0.25\pi t^2$, where t represents the number of days since the rig exploded.

(a) Identify the dependent and independent variables.
(b) Evaluate $A(30)$ and explain what $A(30)$ means.

A.IR2 Exercises — MyLab Statistics

Underlined exercises have complete video solutions in MyLab.

*Problems **1–24** are the Quick ✓ s that follow the EXAMPLES.*

Building Skills

In Problems 25–34, determine whether each relation represents a function. State the domain and the range of each relation. See Objective 1.

25.
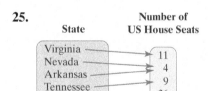

26.
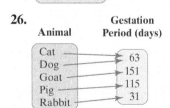

27.
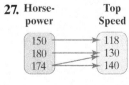

28.
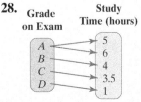

29. $\{(0, 3), (1, 4), (2, 5), (3, 6)\}$

30. $\{(-1, 4), (0, 1), (1, -2), (2, -5)\}$

31. $\{(-3, 5), (1, 5), (4, 5), (7, 5)\}$

32. $\{(-2, 3), (-2, 1), (-2, -3), (-2, 9)\}$

33. $\{(-10, 1), (-5, 4), (0, 3), (-5, 2)\}$

34. $\{(-5, 3), (-2, 1), (5, 1), (7, -3)\}$

In Problems 35–44, determine whether each equation shows y as a function of x. See Objective 2.

35. $y = 2x + 9$

36. $y = -6x + 3$

37. $2x + y = 10$

38. $6x - 3y = 12$

39. $y = \pm 5x$

40. $y = \pm 2x^2$

41. $y = x^2 + 2$

42. $y = x^3 - 3$

43. $x + y^2 = 10$

44. $y^2 = x$

In Problems 45–52, determine whether the graph is that of a function. See Objective 3.

45.
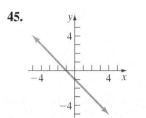

46.
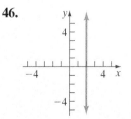

47.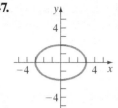
48.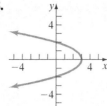
49.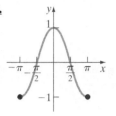
50.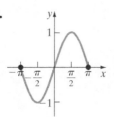
51.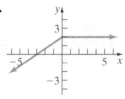
52.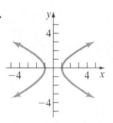

In Problems 53–60, find the indicated value of each function. See Objective 4.

(a) $f(0)$ (b) $f(3)$ (c) $f(-2)$

53. $f(x) = 2x + 3$
54. $f(x) = 3x + 1$
55. $f(x) = -5x + 2$
56. $f(x) = -2x - 3$
57. $f(x) = x^2 - 3x$
58. $f(x) = 2x^2 + 5x$
59. $f(x) = -x^2 + x + 3$
60. $f(x) = -x^2 + 2x - 5$

In Problems 61–64, find the indicated value of each function. See Objective 4.

(a) $f(-x)$ (b) $f(x + 2)$ (c) $f(2x)$
(d) $-f(x)$ (e) $f(x + h)$

61. $f(x) = 2x - 5$
62. $f(x) = 4x + 3$
63. $f(x) = 7 - 5x$
64. $f(x) = 8 - 3x$

In Problems 65–72, evaluate each function. See Objective 4.

65. $f(x) = x^2 + 3; f(2)$
66. $f(x) = -2x^2 + x + 1; f(-3)$
67. $s(t) = -t^3 - 4t; s(-2)$
68. $g(h) = -h^2 + 5h - 1; g(4)$
69. $F(x) = |x - 2|; F(-3)$
70. $G(z) = 2|z + 5|; G(-6)$
71. $F(z) = \dfrac{z + 2}{z - 5}; F(4)$
72. $h(q) = \dfrac{3q^2}{q + 2}; h(2)$

In Problems 73–80, find the domain of each function. See Objective 5.

73. $f(x) = 4x + 7$
74. $G(x) = -8x + 3$
75. $F(z) = \dfrac{2z + 1}{z - 5}$
76. $H(x) = \dfrac{x + 5}{2x + 1}$
77. $f(x) = 3x^4 - 2x^2$
78. $s(t) = 2t^2 - 5t + 1$
79. $G(x) = \dfrac{3x - 5}{3x + 1}$
80. $H(q) = \dfrac{1}{6q + 5}$

Applying the Concepts

81. If $f(x) = 3x^2 - x + C$ and $f(3) = 18$, what is the value of C?

82. If $f(x) = -2x^2 + 5x + C$ and $f(-2) = -15$, what is the value of C?

83. If $f(x) = \dfrac{2x + 5}{x - A}$ and $f(0) = -1$, what is the value of A?

84. If $f(x) = \dfrac{-x + B}{x - 5}$ and $f(3) = -1$, what is the value of B?

△ **85. Geometry** Express the area A of a circle as a function of its radius, r. Determine the area of a circle whose radius is 4 inches. That is, find $A(4)$.

△ **86. Geometry** Express the area A of a triangle as a function of its height h, assuming that the length of the base is 8 centimeters. Determine the area of this triangle if its height is 5 centimeters. That is, find $A(5)$.

87. Salary Express the gross salary G of Jackie, who earns $15 per hour, as a function of the number of hours worked, h. Then determine the gross salary of Jackie if she works 25 hours. That is, find $G(25)$.

88. Commissions Roberta is a commissioned salesperson. She earns a base weekly salary of $250 per week plus 15% of the sales price of items sold. Express her gross salary G as a function of the price p of items sold. Determine the weekly gross salary of Roberta if the value of items sold is $10,000. That is, find $G(10,000)$.

89. Population as a Function of Age The function
$$P(a) = -1.02a^2 + 57.36a + 3694.1$$
represents the population (in thousands) of U.S. residents in 2014, P, that are a years of age.
SOURCE: *United States Census Bureau*

(a) Identify the dependent and independent variables.

(b) Evaluate $P(20)$. Provide a verbal explanation of the meaning of $P(20)$.

(c) Evaluate $P(0)$. Provide a verbal explanation of the meaning of $P(0)$.

90. Number of Rooms The function
$$N(r) = -1.33r^2 + 14.68r - 17.09$$
represents the number of housing units (in millions), N, in 2015 that have r rooms, where $1 \leq r \leq 9$.
SOURCE: *United States Census Bureau*

(a) Identify the dependent and independent variables.

(b) Evaluate $N(3)$. Provide a verbal explanation of the meaning of $N(3)$.

(c) Why is it unreasonable to evaluate $N(0)$?

91. Revenue Function The function
$$R(p) = -p^2 + 200p$$
represents the daily revenue R earned from selling MP4 players at p dollars for $0 \leq p \leq 200$.

(a) Identify the dependent and independent variables.

(b) Evaluate $R(50)$. Provide a verbal explanation of the meaning of $R(50)$.

(c) Evaluate $R(120)$. Provide a verbal explanation of the meaning of $R(120)$.

92. Average Trip Length The function
$$T(x) = 0.01x^2 - 0.12x + 8.89$$
represents the average vehicle trip length T (in miles) x years since 1969.

(a) Identify the dependent and independent variables.

(b) Evaluate $T(35)$. Provide a verbal explanation of the meaning of $T(35)$.

(c) Evaluate $T(0)$. Provide a verbal explanation of the meaning of $T(0)$.

△ **93. Geometry** The volume V of a sphere as a function of its radius r is given by $V(r) = \frac{4}{3}\pi r^3$. What is the domain of this function?

△ **94. Geometry** The area A of a triangle as a function of its height h, assuming that the length of the base is 5 centimeters, is $A = \frac{5}{2}h$. What is the domain of the function?

95. Salary The gross salary G of Kevin as a function of the number of hours worked, h, is given by $G(h) = 22.5h$. What is the domain of the function if he can work up to 60 hours per week?

96. Commissions Carlos is a commissioned salesperson. He earns a base weekly salary of $350 plus 12% of the sales price of items sold. His gross salary G as a function of the price p of items sold is given by $G(p) = 350 + 0.12p$. What is the domain of the function?

97. Demand for Hot Dogs Suppose the function $D(p) = 1200 - 10p$ represents the demand for hot dogs, whose price is p, at a baseball game. Find the domain of the function.

98. Revenue Function The function $R(p) = -p^2 + 200p$ represents the daily revenue earned from selling MP4 players at p dollars for $0 \leq p \leq 200$. Explain why any p greater than $200 is not in the domain of the function.

Extending the Concepts

99. Math for the Future: College Algebra A **piecewise-defined function** is a function defined by more than one equation. For example, the absolute value function $f(x) = |x|$ is actually defined by two equations: $f(x) = x$ if $x \geq 0$ and $f(x) = -x$ if $x < 0$. We can combine these equations into one expression as
$$f(x) = \begin{cases} -x & x < 0 \\ x & x \geq 0 \end{cases}$$
To evaluate $f(3)$, we recognize that $3 \geq 0$, so we use the rule $f(x) = x$ and obtain $f(3) = 3$. To evaluate $f(-4)$, we recognize that $-4 < 0$, so we use the rule $f(x) = -x$ and obtain $f(-4) = -(-4) = 4$.

(a) $f(x) = \begin{cases} x + 3 & x < 0 \\ -2x + 1 & x \geq 0 \end{cases}$

(i) Find $f(3)$. (ii) Find $f(-2)$.
(iii) Find $f(0)$.

(b) $f(x) = \begin{cases} -3x + 1 & x < -2 \\ x^2 & x \geq -2 \end{cases}$

(i) Find $f(-4)$. (ii) Find $f(2)$.
(iii) Find $f(-2)$.

100. Math for the Future: Calculus

(a) If $f(x) = 3x + 7$, find $\dfrac{f(x+h) - f(x)}{h}$.

(b) If $f(x) = -2x + 1$, find $\dfrac{f(x+h) - f(x)}{h}$.

Explaining the Concepts

101. Investigate when the use of function notation $y = f(x)$ first appeared. Start by researching Lejeune Dirichlet.

102. Are all relations functions? Are all functions relations? Explain your answers.

103. Explain what a function is. Be sure to include the terms *domain* and *range* in your explanation.

104. Explain why the vertical line test can be used to identify the graph of a function.

105. What are the four forms of a function presented in this section?

106. Explain why the terms *independent variable* for x and *dependent variable* for y make sense in the function $y = f(x)$.

Technology Exercises

Graphing calculators have the ability to evaluate any function you wish. Figure 10 shows the results obtained in Example 6(c) and 6(d) using a graphing calculator.

Figure 10

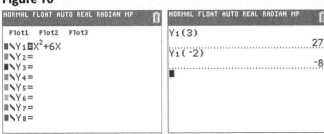

In Problems 107–114, use technology to find the value of each function.

107. $f(x) = x^2 + 3; f(2)$

108. $f(x) = -2x^2 + x + 1; f(-3)$

109. $F(x) = |x - 2|; F(-3)$

110. $G(z) = 2|z + 5|; G(-6)$

111. $H(x) = \sqrt{4x - 3}; H(7)$

112. $g(h) = \sqrt{2h + 1}; g(4)$

113. $F(z) = \dfrac{0.1z + 2.2}{0.2z - 4.8}; F(4)$

114. $h(q) = \dfrac{1.2q^2}{q + 2.8}; h(2)$

A.IR3 Functions and Their Graphs

Objectives
1. Graph a Function
2. Obtain Information from the Graph of a Function
3. Know Properties and Graphs of Basic Functions
4. Interpret Graphs of Functions

Are You Prepared for This Section?

Before getting started, complete the following problems. If you get a problem wrong, go back to the section cited and review the material.

P1. Solve: $3x - 12 = 0$ [Section 2.IR4, pp. IR-91–IR-95]

P2. Graph $y = x^2$ by point-plotting.   [Section 3.IR2, pp. IR-124–IR-128]

▶ 1 Graph a Function

The graph of the linear equation $y = -2x + 4$ (see Figure 11) passes the vertical line test, so the equation is a function. This can be written as $f(x) = -2x + 4$. The graph of a function is the same as the graph of the equation that defines the function. The horizontal axis represents the independent variable and the vertical axis represents the dependent variable. When graphing functions, label the horizontal axis x and the vertical axis either y or by the name of the function.

Figure 11

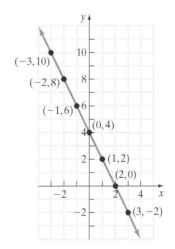

Prepared?... Answers P1. $\{4\}$

P2.

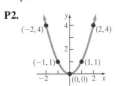

Definition

When a function is defined by an equation in x and y, the **graph of the function** is the set of *all* ordered pairs (x, y) such that $y = f(x)$.

So, if $f(3) = 8$, the point whose ordered pair is $(3, 8)$ is on the graph of $y = f(x)$.

EXAMPLE 1 **Graphing a Function**

Graph the function $f(x) = |x|$.

Solution
To graph $f(x) = |x|$, first determine some ordered pairs $(x, f(x)) = (x, y)$ such that $y = |x|$. See Table 2. Now plot the ordered pairs (x, y) from Table 2 and connect the points as shown in Figure 12.

Table 2

x	$f(x) = \|x\|$	$(x, f(x))$
-3	$\|-3\| = 3$	$(-3, 3)$
-2	$\|-2\| = 2$	$(-2, 2)$
-1	$\|-1\| = 1$	$(-1, 1)$
0	$\|0\| = 0$	$(0, 0)$
1	$\|1\| = 1$	$(1, 1)$
2	$\|2\| = 2$	$(2, 2)$
3	$\|3\| = 3$	$(3, 3)$

Figure 12
$f(x) = |x|$

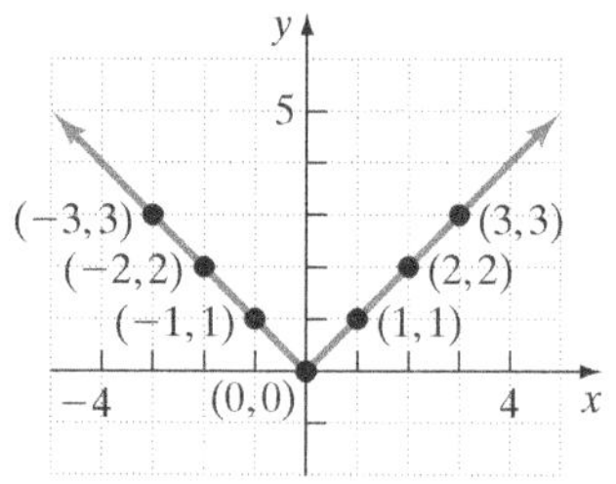

Quick ✓

1. When a function is defined by an equation in x and y, the _____ of the _____ is the set of all ordered pairs (x, y) such that $y = f(x)$.

2. If $f(4) = -7$, then the point whose ordered pair is (__ , __) is on the graph of $y = f(x)$.

In Problems 3–5, graph each function.

3. $f(x) = -2x + 9$ 4. $f(x) = x^2 + 2$ 5. $f(x) = |x - 2|$

❷ Obtain Information from the Graph of a Function

Remember, the domain of a function is the set of all inputs, and the range is the set of all outputs of the function. The domain and the range of a function can be found from its graph.

Section A.IR3 Functions and Their Graphs IR-245

EXAMPLE 2 **Finding the Domain and Range of a Function from Its Graph**

Figure 13 shows the graph of a function.

(a) Determine the function's domain and range.
(b) Identify the intercepts.

Figure 13

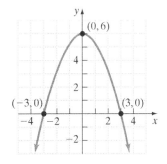

> **In Other Words**
> When the graph of a function is given, its domain may be viewed as the shadow created by the graph on the x-axis by vertical beams of light. Its range can be viewed as the shadow created by the graph on the y-axis by horizontal beams of light.

Solution

(a) The domain of the function consists of all of the graph's x-coordinates. Because the graph exists for all real numbers x, the domain is $\{x \mid x$ is any real number$\}$, or the interval, $(-\infty, \infty)$.

The range of the function consists of all of the graph's y-coordinates. Because the graph exists for all real numbers y less than or equal to 6, the range is $\{y \mid y \leq 6\}$, or the interval, $(-\infty, 6]$.

(b) The x-intercepts are $(-3, 0)$ and $(3, 0)$. The y-intercept is $(0, 6)$.

Quick ✓

6. Use the graph of the function to answer parts (a) and (b).

   (a) Determine the domain and range of the function.

   (b) Identify the intercepts.

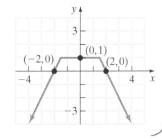

The next example illustrates how to obtain information about a function from its graph. Remember, if $(1, 5)$ is a point on the graph of f, then $f(1) = 5$.

EXAMPLE 3 **Obtaining Information from the Graph of a Function**

Let f be the distance (in feet) above the ground of a person riding in a car on a ferris wheel as a function of time x (in minutes) as shown in Figure 14. Use the graph of f in Figure 14 to answer the following questions.

Figure 14

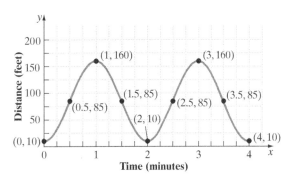

(a) Find $f(1.5)$ and $f(3)$. Interpret these values.

(b) What is the domain of f?

(c) What is the range of f?

(d) For what values of x does $f(x) = 85$? That is, solve $f(x) = 85$.

Solution

(a) Since $(1.5, 85)$ is on the graph of f, then $f(1.5) = 85$. After 1.5 minutes, a rider is 85 feet above the ground. Similarly, since $(3, 160)$ is on the graph, $f(3) = 160$. After 3 minutes, a rider is 160 feet above the ground.

(b) To determine the domain of f, notice that the graph exists for each number x between 0 and 4, inclusive. Therefore, the domain of f is $\{x \mid 0 \leq x \leq 4\}$, or the interval $[0, 4]$.

(c) The points on the graph have y-coordinates between 10 and 160, inclusive. Therefore, the range of f is $\{y \mid 10 \leq y \leq 160\}$, or the interval $[10, 160]$.

(d) Since $(0.5, 85)$, $(1.5, 85)$, $(2.5, 85)$, and $(3.5, 85)$ are the only points on the graph for which $y = f(x) = 85$, the solution set to the equation $f(x) = 85$ is $\{0.5, 1.5, 2.5, 3.5\}$.

Quick ✓

7. If the point $(3, 8)$ is on the graph of a function f, then $f(\underline{}) = \underline{}$. If $g(-2) = 4$, then $(\underline{}, \underline{})$ is a point on the graph of g.

8. Use the graph of $y = f(x)$ to answer the following questions.

   (a) Find $f(-3)$ and $f(1)$.

   (b) What is the domain of f?

   (c) What is the range of f?

   (d) Identify the intercepts.

   (e) For what value of x does $f(x) = 15$? That is, solve $f(x) = 15$.

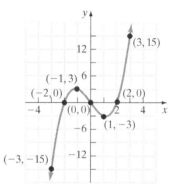

EXAMPLE 4 Obtaining Information about the Graph of a Function

Consider the function $f(x) = 2x - 5$.

(a) Is the point $(3, -1)$ on the graph of the function?

(b) If $x = 1$, what is $f(x)$? Based on this result, what point is on the graph of the function?

(c) If $f(x) = 3$, what is x? Based on this result, what point is on the graph of f?

Solution

(a) If $x = 3$, then

$$f(x) = 2x - 5$$
$$f(3) = 2(3) - 5 = 6 - 5 = 1$$

Since $f(3) = 1$, the point $(3, 1)$ is on the graph but $(3, -1)$ is not.

(b) If $x = 1$, then
$$f(1) = 2(1) - 5 = 2 - 5 = -3$$
Therefore, the point $(1, -3)$ is on the graph of f.

(c) If $f(x) = 3$, then
$$f(x) = 3$$
$$2x - 5 = 3$$
Add 5 to both sides: $\quad 2x = 8$
Divide both sides by 2: $\quad x = 4$

If $f(x) = 3$, then $x = 4$. Therefore, the point $(4, 3)$ is on the graph of f.

Work Smart: Study Skills
Do not confuse the directions "Find $f(3)$" with "If $f(x) = 3$, what is x?" Write down and study errors that you commonly make so that you can avoid them.

Quick ✓

9. Consider the function $f(x) = -3x + 7$.

    (a) Is the point $(-2, 1)$ on the graph of the function?

    (b) If $x = 3$, what is $f(x)$? Based on this result, what point is on the graph of the function?

    (c) If $f(x) = -8$, what is x? What point is on the graph of f?

▶ The Zero of a Function

If $f(r) = 0$ for some number r, then r is a **zero** of f. For example, if $f(x) = x^2 - 4$, then -2 and 2 are zeros of f because $f(-2) = 0$ and $f(2) = 0$. The zeros of a function can be identified from its graph by identifying the x-intercepts of the graph. Why? If $f(r) = 0$, then the point $(r, 0)$ is on the graph of f, and any point with coordinates $(r, 0)$ is an x-intercept of the graph.

In Other Words
A zero of a function is an input that makes the function equal 0.

EXAMPLE 5 Finding the Zeros of a Function from Its Graph

Find the zeros of the function f whose graph is shown in Figure 15.

Figure 15

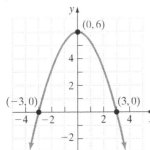

Solution
The x-intercepts of the graph are $(-3, 0)$ and $(3, 0)$. Therefore, the zeros of f are -3 and 3.

Quick ✓

In Problems 10–12, determine whether the value is a zero of the function.

10. $f(x) = 2x + 6; -3$
11. $g(x) = x^2 - 2x - 3; 1$
12. $h(z) = -z^3 + 4z; 2$
13. Find the zeros of the function f whose graph is shown.

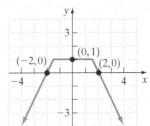

▶ ❸ Know Properties and Graphs of Basic Functions

In Table 3, we list a number of basic functions, their properties, and their graphs.

Table 3

Function	Properties	Graph
Linear Function $f(x) = mx + b$ m and b are real numbers	• Domain and range are all real numbers. • Graph is nonvertical line with slope $= m$ and y-intercept $= (0, b)$.	y, $f(x) = mx + b$, $m > 0$; $(0, b)$
Identity Function (special type of linear function) $f(x) = x$	• Domain and range are all real numbers. • Graph is a line with slope of $m = 1$ and y-intercept $= (0, 0)$. • The line consists of all points for which the x-coordinate equals the y-coordinate.	$f(x) = x$; $(1, 1)$, $(0, 0)$, $(-1, -1)$
Constant Function (special type of linear function) $f(x) = b$ b is a real number	• Domain is the set of all real numbers, and range is the set consisting of a single number b. • Graph is a horizontal line with slope $m = 0$ and y-intercept of $(0, b)$.	$f(x) = b$; $(0, b)$
Square Function $f(x) = x^2$	• Domain is the set of all real numbers, and range is the set of nonnegative real numbers. • The intercept of the graph is $(0, 0)$.	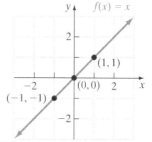
Cube Function $f(x) = x^3$	• Domain and range are the set of all real numbers. • The intercept of the graph is $(0, 0)$.	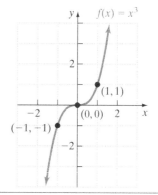
Absolute Value Function $f(x) = \|x\|$	• Domain is the set of all real numbers, and range is the set of nonnegative real numbers. • The intercept of the graph is $(0, 0)$. • If $x \geq 0$, then $f(x) = x$, and the graph of f is part of the line $y = x$; if $x < 0$, then $f(x) = -x$, and the graph of f is part of the line $y = -x$.	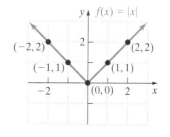

4 Interpret Graphs of Functions

The graph of a function can be used to give a visual description of many different scenarios. Consider the following example.

EXAMPLE 6   Graphing a Verbal Description

Maria decides to take a walk. She leaves her house and walks 3 blocks in 2 minutes at a constant speed. She realizes that she left her front door unlocked, so she runs home in 1 minute. It takes Maria 1 minute to find her keys and lock the door. She next runs 10 blocks in 3 minutes and then rests for 1 minute. She walks 4 more blocks in 10 minutes, and finally hitches a ride home with her neighbor, who happens to drive by, and gets home in 2 minutes. Draw a graph of Maria's distance from home (in blocks) as a function of time.

Solution
Because distance from home is a function of time, draw a Cartesian plane with the horizontal axis representing the independent variable, time, and the vertical axis representing the dependent variable, distance from home.

The ordered pair $(0, 0)$ corresponds to starting the walk. The ordered pair $(2, 3)$ represents being 3 blocks from home after 2 minutes. Start the graph at $(0, 0)$ and then draw a straight line from $(0, 0)$ to $(2, 3)$. Next, draw a straight line to $(3, 0)$, which represents the return trip home to lock the door. Draw a line segment from $(3, 0)$ to $(4, 0)$ to represent the time it takes to lock the door. Draw a line segment from $(4, 0)$ to $(7, 10)$, which represents the 10 block run in 3 minutes. Now draw a horizontal line from $(7, 10)$ to $(8, 10)$. This represents the resting period. Draw a line from $(8, 10)$ to $(18, 14)$ to represent the 4-block walk in 10 minutes. Finally, draw a line segment from $(18, 14)$ to $(20, 0)$ to represent the ride home. See Figure 16.

Figure 16

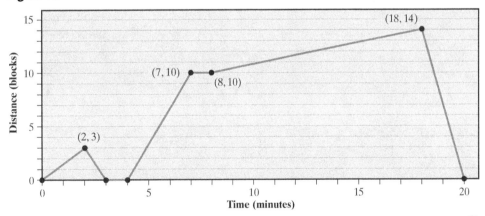

Quick ✓

14. Clara decides to take a walk. She leaves her house and walks 5 blocks in 5 minutes at a constant speed. She realizes that she left her front door unlocked, so she runs home in 2 minutes. It takes her 1 minute to find her keys and lock the door. She next jogs 8 blocks in 5 minutes, and then runs 3 blocks in 1 minute. After resting for 2 minutes she walks home in 10 minutes. Draw a graph of Clara's distance from home (in blocks) as a function of time (in minutes).

A.IR3 Exercises

Problems **1–14** are the Quick ✓s that follow the EXAMPLES.

Building Skills

In Problems 15–22, graph each function. See Objective 1.

15. $f(x) = 4x - 6$
16. $g(x) = -3x + 5$
17. $h(x) = x^2 - 2$
18. $F(x) = x^2 + 1$
19. $G(x) = |x - 1|$
20. $H(x) = |x + 1|$
21. $g(x) = x^3$
22. $h(x) = x^3 - 3$

In Problems 23–32, for each graph of a function, find (a) the domain and the range, (b) the intercepts, if any, and (c) the zeros, if any. See Objective 2.

23.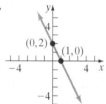
24.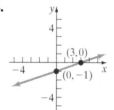
25.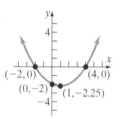
26.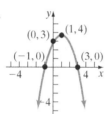
27.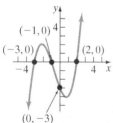
28.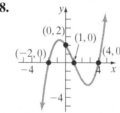
29.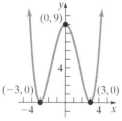
30.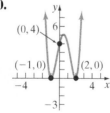
31.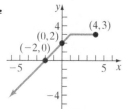
32.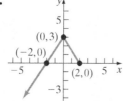

33. Use the graph of the function f shown to answer parts (a) – (l).

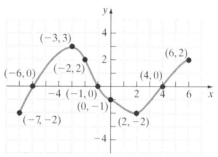

(a) Find $f(-7)$.
(b) Find $f(-3)$.
(c) Find $f(6)$.
(d) Is $f(2)$ positive or negative?
(e) For what numbers x is $f(x) = 0$?
(f) What is the domain of f?
(g) What is the range of f?
(h) What are the x-intercepts?
(i) What is the y-intercept?
(j) For what numbers x is $f(x) = -2$?
(k) For what number x is $f(x) = 3$?
(l) What are the zeros of f?

34. Use the graph of the function g shown to answer parts (a) – (l).

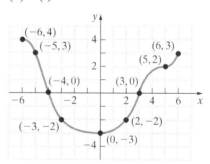

(a) Find $g(-3)$.
(b) Find $g(5)$.
(c) Find $g(6)$.
(d) Is $g(-5)$ positive or negative?
(e) For what numbers x is $g(x) = 0$?
(f) What is the domain of g?
(g) What is the range of g?
(h) What are the x-intercepts?
(i) What is the y-intercept?
(j) For what numbers x is $g(x) = -2$?
(k) For what number x is $g(x) = 3$?
(l) What are the zeros of g?

35. Use the table of values for the function F to answer questions (a)–(e).

x	$F(x)$
-4	0
-2	3
-1	5
0	2
3	-6

(a) What is $F(-2)$?
(b) What is $F(3)$?
(c) For what number(s) x is $F(x) = 5$?
(d) What is the x-intercept of the graph of F?
(e) What is the y-intercept of the graph of F?

36. Use the table of values for the function G to answer questions (a)–(e).

x	$G(x)$
-7	-3
-4	0
0	5
3	8
7	5

(a) What is $G(3)$?
(b) What is $G(7)$?
(c) For what number(s) x is $G(x) = 5$?
(d) What is the x-intercept of the graph of G?
(e) What is the y-intercept of the graph of G?

In Problems 37–40, answer the questions about the given function. See Objective 2.

37. $f(x) = 4x - 9$
(a) Is the point $(2, 1)$ on the graph of the function?
(b) If $x = 3$, what is $f(x)$? What point is on the graph of the function?
(c) If $f(x) = 7$, what is x? What point is on the graph of f?
(d) Is 2 a zero of f?

38. $f(x) = 3x + 5$
(a) Is the point $(-2, 1)$ on the graph of the function?
(b) If $x = 4$, what is $f(x)$? What point is on the graph of the function?
(c) If $f(x) = -4$, what is x? What point is on the graph of f?
(d) Is -2 a zero of f?

39. $g(x) = -\dfrac{1}{2}x + 4$
(a) Is the point $(4, 2)$ on the graph of the function?
(b) If $x = 6$, what is $g(x)$? What point is on the graph of the function?
(c) If $g(x) = 10$, what is x? What point is on the graph of g?
(d) Is 8 a zero of g?

40. $H(x) = \dfrac{2}{3}x - 4$
(a) Is the point $(3, -2)$ on the graph of the function?
(b) If $x = 6$, what is $H(x)$? What point is on the graph of the function?
(c) If $H(x) = -4$, what is x? What point is on the graph of H?
(d) Is 6 a zero of H?

In Problems 41–46, match each graph to the function listed whose graph most resembles the one given. See Objective 3.

(a) Constant function (b) Linear function
(c) Square function   (d) Cube function
(e) Absolute value function (f) Identity function

41.    **42.**

43.    **44.**

45. 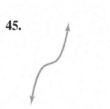   **46.**

In Problems 47–50, sketch the graph of each function. Label at least three points. See Objective 3.

47. $f(x) = x^2$ **48.** $f(x) = x^3$
49. $f(x) = |x|$ **50.** $f(x) = 4$

Applying the Concepts

51. Match each of the following functions with the graph on the following page that best describes the situation.
(a) The distance from ground level of a person who is jumping on a trampoline as a function of time
(b) The cost of an Uber ride as a function of distance
(c) The height of a human as a function of time
(d) The height of a rocket as a function of time
(e) The book value of a machine that is depreciated by equal amounts each year as a function of the year

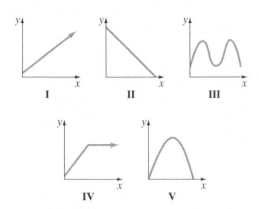

52. Match each of the following functions with the graph that best describes the situation.
    (a) The average high temperature each day as a function of the day of the year
    (b) The height of a human as a function of their age
    (c) The distance that a person rides her bicycle at a constant speed as a function of time
    (d) The temperature of a pizza after it is removed from the oven as a function of time
    (e) The value of a car as a function of time

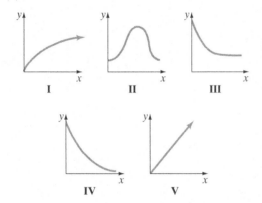

53. **Pulse Rate** Consider the following scenario: Zach starts jogging on a treadmill. His resting pulse rate is 70 beats per minute (b.p.m.). As he continues to jog on the treadmill, his pulse increases at a constant rate until, after 10 minutes, his pulse is 120 b.p.m. He then starts jogging faster and his pulse increases at a constant rate for 2 minutes, at which time his pulse is up to 150 b.p.m. He then begins a cooling-off period for 7 minutes until his pulse goes down to 110 b.p.m. He then gets off the treadmill and his pulse returns to 70 b.p.m. after 12 minutes. Draw a graph of Zach's pulse as a function of time.

54. **Altitude of an Airplane** Suppose that a plane is flying from Chicago to New Orleans. The plane leaves the gate and taxis for 5 minutes. The plane takes off and gets up to 10,000 feet after 5 minutes. The plane continues to ascend at a constant rate until it reaches its cruising altitude of 35,000 feet after another 25 minutes. For the next 80 minutes, the plane maintains a constant height of 35,000 feet. The plane then descends at a constant rate until it lands after 20 minutes. It requires 5 minutes to taxi to the gate. Draw a graph of the height of the plane as a function of time.

55. **Height of a Swing** An 8-year-old girl gets on a swing and starts swinging for 10 minutes. Draw a graph that represents the height of the child from the ground as a function of time.

56. **Temperature of Pizza** Marissa is hungry and would like a pizza. Her mother pulls a frozen pizza out of the freezer and puts it in the oven. After 12 minutes the pizza is done, but Mom lets the pizza cool for 5 minutes before serving it to Marissa. Draw a graph that represents the temperature of the pizza as a function of time.

57. The graph below shows the weight in pounds of a person as a function of his age in years. Describe the weight of the individual over the course of his life.

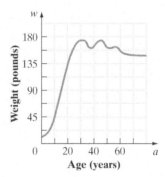

58. The following graph shows the depth of a lake (in feet) as a function of time (in days). Describe the depth of the lake over the course of the 400 days.

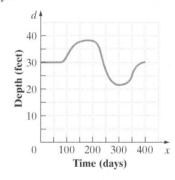

Extending the Concepts

59. Draw a graph of a function f with the following characteristics: x-intercepts: $(-4, 0)$, $(-1, 0)$, and $(2, 0)$; y-intercept: $(0, -2)$; $f(-3) = 7$ and $f(3) = 8$.

60. Draw a graph of a function f with the following characteristics: x-intercepts: $(-3, 0)$, $(2, 0)$, and $(5, 0)$; y-intercept: $(0, 3)$; $f(3) = -2$.

Explaining the Concepts

61. Using the definition of a function, explain why the graph of a function can have at most one y-intercept.

62. Explain what the domain of a function is. In your explanation, discuss how domains are determined in applications.

63. Explain what the range of a function is.

64. Explain the relationship between the x-intercepts of a function and its zeros.

65. Why are y-intercepts not considered zeros of a function?

A.IR4 Linear Functions and Models

Objectives

1. Graph Linear Functions
2. Find the Zero of a Linear Function
3. Build Linear Models from Verbal Descriptions

Are You Prepared for This Section?

Before getting started, complete the following problems. If you get a problem wrong, go back to the section cited and review the material.

P1. Graph: $y = 2x - 3$ [Section 3.IR2, pp. IR-124–IR-128]

P2. Graph: $\frac{1}{2}x + y = 2$ [Section 3.IR2, pp. IR-129–IR-132]

P3. Graph: $y = -4$ [Section 3.IR2, pp. IR-132–IR-134]

P4. Graph: $x = 5$ [Section 3.IR2, pp. IR-132–IR-134]

P5. Find and interpret the slope of the line through $(-1, 3)$ and $(3, -4)$. [Section 3.IR3, pp. IR-137–IR-140]

P6. Find the equation of the line through $(1, 3)$ and $(4, 9)$. [Section 3.IR5, pp. IR-158–IR-159]

P7. Solve: $0.5(x - 40) + 100 = 84$ [Section 2.IR4, pp. IR-95–IR-97]

P8. Solve: $4x + 20 \geq 32$ [Section 4.IR3, pp. IR-179–IR-184]

▶ 1 Graph Linear Functions

Recall from Section 3.IR2 that a linear equation in two variables has the form $Ax + By = C$, where $A, B,$ and C are real numbers, and A and B are not both zero.

Consider the graphs of the four lines in Figure 17. Notice the lines that rise from left to right have positive slope, and lines that fall from left to right have negative slope. Lines that have zero slope are horizontal lines, and lines that have undefined slope are vertical lines. Remember, just as we read a text from left to right, we also read graphs from left to right.

Figure 17

Prepared?...Answers

P1. 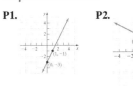 P2.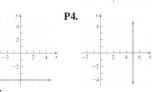

P3. 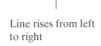 P4.

P5. $-\frac{7}{4}$; y decreases by 7 when x increases by 4. P6. $y = 2x + 1$ P7. $\{8\}$
P8. $\{x \mid x \geq 3\}$; $[3, \infty)$

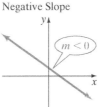

Positive Slope $m > 0$

Line rises from left to right

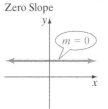

Negative Slope $m < 0$

Line falls from left to right

Zero Slope $m = 0$

Horizontal line

Undefined Slope m is undefined

Vertical line

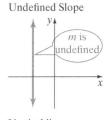

In Figure 17, all the graphs except one pass the vertical line test for identifying graphs of functions. This leads to the conclusion that **all linear equations except those of the form $x = a$, vertical lines, are functions.**

Therefore any linear equation that is in the form $Ax + By = C$ can be written using function notation, provided $B \neq 0$, as follows:

$$Ax + By = C \quad B \neq 0$$

Subtract Ax from both sides: $\quad By = -Ax + C$

Divide both sides by B: $\quad \dfrac{By}{B} = \dfrac{-Ax + C}{B}$

Simplify: $\quad y = -\dfrac{A}{B}x + \dfrac{C}{B}$

$$\updownarrow \quad \updownarrow \quad \updownarrow$$
$$f(x) = mx + b$$

This leads to the following definition:

Definition
A **linear function** is a function of the form

$$f(x) = mx + b$$

where m is the slope and $(0, b)$ is the y-intercept. The graph of a linear function is a line.

Linear functions can be graphed using the same techniques used to graph linear equations written in slope-intercept form, $y = mx + b$ (see Section 3.IR4).

EXAMPLE 1 **Graphing a Linear Function**

Figure 18

Graph the linear function: $f(x) = 3x - 5$

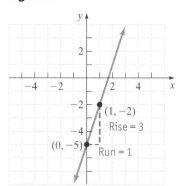

Solution
Since $f(x) = 3x - 5$ is in the form $f(x) = mx + b$, the y-intercept is $(0, -5)$, so plot that point in the coordinate plane. Because the slope $m = 3 = \dfrac{3}{1} = \dfrac{\Delta y}{\Delta x} = \dfrac{\text{rise}}{\text{run}}$, go to the right 1 unit and up 3 units and end up at $(1, -2)$. The line through these two points is the graph of $f(x) = 3x - 5$, shown in Figure 18. ●

Work Smart
Because $\dfrac{3}{1}$ is equivalent to $\dfrac{-3}{-1}$, we could go to the left 1 and down 3 and still remain on the line.

Quick ✓

1. For the graph of a linear function $f(x) = mx + b$, m is the _____ and $(0, b)$ is the _____.

2. The graph of a linear function is called a _____.

3. *True or False* All linear equations are functions.

4. For the linear function $G(x) = -2x + 3$, the slope is _____ and the y-intercept is _____.

In Problems 5–8, graph each linear function.

5. $f(x) = 2x - 3$
6. $G(x) = -5x + 4$
7. $h(x) = \dfrac{3}{2}x + 1$
8. $f(x) = 4$

▶ ❷ Find the Zero of a Linear Function

In Section A.IR3 we stated that if r is a zero of a function f, then $f(r) = 0$. To find the zero of any function f, solve the equation $f(x) = 0$.

EXAMPLE 2   Finding the Zero of a Linear Function

Find the zero of $f(x) = -4x + 12$.

Solution
Find the zero by solving $f(x) = 0$.

$$f(x) = 0$$
$$-4x + 12 = 0$$

Subtract 12 from both sides of the equation: $-4x = -12$

Divide both sides of the equation by -4: $x = 3$

Check: Since $f(3) = -4(3) + 12 = 0$, the zero of f is 3.

Quick ✓
In Problems 9–11, find the zero of each linear function.

9. $f(x) = 3x - 15$

10. $G(x) = \dfrac{1}{2}x + 4$

11. $F(p) = -\dfrac{2}{3}p + 8$

Applications of Linear Functions

Linear functions have many applications. For example, the cost of cab fare, sales commissions, and the cost of breakfast as a function of the number of eggs ordered can be modeled by linear functions.

EXAMPLE 3   Sales Commissions

Tony's weekly salary at Apple Chevrolet is 0.75% of his weekly sales plus $450, so $S(x) = 0.0075x + 450$ describes Tony's weekly salary S as a linear function of his weekly sales x.

(a) What is the implied domain of the function?
(b) If Tony sells cars worth a total of $50,000 one week, what is his salary?
(c) If Tony earned $600 one week, what was the value of the cars that he sold?
(d) Draw a graph of the function.
(e) For what value of cars sold will Tony's weekly salary exceed $1200?

Solution

(a) The independent variable is weekly sales, x. Because negative weekly sales do not make sense, the function's domain is $\{x \mid x \geq 0\}$ or, using interval notation, $[0, \infty)$.

(b) If Tony's weekly sales are $x = \$50,000$, then he earns

$$S(50{,}000) = 0.0075(50{,}000) + 450$$
$$= \$825$$

Tony will earn $825 for the week, if he sells $50,000 worth of cars.

(c) Here, solve the equation $S(x) = 600$.

$$0.0075x + 450 = 600$$

Subtract 450 from both sides: $0.0075x = 150$

Divide both sides by 0.0075: $x = \$20{,}000$

If Tony earned $600, then he sold $20,000 worth of cars in a week.

(d) Plot the independent variable, weekly sales, along the horizontal axis and the dependent variable, salary, along the vertical axis. Graph the equation by plotting points. Part (b) shows that $(50{,}000, 825)$ is on the graph. Part (c) indicates that $(20{,}000, 600)$ is on the graph. Why do we also know that $(0, 450)$ is on the graph? Plot these points to get the graph in Figure 19.

Figure 19

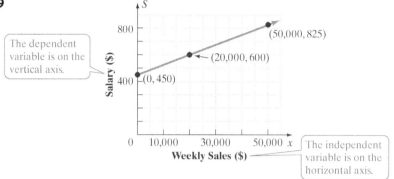

(e) Solve the inequality $S(x) > 1200$.

$$0.0075x + 450 > 1200$$

Subtract 450 from both sides: $\quad 0.0075x > 750$

Divide both sides of the inequality by 0.0075: $\quad x > 100{,}000$

If Tony sells more than \$100,000 worth of cars for the week, his salary will exceed \$1200.

Notice that in Figure 19 the function was graphed only over its domain, $[0, \infty)$ — that is, only in quadrant I. Also notice that the horizontal axis is labeled x for the independent variable, weekly sales, and the vertical axis is labeled S for the dependent variable, salary. For this reason, the intercept on the vertical axis is the S-intercept, not the y-intercept. It is also indicated on the axes what x and S represent. Labeling your axes is always a good practice.

Quick ✓

12. The cost, C, of renting a 12-foot moving truck for a day is \$40 plus \$0.35 times the number of miles driven. The linear function $C(x) = 0.35x + 40$ describes the cost C of driving the truck x miles.

(a) What is the domain of this linear function?

(b) Determine the C-intercept of the graph of the linear function.

(c) What is the rental cost if the truck is driven 80 miles?

(d) How many miles was the truck driven if the rental cost is \$85.50?

(e) Graph the linear function.

(f) How many miles can you drive if you can spend up to \$127.50?

❸ Build Linear Models from Verbal Descriptions

A linear function has the form $f(x) = mx + b$, where m is the slope of the linear function and $(0, b)$ is its y-intercept. In Section 3.IR4, we learned that slope can be thought of as an average rate of change. Slope describes how much a dependent variable changes for a given change in the independent variable. For example, in the linear function $f(x) = 4x + 3$, the slope is $4 = \dfrac{4}{1} = \dfrac{\Delta y}{\Delta x}$, so the dependent variable y increases by 4 units for every 1-unit increase in x (the independent variable). When the average rate of change of a function is constant, a linear function is used to model the situation. For example, if your phone company charges $0.05 per minute to talk regardless of the number of minutes on the phone, then a linear function can be used to model the cost of talking, with slope $m = \dfrac{0.05 \text{ dollar}}{1 \text{ minute}}$.

EXAMPLE 4 Cost Function

In the linear cost function $C(x) = ax + b$, b represents the fixed costs of operating a business, and a represents the costs associated with manufacturing one additional item. Suppose that a bicycle manufacturer has daily fixed costs of $2000 and each bicycle costs $80 to manufacture.

(a) Write a linear function showing the cost to manufacture x bicycles in a day.

(b) What is the cost to manufacture 5 bicycles in a day?

(c) How many bicycles can be manufactured for $2800?

(d) Graph the linear function.

Solution

(a) Because each bicycle costs $80 to manufacture, $a = 80$. The fixed costs are $2000, so $b = 2000$. Therefore, the cost function is

$$C(x) = 80x + 2000$$

(b) Evaluate the function for $x = 5$.

$$C(5) = 80(5) + 2000$$
$$= \$2400$$

It will cost $2400 to manufacture 5 bicycles.

(c) Solve $C(x) = 2800$.

$$C(x) = 2800$$
$$80x + 2000 = 2800$$

Subtract 2000 from both sides: $\quad 80x = 800$

Divide both sides by 80: $\quad x = 10$

Ten bicycles can be manufactured for $2800.

(d) Label the horizontal axis x and the vertical axis C. See Figure 20 for the graph of the cost function.

Figure 20

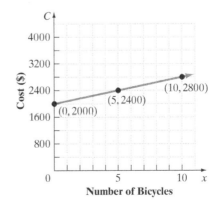

Quick ✓

13. Suppose the business presented in Example 4 must pay a tax of $1 per bicycle produced.

(a) Write a linear function that expresses the cost C of producing x bicycles in a day.

(b) What is the cost of manufacturing 5 bicycles in a day?

(c) How many bicycles can be manufactured for $2810?

(d) Graph the linear function.

EXAMPLE 5   **Straight-Line Depreciation**

The *book value* of an asset such as a building or piece of machinery is the value that the company uses to create its balance sheet. Some companies use *straight-line depreciation* so that the book value of the asset declines by a constant amount each year. The amount of the decline depends on the useful life that the company places on the asset. Suppose Pearson Publishing Company just purchased a new fleet of cars for its sales force at a cost of $29,400 per car. The company uses the straight-line depreciation method for 7 years.

(a) Write a linear function that expresses the book value V of each car as a function of its age, x.

(b) What is the domain of this linear function?

(c) What is the book value of each car after 3 years?

(d) When will the book value of each car be $12,600?

(e) Graph the linear function.

Solution

(a) Let the linear function $V(x) = mx + b$ represent the book value of each car after x years. The original value of each car is $29,400, so $V(0) = 29,400$. Thus the V-intercept of the function is $(0, 29,400)$. After 7 years, the book value of the car is $0. Use the ordered pairs $(0, 29,400)$ and $(7, 0)$ to find the slope, or amount of yearly (annual) depreciation, of V:

$$m = \frac{0 - 29{,}400}{7 - 0} = \frac{-29{,}400}{7} = -4200$$

So each car depreciates by $4200 per year. The linear function that represents the book value of each car after x years is

$$V(x) = -4200x + 29{,}400$$

(b) A car cannot have a negative age, so the age x must be greater than or equal to zero. In addition, each car is depreciated over 7 years, at which time $V(7) = 0$. Therefore, the domain of the function is $\{x \mid 0 \leq x \leq 7\}$, or $[0, 7]$ using interval notation.

(c) The book value of each car after $x = 3$ years is given by $V(3)$.

$$V(3) = -4200(3) + 29{,}400$$
$$= \$16{,}800$$

(d) To find when the book value is $12,600, solve the equation
$$V(x) = 12,600$$
$$-4200x + 29,400 = 12,600$$
Subtract 29,400 from both sides: $\quad -4200x = -16,800$
Divide both sides by -4200: $\quad x = 4$

Each car will have a book value of $12,600 in 4 years.

(e) Label the horizontal axis x and the vertical axis V. Since $V(0) = 29,400$, the point $(0, 29,400)$ is on the graph. Since $V(7) = 0$, the point $(7, 0)$ is on the graph. To graph the function, we use these intercepts, along with the points $(3, 16,800)$ and $(4, 12,600)$ from parts (c) and (d). See Figure 21.

Figure 21

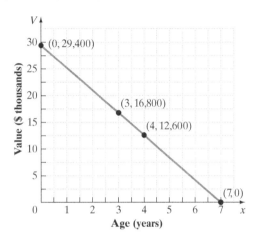

Quick ✓

14. Roberta's monthly payment for her new car is $250 per month. She estimates that maintenance and gas cost her $0.18 per mile.
  (a) Write a linear function that expresses the monthly cost C of operating the car as a function of miles driven, x.
  (b) What is the domain of this linear function?
  (c) What is the monthly cost of driving 320 miles?
  (d) How many miles can Roberta drive each month if she can afford the monthly cost to be $282.40?
  (e) Graph the linear function.

A.IR4 Exercises   MyLab Statistics

Underlined exercises have complete video solutions in MyLab.

Problems 1–14 are the Quick ✓ s that follow the EXAMPLES.

Building Skills

For Problems 15–26, graph each linear function. See Objective 1.

15. $F(x) = 5x - 2$
16. $F(x) = 4x + 1$
17. $G(x) = -3x + 7$
18. $G(x) = -2x + 5$
19. $H(x) = -2$
20. $P(x) = 5$
21. $f(x) = \frac{1}{2}x - 4$
22. $f(x) = \frac{1}{3}x - 3$
23. $F(x) = -\frac{5}{2}x + 5$
24. $P(x) = -\frac{3}{5}x - 1$
25. $G(x) = -\frac{3}{2}x$
26. $f(x) = \frac{4}{5}x$

In Problems 27–34, find the zero of the linear function. See Objective 2.

27. $f(x) = 2x + 10$
28. $f(x) = 3x + 18$
29. $G(x) = -5x + 40$
30. $H(x) = -4x + 36$
31. $s(t) = \frac{1}{2}t - 3$
32. $p(q) = \frac{1}{4}q + 2$
33. $P(z) = -\frac{4}{3}z + 12$
34. $F(t) = -\frac{3}{2}t + 6$

In Problems 35–38, determine whether the scatter diagram indicates that a linear relation may exist between the two variables. If a linear relation does exist, indicate whether the slope is positive or negative. See Objective 4.

35.
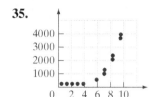

36.
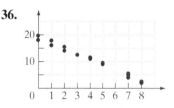

37.
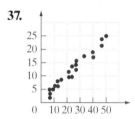

38.
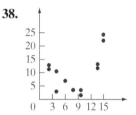

Mixed Practice

39. Suppose that $f(x) = 3x + 2$.
 (a) What is the slope?
 (b) What is the y-intercept?
 (c) What is the zero of f?
 (d) Solve $f(x) = 5$. What point is on the graph of f?
 (e) Solve $f(x) \leq -1$.
 (f) Graph f.

40. Suppose that $g(x) = 8x + 3$.
 (a) What is the slope?
 (b) What is the y-intercept?
 (c) What is the zero of g?
 (d) Solve $g(x) = 19$. What point is on the graph of g?
 (e) Solve $g(x) > -5$.
 (f) Graph g.

41. Suppose that $f(x) = x - 5$ and $g(x) = -3x + 7$.
 (a) Solve $f(x) = g(x)$. What is the value of f at the solution? What point is on the graph of f? What point is on the graph of g?
 (b) Solve $f(x) > g(x)$.
 (c) Graph f and g in the same Cartesian plane. Label the intersection point.

42. Suppose that $f(x) = \dfrac{4}{3}x + 5$ and $g(x) = \dfrac{1}{3}x + 1$.
 (a) Solve $f(x) = g(x)$. What is the value of f at the solution? What point is on the graph of f? What point is on the graph of g?
 (b) Solve $f(x) \leq g(x)$.
 (c) Graph f and g in the same Cartesian plane. Label the intersection point.

43. Find a linear function f such that $f(2) = 6$ and $f(5) = 12$. What is $f(-2)$?

44. Find a linear function g such that $g(1) = 5$ and $g(5) = 17$. What is $g(-3)$?

45. Find a linear function h such that $h(3) = 7$ and $h(-1) = 14$. What is $h\left(\dfrac{1}{2}\right)$?

46. Find a linear function F such that $F(2) = 5$ and $F(-3) = 9$. What is $F\left(-\dfrac{3}{2}\right)$?

47. In parts (a)–(e), use the figure shown below.

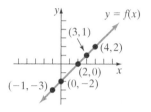

 (a) Solve $f(x) = 1$.
 (b) Solve $f(x) = -3$.
 (c) What is $f(4)$?
 (d) What are the intercepts of the function $y = f(x)$?
 (e) Write the equation of the function whose graph is given in the form $f(x) = mx + b$.

48. In parts (a)–(e), use the figure shown below.

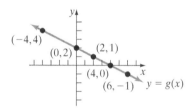

 (a) Solve $g(x) = 1$.
 (b) Solve $g(x) = -1$.
 (c) What is $g(-4)$?
 (d) What are the intercepts of the function $y = g(x)$?
 (e) Write the equation of the function whose graph is given in the form $g(x) = mx + b$.

Applying the Concepts

49. Cab Fare The linear function $C(m) = 1.5m + 2$ describes the cab fare C for a ride of m miles.
 (a) What is the domain of this linear function?
 (b) What is $C(0)$? Explain what this result means.
 (c) What is cab fare for a 5-mile ride?
 (d) Graph the linear function.

(e) How many miles can you ride in a cab if you have $13.25?

(f) Over what range of miles can you ride if you can spend no more than $39.50?

50. Sales Commissions Tanya works for Pearson Education as a book representative. The linear function $I(s) = 0.01s + 20{,}000$ describes the annual income I of Tanya when she has total sales s.

(a) What is the domain of this linear function?

(b) What is $I(0)$? Explain what this result means.

(c) What is Tanya's salary if she sells $500,000 in books for the year?

(d) Graph the linear function.

(e) At what level of sales will Tanya's income be $45,000?

51. Taxes The function $T(x) = 0.15(x - 9226) + 922.50$ represents the federal income tax bill T of a single person whose adjusted gross income in 2015 was x dollars for income between $9226 and $37,450, inclusive. SOURCE: *Internal Revenue Service*

(a) What is the domain of this linear function?

(b) What was a single filer's tax bill if adjusted gross income was $20,000?

(c) Which variable is independent and which is dependent?

(d) Graph the linear function over the domain specified in part (a).

(e) What was a single filer's adjusted gross income if his or her tax bill was $2996.25?

52. Luxury Tax In 2002, Major League Baseball signed a labor agreement with the players. In this agreement, any team whose payroll exceeds $189 million in 2015 will have to pay a luxury tax of 17.5% (for first offenses). The linear function $T(p) = 0.175(p - 189)$ describes the luxury tax T of a team whose payroll was p (in millions).

(a) What is the domain of this linear function?

(b) What was the luxury tax for the Boston Red Sox whose payroll was $200 million in 2015?

(c) Graph the linear function.

(d) What was the payroll of the San Francisco Giants, who paid a luxury tax of $1.3 million in 2015?

53. Health Costs The annual cost of health insurance H as a function of age a is given by the function $H(a) = 22.8a - 117.5$ for $15 \leq a \leq 90$. SOURCE: *Statistical Abstract*

(a) What are the independent and dependent variables?

(b) What is the domain of this linear function?

(c) What is the health insurance premium of a 30-year-old?

(d) Graph the linear function over its domain.

(e) What is the age of an individual whose health insurance premium is $976.90?

54. Birth Rate A multiple birth is any birth with 2 or more children born. The birth rate is the number of births per 1000 women. The birth rate B of multiple births as a function of age a is given by the function $B(a) = 1.73a - 14.56$ for $15 \leq a \leq 44$. SOURCE: *Centers for Disease Control*

(a) What are the independent and dependent variables?

(b) What is the domain of this linear function?

(c) What is the multiple birth rate of women who are 22 years of age, according to the model?

(d) Graph the linear function over its domain.

(e) What is the age of women whose multiple birth rate is 49.45?

55. Phone Charges Sprint has a long-distance phone plan that charges a monthly fee of $5.95 plus $0.05 per minute. SOURCE: Sprint.com

(a) Find a linear function that expresses the monthly bill B as a function of minutes used m.

(b) What are the independent and dependent variables?

(c) What is the domain of this linear function?

(d) What is the monthly bill if 300 minutes are used for long-distance phone calls?

(e) How many minutes were used for long distance if the long-distance phone bill was $17.95?

(f) Graph the linear function.

(g) Over what range of minutes can you talk each month if you don't want to spend more than $18.45?

56. RV Rental The weekly rental cost R of a class C 20-foot recreational vehicle is $129 plus $0.32 per mile, up to a maximum of 500 miles. SOURCE: *westernrv.com*

(a) Find a linear function that expresses the cost R as a function of miles driven m.

(b) What are the independent and dependent variables?

(c) What is the domain of this linear function?

(d) What is the rental cost if 360 miles are driven?

(e) How many miles were driven if the rental cost is $275.56?

(f) Graph the linear function.

(g) Over what range of miles can you drive if you have a budget of $273?

57. Depreciation Suppose that a company has just purchased a new computer for $2700. The company chooses to depreciate the computer using the straight-line method over 3 years.

(a) Find a linear function that expresses the book value V of the computer as a function of its age x.

(b) What is the domain of this linear function?

(c) What is the book value of the computer after the first year?

(d) What are the intercepts of the graph of the linear function?

(e) When will the book value of the computer be $900?

(f) Graph the linear function.

58. Depreciation Suppose that a company just purchased a new machine for its manufacturing facility for $1,200,000. The company chooses to depreciate the machine using the straight-line method over 20 years.

(a) Find a linear function that expresses the book value V of the machine as a function of its age x.

(b) What is the domain of this linear function?

(c) What is the book value of the machine after three years?

(d) What are the intercepts of the graph of the linear function?

(e) When will the book value of the machine be $480,000?

(f) Graph the linear function.

Extending the Concepts

59. Math for the Future: Calculus The **average rate of change** of a function $y = f(x)$ from c to x is defined as

$$\text{Average rate of change} = \frac{\Delta y}{\Delta x} = \frac{f(x) - f(c)}{x - c}, \quad x \neq c$$

provided that c is in the domain of f. The average rate of change of a function is simply the slope of the line joining the points $(c, f(c))$ and $(x, f(x))$. The line joining these points is called a **secant line**. The slope of the secant line is

$$m_{\text{sec}} = \frac{f(x) - f(c)}{x - c}$$

The following figure illustrates the idea.

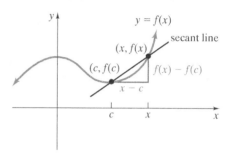

Below, we show the graph of the function $f(x) = 2x^2 - 4x + 1$.

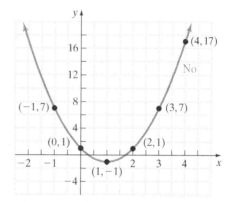

(a) On the graph of $f(x) = 2x^2 - 4x + 1$, draw a line through the points $(1, f(1))$ and $(x, f(x))$, where $x = 4$.

(b) Find the slope of the secant line through $(1, f(1))$ and $(x, f(x))$, where $x = 4$.

(c) Find the equation of the secant line through $(1, f(1))$ and $(x, f(x))$, where $x = 4$.

(d) Repeat parts (a)–(c) for $x = 3, x = 2, x = 1.5$, and $x = 1.1$.

(e) What happens to the slope of the secant line as x gets closer to 1?

60. A strain of *E. coli* Beu 397-recA441 is placed into a Petri dish at 30° Celsius and allowed to grow. The population is estimated by means of an optical device in which the amount of light that passes through the Petri dish is measured. The data below are collected. Do you think that a linear function could be used to describe the relation between the two variables? Why or why not?

Time, x	Population, y
0	0.09
2.5	0.18
3.5	0.26
4.5	0.35
6	0.50

SOURCE: Dr. Polly Lavery, Joliet Junior College

Getting Ready for Exponential and Logarithmic Functions
Laws of Exponents

Objectives

1. Simplify Exponential Expressions Using the Product Rule
2. Simplify Exponential Expressions Using the Quotient Rule
3. Evaluate Exponential Expressions with a Zero or Negative Exponent
4. Simplify Exponential Expressions Using the Power Rule
5. Simplify Exponential Expressions Containing Products or Quotients

Work Smart

Read a^n as "a raised to the power of n" or "a raised to the nth power." We usually read a^2 as "a squared" and a^3 as "a cubed."

Before we can discuss exponential and logarithmic functions, we must discuss some laws of exponents.

Recall that if a is a real number and n is a positive integer, then the symbol a^n means that a should be used as a factor n times. That is,

$$a^n = \underbrace{a \cdot a \cdot \cdots \cdot a}_{n \text{ factors}}$$

For example,

$$4^3 = \underbrace{4 \cdot 4 \cdot 4}_{3 \text{ factors}}$$

In the notation a^n, a is called the **base** and n the **power** or **exponent**.

① Simplify Exponential Expressions Using the Product Rule

Several general rules can be discovered for simplifying expressions involving *positive* integer exponents. The first rule that we introduce is used when multiplying two exponential expressions that have the same base. Consider the following:

$$\underset{\text{Same base}}{x^2} \cdot x^4 = \underbrace{(x \cdot x)}_{2 \text{ factors}} \underbrace{(x \cdot x \cdot x \cdot x)}_{4 \text{ factors}} = \underbrace{x \cdot x \cdot x \cdot x \cdot x \cdot x}_{6 \text{ factors}} = \underset{\text{Same base}}{x^6}$$

Sum of powers 2 and 4

Based on the above, the following results:

In Other Words

When multiplying two exponential expressions with the same base, add the exponents. Then write the common base to the power of this sum.

Product Rule for Exponents (Positive Integer Exponents)
If a is a real number and m and n are positive integers, then
$$a^m a^n = a^{m+n}$$

EXAMPLE 1 | **Using the Product Rule to Simplify Expressions Involving Exponents**

Simplify each expression. All answers should contain only positive integer exponents.

▶ (a) $2^2 \cdot 2^3$ ▶ (b) $3z^2 \cdot 4z^4$

Solution

(a) $2^2 \cdot 2^3 = 2^{2+3}$
$\qquad = 2^5$
$\qquad = 32$

(b) $3z^2 \cdot 4z^4 = 3 \cdot 4 \cdot z^2 \cdot z^4$
$\qquad = 12z^{2+4}$
$\qquad = 12z^6$

Quick ✓

1. In the notation a^n we call a the _____ and n the _____ or _____.
2. If a is a real number, and m and n are positive integers, then $a^m a^n =$ _____.

In Problems 3–7, simplify each expression. All answers should contain only positive integer exponents.

3. $5^2 \cdot 5$
4. $(-3)^2(-3)^3$
5. $y^4 y^3$
6. $(5x^2)(-2x^5)$
7. $6y^3(-y^2)$

❷ Simplify Exponential Expressions Using the Quotient Rule

To find a general rule for the quotient of two exponential expressions with *positive* integer exponents, use the Reduction Property to divide out common factors. Consider the following:

$$\frac{y^6}{y^2} = \frac{\overbrace{y \cdot y \cdot y \cdot y \cdot y \cdot y}^{6 \text{ factors}}}{\underbrace{y \cdot y}_{2 \text{ factors}}} = \underbrace{y \cdot y \cdot y \cdot y}_{4 \text{ factors}} = y^{\overset{\text{Difference of powers}}{\overset{6 \text{ and } 2 \downarrow}{4}}}$$

Conclude from this result that

$$\frac{y^6}{y^2} = y^{6-2} = y^4$$

This result is true in general.

In Other Words
When dividing two exponential expressions with a common base, subtract the exponent in the denominator from the exponent in the numerator. Then write the common base to the power of this difference.

Quotient Rule for Exponents (Positive Integer Exponents)
If a is a real number and m and n are positive integers, then

$$\frac{a^m}{a^n} = a^{m-n} \quad \text{if } a \neq 0$$

EXAMPLE 2 **Using the Quotient Rule to Simplify Expressions Involving Exponents**

Simplify each expression. Answers should contain only positive integer exponents.

(a) $\dfrac{8^5}{8^3}$ 	(b) $\dfrac{27z^9}{12z^4}$

Solution

(a) $\dfrac{8^5}{8^3} = 8^{5-3}$ 	(b) $\dfrac{27z^9}{12z^4} = \dfrac{9 \cdot 3}{4 \cdot 3} z^{9-4}$

$\phantom{(a) \dfrac{8^5}{8^3}} = 8^2$ 	$\phantom{(b) \dfrac{27z^9}{12z^4}} = \dfrac{9}{4} z^5$

$\phantom{(a) \dfrac{8^5}{8^3}} = 64$

Quick ✓

8. *True or False* To divide two exponential expressions having the same base, keep the base and subtract the exponents.

In Problems 9–12, simplify each expression. All answers should contain only positive integer exponents.

9. $\dfrac{5^6}{5^4}$ 	10. $\dfrac{y^8}{y^6}$ 	11. $\dfrac{16a^6}{10a^5}$ 	12. $\dfrac{-24b^5}{16b^3}$

❸ Evaluate Exponential Expressions with a Zero or Negative Exponent

Now extend the definition of exponential expressions to *all* integer exponents. That is, evaluate exponential expressions where the exponent can be a positive integer, zero, or a negative integer. Begin with raising a real number to the 0 power.

▶ Zero-Exponent Rule
If a is a nonzero real number, then

$$a^0 = 1 \quad \text{if } a \neq 0$$

This rule is based on the Product Rule and the Identity Property of Multiplication. From the Product Rule for Exponents, it is known that

$$a^0 a^n = a^{0+n}$$
$$= a^n$$
$$= 1 \cdot a^n$$

From the Identity Property of Multiplication, this means that $a^0 = 1$.

Suppose we want to simplify $\dfrac{z^3}{z^5}$. If the Quotient Rule for Exponents is used,

$$\frac{z^3}{z^5} = z^{3-5} = z^{-2}$$

This expression could also be simplified using the Reduction Property.

$$\frac{z^3}{z^5} = \frac{\cancel{z} \cdot \cancel{z} \cdot \cancel{z}}{\cancel{z} \cdot \cancel{z} \cdot \cancel{z} \cdot z \cdot z} = \frac{1}{z^2}$$

This implies that $z^{-2} = \dfrac{1}{z^2}$, which suggests that a raised to a negative power is defined as follows:

Negative-Exponent Rule

If n is a positive integer and if a is a nonzero real number, then

$$a^{-n} = \frac{1}{a^n} \quad \text{or} \quad \frac{1}{a^{-n}} = a^n \text{ if } a \neq 0$$

EXAMPLE 3 **Evaluating Exponential Expressions Containing Integer Exponents**

Simplify each expression. All exponents should be positive integers.

(a) 3^{-4} (b) $\dfrac{1}{3^{-2}}$ (c) $5x^0$ (d) $4x^{-5}$

Solution

(a) $3^{-4} = \dfrac{1}{3^4}$ (b) $\dfrac{1}{3^{-2}} = 3^2$ (c) $5x^0 = 5 \cdot 1$ (d) $4x^{-5} = \dfrac{4}{x^5}$

$\qquad = \dfrac{1}{81}$ $\qquad = 9$ $\qquad = 5$

Work Smart

Remember, $3^{-4} = \dfrac{1}{3^4}$ (not -3^4).
The negative applies only to the exponent. Similarly,
$4x^{-5} = \dfrac{4}{x^5}$ $\left(\text{not } \dfrac{1}{4x^5}\right)$. The negative exponent applies only to the base "x" not to the coefficient.

Quick ✓

13. $a^0 =$ ___, provided $a \neq$ ___.

14. $a^{-n} =$ ___, provided $a \neq$ ___.

In Problems 15–20, simplify each expression. All exponents should be positive integers.

15. 5^{-3}

16. $5z^{-7}$

17. $\dfrac{1}{x^{-4}}$

18. $\dfrac{5}{y^{-3}}$

19. -4^0

20. $(-10)^0$

EXAMPLE 4 **Evaluating Exponential Expressions Containing Integer Exponents**

Simplify each expression. All exponents should be positive integers.

(a) $\left(\dfrac{2}{3}\right)^{-3}$

(b) $\left(\dfrac{1}{7}\right)^{-2}$

Solution

(a) $\left(\dfrac{2}{3}\right)^{-3} = \dfrac{1}{\left(\dfrac{2}{3}\right)^3}$

$= \dfrac{1}{\dfrac{2}{3} \cdot \dfrac{2}{3} \cdot \dfrac{2}{3}}$

$= \dfrac{1}{\dfrac{8}{27}}$

$= \dfrac{27}{8}$

(b) $\left(\dfrac{1}{7}\right)^{-2} = \dfrac{1}{\left(\dfrac{1}{7}\right)^2}$

$= \dfrac{1}{\dfrac{1}{7} \cdot \dfrac{1}{7}}$

$= \dfrac{1}{\dfrac{1}{49}}$

$= 49$

> **In Other Words**
> To evaluate $\left(\dfrac{a}{b}\right)^{-n}$, determine the reciprocal of the base and then raise it to the *n*th power.

The following shortcut is based on the results of Example 4:

If a and b are real numbers and n is an integer, then

$$\left(\dfrac{a}{b}\right)^{-n} = \left(\dfrac{b}{a}\right)^n \quad \text{if } a \neq 0, b \neq 0$$

Quick ✓

In Problems 21–24, simplify each expression. All exponents should be positive integers.

21. $\left(\dfrac{4}{3}\right)^{-2}$ **22.** $\left(-\dfrac{1}{4}\right)^{-3}$ **23.** $\left(\dfrac{3}{x}\right)^{-2}$ **24.** $\dfrac{5}{2^{-2}}$

Now that we have definitions for 0 as an exponent and negative exponents, the Product Rule and Quotient Rule for Exponents can be restated, assuming that the exponent is any integer (positive, negative, or zero).

Product Rule for Exponents

If a is a real number and m and n are integers, then

$$a^m a^n = a^{m+n}$$

If m, n, or $m + n$ is 0 or negative, then a cannot be 0.

Quotient Rule for Exponents

If a is a real number and if m and n are integers, then

$$\dfrac{a^m}{a^n} = a^{m-n} \text{ if } a \neq 0$$

Notice that allowing the exponents to be any integer (not just any positive integer) requires restrictions on the value of the base.

EXAMPLE 5 **Using the Product Rule to Simplify Expressions Containing Exponents**

Simplify each expression. All exponents should be positive integers.

(a) $(-3)^2(-3)^{-4}$ (b) $\dfrac{3}{4}y^5 \cdot \dfrac{20}{9}y^{-2}$

Solution

(a) $(-3)^2(-3)^{-4} = (-3)^{2+(-4)}$
$= (-3)^{-2}$
$= \dfrac{1}{(-3)^2}$
$= \dfrac{1}{9}$

(b) $\dfrac{3}{4}y^5 \cdot \dfrac{20}{9}y^{-2} = \dfrac{3}{4} \cdot \dfrac{20}{9} y^{5+(-2)}$
$= \dfrac{5}{3}y^3$

EXAMPLE 6 **Using the Quotient Rule to Simplify Expressions Containing Exponents**

Simplify each expression. All exponents should be positive integers.

(a) $\dfrac{w^{-2}}{w^{-5}}$

(b) $\dfrac{20a^3b}{4ab^4}$

Solution

(a) $\dfrac{w^{-2}}{w^{-5}} = w^{-2-(-5)}$
$= w^{-2+5}$
$= w^3$

(b) $\dfrac{20a^3b}{4ab^4} = 5a^{3-1}b^{1-4}$
$= 5a^2b^{-3}$
$a^{-n} = \dfrac{1}{a^n} = \dfrac{5a^2}{b^3}$

Quick ✓

In Problems 25–30, simplify each expression. All exponents should be positive integers.

25. $6^3 \cdot 6^{-5}$

26. $\dfrac{10^{-3}}{10^{-5}}$

27. $(4x^2y^3)(5xy^{-4})$

28. $\left(\dfrac{3}{4}a^3b\right)\left(\dfrac{8}{9}a^{-2}b^3\right)$

29. $\dfrac{-24b^5}{16b^{-3}}$

30. $\dfrac{50s^2t}{15s^5t^{-4}}$

▶ ④ Simplify Exponential Expressions Using the Power Rule

Another law of exponents applies when an exponential expression containing a power is itself raised to a power.

$(3^2)^4 = \underbrace{3^2 \cdot 3^2 \cdot 3^2 \cdot 3^2}_{\text{4 factors}} = \underbrace{(3\cdot3)}_{\text{2 factors}}\underbrace{(3\cdot3)}_{\text{2 factors}}\underbrace{(3\cdot3)}_{\text{2 factors}}\underbrace{(3\cdot3)}_{\text{2 factors}} = 3^8$

$2 \cdot 4 = 8$ factors

The following results:

In Other Words

If an exponential expression contains a power raised to a power, keep the base and multiply the powers.

Power Rule for Exponential Expressions

If a is a real number and m and n are integers, then

$$(a^m)^n = a^{mn}$$

If m or n is 0 or negative, then a must not be 0.

EXAMPLE 7 **Using the Power Rule to Simplify Exponential Expressions**

Simplify each expression. All exponents should be positive integers.

(a) $(y^3)^5$

(b) $[(-3)^3]^2$

(c) $(6^3)^0$

Solution

(a) $(y^3)^5 = y^{3\cdot 5}$
$= y^{15}$

(b) $[(-3)^3]^2 = (-3)^{3\cdot 2}$
$= (-3)^6$
$= 729$

(c) $(6^3)^0 = 6^{3\cdot 0}$
$= 6^0$
$= 1$

Quick ✓

In Problems 31–36, simplify each expression. All exponents should be positive integers.

31. $(2^2)^3$ **32.** $(5^8)^0$ **33.** $[(-4)^3]^2$

34. $(a^3)^5$ **35.** $(z^3)^{-6}$ **36.** $(s^{-3})^{-7}$

⑤ Simplify Exponential Expressions Containing Products or Quotients

There are two additional laws of exponents. The first deals with raising a product to a power, and the second deals with raising a quotient to a power. Consider the following product to a power:

$$(xy)^3 = (xy)(xy)(xy)$$
$$= (x \cdot x \cdot x)(y \cdot y \cdot y)$$
$$= x^3 y^3$$

The following results:

Work Smart

Do not use this rule to try to simplify $(a + b)^2$ as $a^2 + b^2$ or $(a + b)^3$ as $a^3 + b^3$. For this rule to apply, the base must be the *product* of two numbers—not a sum.

Product-to-a-Power Rule

If a and b are real numbers and n is an integer, then

$$(ab)^n = a^n b^n$$

If n is 0 or negative, neither a nor b can be 0.

EXAMPLE 8   Using the Product-to-a-Power Rule to Simplify Exponential Expressions

Simplify each expression. All exponents should be positive integers.

(a) $(3z)^4$ (b) $(-5y^{-2})^{-3}$ (c) $(-4a^2)^{-2}$

Solution

(a) $(3z)^4 = 3^4 z^4$
$= 81 z^4$

(b) $(-5y^{-2})^{-3} = (-5)^{-3}(y^{-2})^{-3}$
$= \dfrac{y^{-2(-3)}}{(-5)^3}$
$= \dfrac{y^6}{-125}$
$= -\dfrac{y^6}{125}$

(c) $(-4a^2)^{-2} = \dfrac{1}{(-4a^2)^2}$
$= \dfrac{1}{(-4)^2(a^2)^2}$
$= \dfrac{1}{16a^4}$

Quick ✓

In Problems 37–40, simplify each expression. All exponents should be positive integers.

37. $(5y)^3$ **38.** $(6y)^0$ **39.** $(3x^2)^4$ **40.** $(4a^3)^{-2}$

Now look at a quotient raised to a power:
$$\left(\frac{2}{3}\right)^4 = \left(\frac{2}{3}\right)\left(\frac{2}{3}\right)\left(\frac{2}{3}\right)\left(\frac{2}{3}\right) = \frac{2^4}{3^4}$$
The following results:

Quotient-to-a-Power Rule

If a and b are real numbers and n is an integer, then
$$\left(\frac{a}{b}\right)^n = \frac{a^n}{b^n} \quad \text{if } b \neq 0$$
If n is negative or 0, then a cannot be 0.

EXAMPLE 9 Using the Quotient-to-a-Power Rule to Simplify Exponential Expressions

Simplify each expression. All exponents should be positive integers.

(a) $\left(\dfrac{w}{4}\right)^3$ (b) $\left(\dfrac{2x^2}{y^3}\right)^4$

Solution

(a) $\left(\dfrac{w}{4}\right)^3 = \dfrac{w^3}{4^3}$

$= \dfrac{w^3}{64}$

(b) $\left(\dfrac{2x^2}{y^3}\right)^4 = \dfrac{(2x^2)^4}{(y^3)^4}$

$= \dfrac{2^4(x^2)^4}{(y^3)^4}$

$= \dfrac{16x^{2\cdot 4}}{y^{3\cdot 4}}$

$= \dfrac{16x^8}{y^{12}}$

Quick ✓

In Problems 41–44, simplify each expression. All exponents should be positive integers.

41. $\left(\dfrac{z}{3}\right)^4$ **42.** $\left(\dfrac{x}{2}\right)^{-5}$ **43.** $\left(\dfrac{x^2}{y^3}\right)^4$ **44.** $\left(\dfrac{3a^{-2}}{b^4}\right)^3$

The Laws of Exponents are now summarized.

Work Smart: Study Skills

When learning new rules, it is helpful to write each rule on a notecard that you can study with regularly. For example, you could put a^0, $a \neq 0$ on one side of a notecard and its value of 1 on the other side. This will allow you to quiz yourself. There is also a free app called Quizlet that may be used for electronic flash cards.

The Laws of Exponents

If a and b are real numbers and if m and n are integers, then assuming the expression is defined,

Zero-Exponent Rule: $\quad a^0 = 1 \quad$ if $a \neq 0$

Negative-Exponent Rule: $\quad a^{-n} = \dfrac{1}{a^n} \quad$ if $a \neq 0$

Product Rule: $\quad a^m a^n = a^{m+n}$

Quotient Rule: $\quad \dfrac{a^m}{a^n} = a^{m-n} \quad$ if $a \neq 0$

Power Rule: $\quad (a^m)^n = a^{mn}$

Product-to-a-Power Rule: $\quad (ab)^n = a^n b^n$

Quotient-to-a-Power Rule: $\quad \left(\dfrac{a}{b}\right)^n = \dfrac{a^n}{b^n} \quad$ if $b \neq 0$

Quotient-to-a-Negative-Power Rule: $\quad \left(\dfrac{a}{b}\right)^{-n} = \left(\dfrac{b}{a}\right)^n \quad$ if $a \neq 0, b \neq 0$

Getting Ready Exercises — MyLab Statistics

Problems 1–44 are the Quick ✓s that follow the EXAMPLES.

Mixed Practice

In Problems 45–76, simplify each expression. All exponents should be positive integers.

45. -5^2
46. 5^{-2}
47. -5^{-2}
48. -5^0
49. $-8^2 \cdot 8^{-2}$
50. $\dfrac{8^7}{8^5} \cdot 8^{-2}$
51. $\left(\dfrac{4}{9}\right)^{-2}$
52. $\left(\dfrac{3}{4}\right)^{-3}$
53. $(-3)^2(-3)^{-5}$
54. $(-4)^{-5}(-4)^3$
55. $\dfrac{(-4)^2}{(-4)^{-1}}$
56. $\dfrac{(-3)^3}{(-3)^{-2}}$
57. $\dfrac{2^3 \cdot 3^{-2}}{2^{-2} \cdot 3^{-4}}$
58. $\dfrac{3^{-2} \cdot 5^3}{3^2 \cdot 5}$
59. $(6x)^3(6x)^{-3}$
60. $(5a^2)^5(5a^2)^{-5}$
61. $(2s^{-2}t^4)(-5s^2 t)$
62. $(6ab)(3a^3 b^{-4})$
63. $\left(\dfrac{1}{4}xy\right) \cdot (20xy^{-2})$
64. $(3xy^3)\left(\dfrac{1}{9}x^2 y\right)$
65. $\dfrac{36x^7 y^3}{9x^5 y^2}$
66. $\dfrac{25a^2 b^3}{5ab^6}$
67. $\dfrac{21a^2 b}{14a^3 b^{-2}}$
68. $\dfrac{25x^{-2} y}{10xy^3}$
69. $(x^{-2})^4$
70. $(z^2)^{-6}$
71. $(3x^2 y)^3$
72. $(5a^2 b^{-1})^2$
73. $\left(\dfrac{z}{4}\right)^{-3}$
74. $\left(\dfrac{x}{y}\right)^{-8}$
75. $(3a^{-3})^{-2}$
76. $(2y^{-2})^{-4}$

Applying the Concepts

△ **77. Cubes** Suppose the length of a side of a cube is x^2. Find the volume of the cube in terms of x.

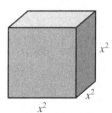

△ **78. Circles** The radius r of a circle is $\dfrac{d}{2}$, where d is the diameter. The area of a circle is given by the formula $A = \pi r^2$. Find the area of a circle in terms of its diameter d.

In Problems 79–82, simplify each algebraic expression by rewriting each factor with a common base. (Hint: Consider that $8 = 2^3$.)

79. $\dfrac{2^{x+3}}{4}$
80. $\dfrac{3^{2x}}{27}$
81. $3^x \cdot 27^{3x+1}$
82. $9^{-x} \cdot 3^{x+1}$
83. If $3^x = 5$, what does 3^{4x} equal?
84. If $4^x = 6$, what does 4^{5x} equal?
85. If $2^x = 7$, what does 2^{-4x} equal?
86. If $5^x = 3$, what does 5^{-3x} equal?

Explaining the Concepts

87. A friend of yours has a homework problem in which he must simplify $(x^4)^3$. He tells you that he thinks the answer is x^7. Is he right? If not, explain where he went wrong.

88. A friend of yours is convinced that x^0 must equal 0. Write an explanation that details why $x^0 = 1$. Include any restrictions that must be placed on x.

89. Explain why a cannot be 0 when $m, n,$ or $m + n$ is negative or 0 in the expression a^{m+n}. Use examples to support your explanation.

90. Explain why a cannot be 0 when n is negative or 0 in the expression $(a^m)^n$. Use examples to support your explanation.

91. Explain why neither a nor b can be 0 when n is 0 or negative in the expression $(ab)^n = a^n b^n$.

92. Provide a justification for the Product Rule for Exponents.

93. Provide a justification for the Quotient Rule for Exponents.

94. Provide a justification for the Power Rule for Exponents.

95. Provide a justification for the Product-to-a-Power Rule for Exponents.

A.IR5 Exponential Functions

Objectives
1. Evaluate Exponential Expressions
2. Graph Exponential Functions
3. Define the Number e
4. Solve Exponential Equations
5. Use Exponential Models That Describe Our World

Are You Prepared for This Section?
Before getting started, complete the following problems. If you get a problem wrong, go back to the section cited and review the material.

P1. Evaluate: **(a)** 2^3 **(b)** 2^{-1} **(c)** 3^4 [Getting Ready, pp. IR-263–IR-266]

P2. Graph: $f(x) = x^2$ [Section A.IR3, p. IR-244]

P3. State the definition of a rational number. [Section 1.IR3, p. IR-17]

P4. State the definition of an irrational number. [Section 1.IR3, p. IR-17]

P5. Write 3.20349193 as a decimal **(a)** rounded to four decimal places **(b)** truncated to four decimal places. [Section 1.IR2, pp. IR-13–IR-14]

P6. Solve: $3x + 2 = 14$ [Section 2.IR4, pp. IR-91–IR-95]

Prepared?...Answers
P1. (a) 8 **(b)** $\frac{1}{2}$ **(c)** 81

P2.

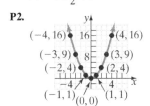

P3. A rational number is a number that can be expressed as a quotient $\frac{p}{q}$ of two integers. The integer p is called the numerator, and the integer q, which cannot be 0, is called the denominator. The set of rational numbers are the numbers $\mathbb{Q} = \left\{ x \mid x = \frac{p}{q}, \text{where } p, q \text{ are integers and } q \neq 0 \right\}$.

P4. An irrational number has a decimal representation that neither repeats nor terminates.

P5. (a) 3.2035 **(b)** 3.2034

P6. (a) m^8 **(b)** a^5 **(c)** z^{12}

P7. {4}

Suppose editor Suzanna Bainbridge has just hired you as a proofreader for Pearson Publishing. Suzanna offers you two options: Option A states that you will be paid $100 for each error you find in the final page proofs of a text. Option B states that you will start with $1 and your payment will double for each error you find in the final page proofs. You know from experience that the final page proofs of a text typically have about 15–20 errors. Which option will you choose?

If there is one error, Option A pays $100, while Option B pays $2. For two errors, Option A pays 2($100) = $200, while Option B pays 2^2 = $4. For three errors, Option A pays 3($100) = $300, while Option B pays 2^3 = $8. Option A seems to be the way to go. To complete the analysis, set up Table 5, which lists the payment amount as a function of the number of errors in the page proofs. Remember, in Option B, the payment amount doubles each time you find an error.

Table 5

Number of Errors	Option A Payment	Option B Payment	Number of Errors	Option A Payment	Option B Payment
0	$0	$1	11	$1100	$2048
1	$100	$2	12	$1200	$4096
2	$200	$4	13	$1300	$8192
3	$300	$8	14	$1400	$16,384
4	$400	$16	15	$1500	$32,768
5	$500	$32	16	$1600	$65,536
6	$600	$64	17	$1700	$131,072
7	$700	$128	18	$1800	$262,144
8	$800	$256	19	$1900	$524,288
9	$900	$512	20	$2000	$1,048,576
10	$1000	$1024			

Holy cow! If you find 20 errors, you will get paid over a million dollars! Suzanna Bainbridge better reconsider her offer! If x represents the number of errors, Option A can be expressed as a linear function, $f(x) = 100x$; Option B can be expressed as an *exponential function*, $g(x) = 2^x$.

Definition

An **exponential function** is a function of the form

$$f(x) = a^x$$

where $a \neq 1$ is a positive real number $(a > 0)$. The domain of the exponential function is the set of all real numbers.

The restrictions on the base a will be addressed shortly. The key point is that *the independent variable is in the exponent of the exponential expression.* Contrast this idea with polynomial functions (such as $f(x) = x^2 - 4x$ or $g(x) = 2x^3 + x^2 - 5$), where the independent variable is the base of each exponential expression.

▶ ❶ Evaluate Exponential Expressions

In the 'Getting Ready' section, a^n was defined, where the base a is a positive real number and the exponent n is an integer.

But what if you want to raise the base a to a real number? The answer to this question requires advanced mathematics, so we will use a calculator.

To evaluate expressions of the form a^x using a scientific calculator, enter the base a, press the $\boxed{x^y}$ key, enter the exponent x, and press $\boxed{=}$. To evaluate expressions of the form a^x using a graphing calculator, enter the base a, press the caret $\boxed{\wedge}$ key, enter the exponent x, and press $\boxed{\text{ENTER}}$.

EXAMPLE 1 **Evaluating Exponential Expressions**

Using a calculator, evaluate each of the following expressions. Write as many places as your calculator allows.

(a) $3^{1.4}$ (b) $3^{1.41}$ (c) $3^{1.414}$

(d) $3^{1.4142}$ (e) $3^{\sqrt{2}}$

Solution

(a) $3^{1.4} \approx 4.655536722$ (b) $3^{1.41} \approx 4.706965002$

(c) $3^{1.414} \approx 4.727695035$ (d) $3^{1.4142} \approx 4.72873393$

(e) $3^{\sqrt{2}} \approx 4.728804388$

Quick ✓

1. An exponential function is a function of the form $f(x) = a^x$ where $a \_\_ 0$ and $a \_\_ 1$.

In Problem 2, use a calculator to evaluate each of the following expressions. Write as many places as your calculator allows.

2. (a) $2^{1.7}$ **(b)** $2^{1.73}$ **(c)** $2^{1.732}$ **(d)** $2^{1.7321}$ **(e)** $2^{\sqrt{3}}$

Example 1 illustrates that the value of an exponential expression at any real number can be approximated. This is why the domain of any exponential function is the set of all real numbers.

In the definition of an exponential function, $a = 1$ was ruled out, and it was required that a be positive. The base $a = 1$ is excluded because this function is the constant function $f(x) = 1^x = 1$. Exclude bases that are negative to avoid problems for exponents such as $\dfrac{1}{2}$ or $\dfrac{3}{4}$. For example, suppose $f(x) = (-2)^x$. In the real number system, $f\left(\dfrac{1}{2}\right)$ could not be evaluated because $f\left(\dfrac{1}{2}\right) = (-2)^{\frac{1}{2}} = \sqrt{-2}$, which is not a real number. Finally, exclude a base of 0, because this function is $f(x) = 0^x$, which equals zero for $x > 0$ and is undefined when $x < 0$. When $x = 0, f(x) = 0^x$ is *indeterminate* because its value is not precisely determined.

❷ Graph Exponential Functions

Let's learn about properties of exponential functions from their graphs.

EXAMPLE 2 **Graphing an Exponential Function**

Graph the exponential function $f(x) = 2^x$ using point plotting. From the graph, state the domain and the range of the function.

Solution
Begin by locating some points on the graph of $f(x) = 2^x$ as shown in Table 6. Plot the points in Table 6 and connect them in a smooth curve. Figure 22 shows the graph of $f(x) = 2^x$.

Table 6

x	$f(x) = 2^x$	$(x, f(x))$
-3	$f(-3) = 2^{-3} = \dfrac{1}{2^3} = \dfrac{1}{8}$	$\left(-3, \dfrac{1}{8}\right)$
-2	$f(-2) = 2^{-2} = \dfrac{1}{2^2} = \dfrac{1}{4}$	$\left(-2, \dfrac{1}{4}\right)$
-1	$f(-1) = 2^{-1} = \dfrac{1}{2^1} = \dfrac{1}{2}$	$\left(-1, \dfrac{1}{2}\right)$
0	$f(0) = 2^0 = 1$	$(0, 1)$
1	$f(1) = 2^1 = 2$	$(1, 2)$
2	$f(2) = 2^2 = 4$	$(2, 4)$
3	$f(3) = 2^3 = 8$	$(3, 8)$

Figure 22

The domain of any exponential function is the set of all real numbers. Notice that 2^x is greater than or equal to 0 for all the x-values in Table 6. Based on this and the graph, the range of $f(x) = 2^x$ is the set of all positive real numbers $\{y | y > 0\}$, or $(0, \infty)$ in interval notation.

The graph of $f(x) = 2^x$ in Figure 22 is typical of all exponential functions that have a base larger than 1. Figure 23 shows the graphs of two other exponential functions whose bases are larger than 1: $y = 3^x$ and $y = 6^x$. Notice that the larger the base, the steeper the graph is for $x > 0$ and the closer the graph is to the x-axis for $x < 0$.

The information for $f(x) = a^x$, where $a > 1$, is summarized below.

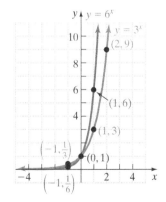

Figure 23

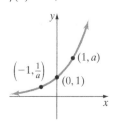

Figure 24
$f(x) = a^x, a > 1$

> **Properties of the Graph of an Exponential Function $f(x) = a^x, a > 1$**
>
> 1. The domain is the set of all real numbers. The range is the set of all positive real numbers.
> 2. There are no x-intercepts; the y-intercept is $(0, 1)$.
> 3. The graph of f contains the points $\left(-1, \dfrac{1}{a}\right)$, $(0, 1)$, and $(1, a)$.
>   See Figure 24.

Quick ✓

3. Graph the exponential function $f(x) = 4^x$ using point plotting. From the graph, state the domain and the range of the function.

▶ Now consider $f(x) = a^x, 0 < a < 1$.

EXAMPLE 3   **Graphing an Exponential Function**

Graph the exponential function $f(x) = \left(\dfrac{1}{2}\right)^x$ using point plotting. From the graph, state the domain and the range of the function.

Solution

Begin by locating some points on the graph of $f(x) = \left(\dfrac{1}{2}\right)^x$ as shown in Table 7. Plot the points in Table 7 and connect them in a smooth curve. Figure 25 shows the graph of $f(x) = \left(\dfrac{1}{2}\right)^x$.

Table 7

x	$f(x) = \left(\dfrac{1}{2}\right)^x$	$(x, f(x))$
-3	$f(-3) = \left(\dfrac{1}{2}\right)^{-3} = 2^3 = 8$	$(-3, 8)$
-2	$f(-2) = \left(\dfrac{1}{2}\right)^{-2} = 2^2 = 4$	$(-2, 4)$
-1	$f(-1) = \left(\dfrac{1}{2}\right)^{-1} = 2^1 = 2$	$(-1, 2)$
0	$f(0) = \left(\dfrac{1}{2}\right)^0 = 1$	$(0, 1)$
1	$f(1) = \left(\dfrac{1}{2}\right)^1 = \dfrac{1}{2}$	$\left(1, \dfrac{1}{2}\right)$
2	$f(2) = \left(\dfrac{1}{2}\right)^2 = \dfrac{1}{4}$	$\left(2, \dfrac{1}{4}\right)$
3	$f(3) = \left(\dfrac{1}{2}\right)^3 = \dfrac{1}{8}$	$\left(3, \dfrac{1}{8}\right)$

Figure 25

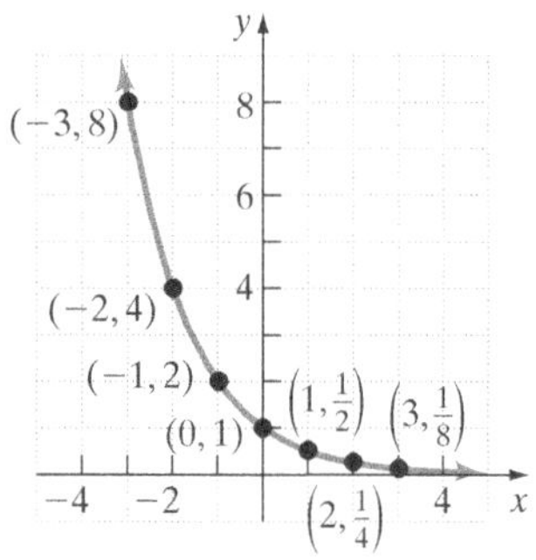

The domain of any exponential function is the set of all real numbers. From the graph, the range of $f(x) = \left(\dfrac{1}{2}\right)^x$ is the set of all positive real numbers $\{y \mid y > 0\}$, or $(0, \infty)$ in interval notation. ●

The graph of $f(x) = \left(\dfrac{1}{2}\right)^x$ in Figure 25 is typical of all exponential functions that have a base between 0 and 1. Figure 26 shows the graphs of two additional exponential functions whose bases are between 0 and 1: $y = \left(\dfrac{1}{3}\right)^x$ and $y = \left(\dfrac{1}{6}\right)^x$. Notice that the smaller the base, the closer the graph is to the x-axis for $x > 0$ and the steeper the graph is for $x < 0$.

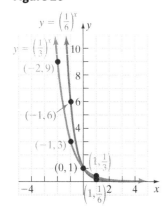

Figure 26

Section A.IR5 Exponential Functions IR-275

The information about $f(x) = a^x$, where $0 < a < 1$, is now summarized.

Figure 27
$f(x) = a^x, 0 < a < 1$

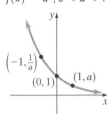

Properties of the Graph of an Exponential Function $f(x) = a^x, 0 < a < 1$

1. The domain is the set of all real numbers. The range is the set of all positive real numbers.
2. There are no x-intercepts; the y-intercept is $(0, 1)$.
3. The graph of f contains the points $\left(-1, \dfrac{1}{a}\right)$, $(0, 1)$, and, $(1, a)$.
   See Figure 27.

Quick ✓

4. The graph of every exponential function $f(x) = a^x$ passes through three points: ___, ___, and ___.
5. *True or False* The domain of the exponential function $f(x) = a^x, a > 0, a \neq 1$, is the set of all real numbers.
6. *True or False* The range of the exponential function $f(x) = a^x, a > 0, a \neq 1$, is the set of all real numbers.
7. Graph the exponential function $f(x) = \left(\dfrac{1}{4}\right)^x$ using point plotting. From the graph, state the domain and the range of the function.

EXAMPLE 4 **Graphing an Exponential Function**

Use point plotting to graph $f(x) = 3^{x+1}$. From the graph, state the domain and the range of the function.

Solution
Choose values of x and find the corresponding values of the function. See Table 8. Then plot the ordered pairs and connect them in a smooth curve. See Figure 28.

Table 8

x	$f(x)$	$(x, f(x))$
-3	$f(-3) = 3^{-3+1} = 3^{-2} = \dfrac{1}{3^2} = \dfrac{1}{9}$	$\left(-3, \dfrac{1}{9}\right)$
-2	$f(-2) = 3^{-2+1} = 3^{-1} = \dfrac{1}{3^1} = \dfrac{1}{3}$	$\left(-2, \dfrac{1}{3}\right)$
-1	$f(-1) = 3^{-1+1} = 3^0 = 1$	$(-1, 1)$
0	$f(0) = 3^{0+1} = 3^1 = 3$	$(0, 3)$
1	$f(1) = 3^{1+1} = 3^2 = 9$	$(1, 9)$

Figure 28

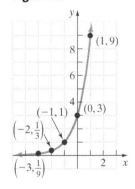

The domain is the set of all real numbers. The range is $\{y | y > 0\}$, or $(0, \infty)$ in interval notation.

Quick ✓

In Problems 8 and 9, graph each function using point plotting. From the graph, state the domain and the range of each function.

8. $f(x) = 2^{x-1}$
9. $f(x) = 3^x + 1$

3 Define the Number e

Many models use an exponential function whose base is an irrational number symbolized by the letter e. The number e can be used to model the growth of a stock's price or to estimate the time of death of a carbon-based life form.

Definition

The **number e** is defined as the number that the expression

$$\left(1 + \frac{1}{n}\right)^n$$

approaches as n increases.

Table 9 shows some values of $\left(1 + \dfrac{1}{n}\right)^n$ as n increases. The last number in column 4 is the number e correct to nine decimal places.

Table 9

n	$\dfrac{1}{n}$	$1 + \dfrac{1}{n}$	$\left(1 + \dfrac{1}{n}\right)^n$
1	1	2	2
2	0.5	1.5	2.25
5	0.2	1.2	2.48832
10	0.1	1.1	2.59374246
100	0.01	1.01	2.704813829
1,000	0.001	1.001	2.716923932
10,000	0.0001	1.0001	2.718145927
100,000	0.00001	1.00001	2.718268237
1,000,000	0.000001	1.000001	2.718280469
1,000,000,000	10^{-9}	$1 + 10^{-9}$	2.718281827

The exponential function $f(x) = e^x$, whose base is the number e, occurs so often in applications that it is sometimes called *the* exponential function. Most calculators have the key $\boxed{e^x}$ or $\boxed{\exp(x)}$, which you can use to evaluate the exponential function $f(x) = e^x$ for a given value of x.

Use your calculator to approximate the values of $f(x) = e^x$ for $x = -1, 0,$ and 1 as was done to create Table 10. The graph of the exponential function is shown in Figure 29(a). Because $2 < e < 3$, the graph of $f(x) = e^x$ lies between the graph of $y = 2^x$ and that of $y = 3^x$. See Figure 29(b).

Table 10

x	$f(x) = e^x$
-2	$e^{-2} \approx 0.135$
-1	$e^{-1} \approx 0.368$
0	$e^0 = 1$
1	$e^1 \approx 2.718$
2	$e^2 \approx 7.389$

Figure 29
$f(x) = e^x$

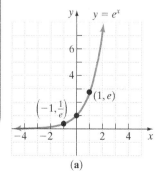
(a)

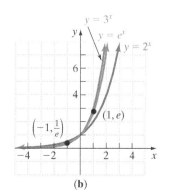
(b)

Quick ✓

10. What is the value of e rounded to five decimal places?
11. Evaluate each of the following rounded to three decimal places:
    (a) e^4
    (b) e^{-4}

④ Solve Exponential Equations

Equations that involve terms of the form a^x, $a > 0$, $a \neq 1$, are called **exponential equations.** Some exponential equations can be solved using the Laws of Exponents and the following property.

Work Smart

In order to use the Property for Solving Exponential Equations, both sides of the equation must have the *same* base.

> **Property For Solving Exponential Equations**
> If $a^u = a^v$, then $u = v$.

This property results from the fact that exponential functions are *one-to-one*. In exponential functions, any output is the result of one (and only one) input. That is, two different inputs cannot yield the same output. Contrast this with $y = x^2$, which is not one-to-one because two different inputs, -2 and 2, correspond to the same output, 4.

EXAMPLE 5 How to Solve an Exponential Equation

Solve: $2^{x-3} = 32$

Step-by-Step Solution

Step 1: Use the Laws of Exponents to write both sides of the equation with the same base.

$32 = 2^5$:
$$2^{x-3} = 32$$
$$2^{x-3} = 2^5$$

Step 2: Set the exponents on each side of the equation equal to each other.

If $a^u = a^v$, then $u = v$: $\quad x - 3 = 5$

Step 3: Solve the equation resulting from Step 2.

Add 3 to both sides: $\quad x - 3 + 3 = 5 + 3$
$$x = 8$$

Step 4: Verify your solution(s).

Work Smart

Did you notice that properties of algebra were used to reduce the exponential equation to a linear equation?

Let $x = 8$:
$$2^{x-3} = 32$$
$$2^{8-3} \stackrel{?}{=} 32$$
$$2^5 \stackrel{?}{=} 32$$
$$32 = 32 \quad \text{True}$$

The solution set is $\{8\}$.

> **Solving Exponential Equations of the Form $a^u = a^v$**
>
> **Step 1:** Use the Laws of Exponents to write both sides of the equation with the same base.
> **Step 2:** Set the exponents on each side of the equation equal to each other.
> **Step 3:** Solve the equation resulting from Step 2.
> **Step 4:** Verify the solution(s).

Quick ✓

In Problems 12 and 13, solve each equation.

12. $5^{x-4} = 5^{-1}$
13. $3^{x+2} = 81$

⑤ Use Exponential Models That Describe Our World

Exponential functions are used in many different disciplines, such as biology (half-life), chemistry (carbon dating), economics (time value of money), and psychology (learning curves). Exponential functions are also used in statistics, as shown in the next example.

EXAMPLE 6 Exponential Probability

The manager of a crisis helpline knows that between 3:00 A.M. and 5:00 A.M., 3 calls per hour occur (that's 0.05 call per minute). The following formula from statistics gives the likelihood that a call will occur within t minutes of 3 A.M.

$$F(t) = 1 - e^{-0.05t}$$

Determine the likelihood that a person will call

(a) within 5 minutes of 3:00 A.M.

(b) within 20 minutes of 3:00 A.M.

Solution

(a) The likelihood that a call will occur within 5 minutes of 3:00 A.M. is found by evaluating the function $F(t) = 1 - e^{-0.05t}$ at $t = 5$.

$$F(5) = 1 - e^{-0.05(5)}$$
$$\approx 0.221$$

The likelihood that a call will occur in this time span is $0.221 = 22.1\%$.

(b) The likelihood that a call will occur within 20 minutes of 3:00 A.M. is found by evaluating the function $F(t) = 1 - e^{-0.05t}$ at $t = 20$, or $F(20)$.

$$F(20) = 1 - e^{-0.05(20)}$$
$$\approx 0.632$$

The likelihood that a call will occur in this time span is $0.632 = 63.2\%$.

Quick ✓

14. A bank manager knows that between 3:00 P.M. and 5:00 P.M., 15 people arrive per hour (that's 0.25 people per minute). The following formula from statistics gives the likelihood that a person will arrive within t minutes of 3:00 P.M.

$$F(t) = 1 - e^{-0.25t}$$

Find the likelihood that a person will arrive

(a) within 10 minutes of 3 P.M.   (b) within 25 minutes of 3 P.M.

EXAMPLE 7 Radioactive Decay

The radioactive **half-life** for a given radioisotope of an element is the time it takes for half the radioactive nuclei in any sample to decay to some other substance. For example, the half-life of plutonium-239 is 24,360 years. Plutonium-239 is particularly dangerous because it emits alpha particles that are absorbed into bone marrow. The maximum amount of plutonium-239 an adult can handle without significant injury is 0.13 microgram ($= 0.000000013$ gram). Suppose a researcher has a 1-gram sample of plutonium-239. The amount A (in grams) of plutonium-239 after t years is given by

$$A(t) = 1\left(\frac{1}{2}\right)^{\frac{t}{24{,}360}}$$

How much plutonium-239 is left in the sample after

(a) 500 years?   (b) 24,360 years?   (c) 73,080 years?

Solution

(a) The amount of plutonium-239 left in the sample after 500 years is found by evaluating A at $t = 500$. That is, determine $A(500)$.

$$A(500) = 1\left(\frac{1}{2}\right)^{\frac{500}{24,360}}$$

Use a calculator: ≈ 0.986 gram

After 500 years, approximately 0.986 gram of plutonium-239 will be left in the sample.

(b) The amount of plutonium-239 remaining after 24,360 years is found by evaluating A at $t = 24,360$. That is, determine $A(24,360)$.

$$A(24,360) = 1\left(\frac{1}{2}\right)^{\frac{24,360}{24,360}}$$
$$= 1\left(\frac{1}{2}\right)^{1}$$
$$= 0.5 \text{ gram}$$

After 24,360 years, there will be 0.5 gram of plutonium-239 left.

(c) To find the amount left after 73,080 years, evaluate $A(73,080)$.

$$A(73,080) = 1\left(\frac{1}{2}\right)^{\frac{73,080}{24,360}}$$
$$= 1\left(\frac{1}{2}\right)^{3}$$
$$= \frac{1}{8} \text{ gram}$$

After 73,080 years, there will be $\frac{1}{8} = 0.125$ gram of plutonium-239 left in the sample.

Quick ✓

15. The half-life of thorium-227 is 18.72 days. Suppose that a researcher possesses a 10-gram sample of thorium-227. The amount A (in grams) of thorium-227 left after t days is given by

$$A(t) = 10 \cdot \left(\frac{1}{2}\right)^{\frac{t}{18.72}}$$

How much thorium-227 is left in the sample after

(a) 10 days? (b) 18.72 days?
(c) 74.88 days? (d) 100 days?

When money is deposited in a bank, the bank pays interest on the balance in the account. When solving interest problems, we use the term **payment period** as shown in Table 11 to indicate how often the bank pays interest.

Table 11

Payment Period	Number of Times Interest Is Paid
Annually	Once per year
Semiannually	Twice per year
Quarterly	4 times per year
Monthly	12 times per year
Daily	360 times per year

When the interest due at the end of a payment period is added to the principal so that the interest computed at the end of the *next* payment period is based on this new principal amount (old principal + interest), the interest has been **compounded**. **Compound interest** is interest paid on the original principal and all previously earned interest.

The following formula can be used to determine the value of an account after a certain period of time.

Compound Interest Formula

The amount A after t years resulting from a principal P invested at an annual interest rate r compounded n times per year is

$$A = P\left(1 + \frac{r}{n}\right)^{nt}$$

Work Smart

When using the compound interest formula, be sure to express the interest rate as a decimal.

For example, if you deposit $500 into an account paying 3% annual interest compounded monthly, then $P = \$500$, $r = 0.03$, and $n = 12$ (twelve compounding periods per year).

EXAMPLE 8 — Future Value of Money

Suppose you deposit $3000 into a Roth IRA today. If the deposit earns 8% interest compounded quarterly, determine its future value A after

(a) 1 year

(b) 10 years

(c) 35 years, when you plan on retiring

Solution

Use the compound interest formula with $P = \$3000$, $r = 0.08$, and $n = 4$ (for quarterly compounding), so

$$A = \$3000\left(1 + \frac{0.08}{4}\right)^{4t}$$
$$= \$3000(1 + 0.02)^{4t}$$
$$= \$3000(1.02)^{4t}$$

(a) The value of the account after $t = 1$ year is

$$A = \$3000(1.02)^{4(1)}$$
$$= \$3000(1.02)^{4}$$

Use a calculator; see Figure 30:
$$= \$3000(1.08243216)$$
$$= \$3247.30$$

Figure 30

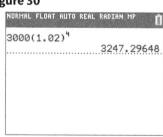

(b) The value after $t = 10$ years is

$$A = \$3000(1.02)^{4(10)}$$
$$= \$3000(1.02)^{40}$$

Use a calculator:
$$= \$3000(2.208039664)$$
$$= \$6624.12$$

(c) The value after $t = 35$ years is

$$A = \$3000(1.02)^{4(35)}$$
$$= \$3000(1.02)^{140}$$

Use a calculator:
$$= \$3000(15.99646598)$$
$$= \$47{,}989.40$$

Quick ✓

16. Suppose that you deposit $2000 into a Roth IRA today. Determine the future value A of the deposit if it earns 5% interest compounded monthly (twelve times per year) after

(a) 1 year (b) 15 years (c) 30 years, when you plan on retiring

A.IR5 Exercises — MyLab Statistics

Underlined exercises have complete video solutions in MyLab.

Problems 1–16 are the Quick ✓s that follow the EXAMPLES.

Building Skills

In Problems 17–20, approximate each number using a calculator. Express your answer rounded to three decimal places. See Objective 1.

17. (a) $3^{2.2}$ (b) $3^{2.23}$ (c) $3^{2.236}$
  (d) $3^{2.2361}$ (e) $3^{\sqrt{5}}$

18. (a) $5^{1.4}$ (b) $5^{1.41}$ (c) $5^{1.414}$
  (d) $5^{1.4142}$ (e) $5^{\sqrt{2}}$

19. (a) $4^{3.1}$ (b) $4^{3.14}$ (c) $4^{3.142}$
  (d) $4^{3.1416}$ (e) 4^{π}

20. (a) $10^{2.7}$ (b) $10^{2.72}$ (c) $10^{2.718}$
  (d) $10^{2.7183}$ (e) 10^{e}

In Problems 21–28, the graph of an exponential function is given. Match each graph to one of the following functions. It may prove useful to create a table of values for each function to help you identify the correct graph. See Objective 2.

(a) $f(x) = 2^x$ (b) $f(x) = 2^{-x}$ (c) $f(x) = 2^{x+1}$
(d) $f(x) = 2^{x-1}$ (e) $f(x) = -2^x$ (f) $f(x) = 2^x + 1$
(g) $f(x) = 2^x - 1$ (h) $f(x) = -2^{-x}$

21.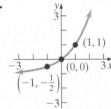

22.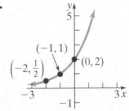

23.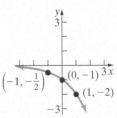

24.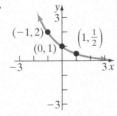

25.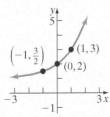

26.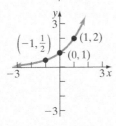

27.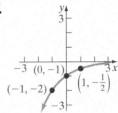

28.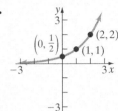

In Problems 29–40, graph each function. State the domain and the range of the function. See Objective 2.

29. $f(x) = 5^x$ **30.** $f(x) = 7^x$

31. $F(x) = \left(\dfrac{1}{5}\right)^x$ **32.** $F(x) = \left(\dfrac{1}{7}\right)^x$

33. $h(x) = 2^{x+2}$ **34.** $H(x) = 2^{x-2}$

35. $f(x) = 2^x + 3$ **36.** $F(x) = 2^x - 3$

37. $F(x) = \left(\dfrac{1}{2}\right)^x - 1$ **38.** $G(x) = \left(\dfrac{1}{2}\right)^x + 2$

39. $P(x) = \left(\dfrac{1}{3}\right)^{x-2}$ **40.** $p(x) = \left(\dfrac{1}{3}\right)^{x+2}$

In Problems 41–48, approximate each number using a calculator. Express your answer rounded to three decimal places. See Objective 3.

41. (a) $3.1^{2.7}$ (b) $3.14^{2.72}$ (c) $3.142^{2.718}$
  (d) $3.1416^{2.7183}$ (e) π^e

42. (a) $2.7^{3.1}$ (b) $2.72^{3.14}$ (c) $2.718^{3.142}$
  (d) $2.7183^{3.1416}$ (e) e^{π}

43. e^2 **44.** e^3 **45.** e^{-2}

46. e^{-3} **47.** $e^{2.3}$ **48.** $e^{1.5}$

In Problems 49–52, graph each function. State the domain and the range of the function. See Objective 3.

49. $g(x) = e^{x-1}$ **50.** $f(x) = e^x - 1$

51. $f(x) = -2e^x$ **52.** $F(x) = \dfrac{1}{2}e^x$

In Problems 53–72, solve each equation. See Objective 4.

53. $2^x = 2^5$

54. $3^x = 3^{-2}$

55. $3^{-x} = 81$

56. $4^{-x} = 64$

57. $\left(\dfrac{1}{2}\right)^x = \dfrac{1}{32}$

58. $\left(\dfrac{1}{3}\right)^x = \dfrac{1}{243}$

59. $5^{x-2} = 125$

60. $2^{x+3} = 128$

61. $4^x = 8$

62. $9^x = 27$

63. $2^{-x+5} = 16^x$

64. $3^{-x+4} = 27^x$

65. $2^x \cdot 8 = 4^{x-3}$

66. $3^x \cdot 9 = 27^x$

67. $\left(\dfrac{1}{5}\right)^x - 25 = 0$

68. $\left(\dfrac{1}{6}\right)^x - 36 = 0$

69. $e^x = e^{3x+4}$

70. $e^{3x} = e^2$

71. $(e^x)^2 = e^{3x-2}$

72. $(e^3)^x = e^2 \cdot e^x$

Mixed Practice

73. Suppose that $f(x) = 2^x$.
 (a) What is $f(3)$? What point is on the graph of f?
 (b) If $f(x) = \dfrac{1}{8}$, what is x? What point is on the graph of f?

74. Suppose that $f(x) = 3^x$.
 (a) What is $f(2)$? What point is on the graph of f?
 (b) If $f(x) = \dfrac{1}{81}$, what is x? What point is on the graph of f?

75. Suppose that $g(x) = 4^x - 1$.
 (a) What is $g(-1)$? What point is on the graph of g?
 (b) If $g(x) = 15$, what is x? What point is on the graph of g?

76. Suppose that $g(x) = 5^x + 1$.
 (a) What is $g(-1)$? What point is on the graph of g?
 (b) If $g(x) = 126$, what is x? What point is on the graph of g?

77. Suppose that $H(x) = 3\left(\dfrac{1}{2}\right)^x$.
 (a) What is $H(-3)$? What point is on the graph of H?
 (b) If $H(x) = \dfrac{3}{4}$, what is x? What point is on the graph of H?

78. Suppose that $F(x) = -2\left(\dfrac{1}{3}\right)^x$.
 (a) What is $F(-1)$? What point is on the graph of F?
 (b) If $F(x) = -18$, what is x? What point is on the graph of F?

Applying the Concepts

79. A Population Model According to the U.S. Census Bureau, the population of the United States in 2016 was 326 million people. In addition, the population of the United States was growing at a rate of 0.99% per year. Assuming that this growth rate continues, the model $P(t) = 326(1.0099)^{t-2016}$ represents the population P (in millions of people) in year t.

 (a) According to this model, what will be the population of the United States in 2021?
 (b) According to this model, what will be the population of the United States in 2050?
 (c) The United States Census Bureau predicts that the United States population will be 439 million in 2050. Compare this estimate to the one obtained in part (b). What might account for any differences?

80. A Population Model According to the U.S. Census Bureau, the population of the world in 2016 was 7563 million people. In addition, the population of the world was growing at a rate of 1.02% per year. Assuming that this growth rate continues, the model $P(t) = 7563(1.0102)^{t-2016}$ represents the population P (in millions of people) in year t.

 (a) According to this model, what will be the population of the world in 2021?
 (b) According to this model, what will be the population of the world in 2025?
 (c) The United States Census Bureau predicts that the world population will be 8000 million (8 billion) in 2025. Compare this estimate to the one obtained in part (b). What might account for any differences?

81. Time Is Money Suppose that you deposit $5000 into a certificate of deposit (CD) today. Determine the future value A of the deposit if it earns 2% interest compounded monthly after
 (a) 1 year.
 (b) 3 years.
 (c) 5 years, when the CD comes due.

82. Time Is Money Suppose that you deposit $8000 into a certificate of deposit (CD) today. Determine the future value A of the deposit if it earns 2% interest compounded quarterly (4 times per year) after
 (a) 1 year.
 (b) 3 years.
 (c) 5 years, when the CD comes due.

83. Do the Compounding Periods Matter? Suppose that you deposit $2000 into an account that pays 3% annual interest. How much will you have after 5 years if interest is compounded

(a) annually? (b) quarterly?
(c) monthly? (d) daily?

(e) Based on the results of parts (a)–(d), what impact does increasing the number of compounding periods have on the future value, all other things equal?

84. Do the Compounding Periods Matter? Suppose that you deposit $1000 into an account that pays 6% annual interest. How much will you have after 3 years if interest is compounded

(a) annually? (b) quarterly?
(c) monthly? (d) daily?

(e) Based on the results of parts (a)–(d), what impact does increasing the number of compounding periods have on the future value, all other things equal?

85. Depreciation Based on data obtained from the *Kelley Blue Book*, the value V of a Ford Focus that is t years old can be modeled by $V(t) = 19{,}841(0.88)^t$.

(a) According to the model, what is the value of a brand-new Focus?

(b) According to the model, what is the value of a 2-year-old Focus?

(c) According to the model, what is the value of a 5-year-old Focus?

86. Depreciation Based on data obtained from the *Kelley Blue Book*, the value V of a Chevy Malibu that is t years old can be modeled by $V(t) = 25{,}258(0.84)^t$.

(a) According to the model, what is the value of a brand-new Chevy Malibu?

(b) According to the model, what is the value of a 2-year-old Chevy Malibu?

(c) According to the model, what is the value of a 5-year-old Chevy Malibu?

87. Radioactive Decay The half-life of beryllium-11 is 13.81 seconds. Suppose that a researcher possesses a 100-gram sample of beryllium-11. The amount A (in grams) of beryllium-11 after t seconds is given by

$$A(t) = 100\left(\frac{1}{2}\right)^{\frac{t}{13.81}}$$

(a) How much beryllium-11 is left in the sample after 1 second?

(b) How much beryllium-11 is left in the sample after 13.81 seconds?

(c) How much beryllium-11 is left in the sample after 27.62 seconds?

(d) How much beryllium-11 is left in the sample after 100 seconds?

88. Radioactive Decay The half-life of carbon-10 is 19.255 seconds. Suppose that a researcher possesses a 100-gram sample of carbon-10. The amount A (in grams) of carbon-10 after t seconds is given by

$$A(t) = 100\left(\frac{1}{2}\right)^{\frac{t}{19.255}}$$

(a) How much carbon-10 is left in the sample after 1 second?

(b) How much carbon-10 is left in the sample after 19.255 seconds?

(c) How much carbon-10 is left in the sample after 38.51 seconds?

(d) How much carbon-10 is left in the sample after 100 seconds?

For Exercises 89 and 90, use Newton's Law of Cooling, which states that the temperature of a heated object decreases exponentially over time toward the temperature of the surrounding medium.

89. Newton's Law of Cooling Suppose a pizza is removed from a 400°F oven and placed in a room whose temperature is 70°F. The temperature u (in °F) of the pizza at time t (in minutes) can be modeled by $u(t) = 70 + 330e^{-0.072t}$.

(a) According to the model, what will be the temperature of the pizza in 5 minutes?

(b) According to the model, what will be the temperature of the pizza in 10 minutes?

(c) If the pizza can be safely consumed when its temperature is 200°F, will it be ready to eat after cooling for 13 minutes?

90. Newton's Law of Cooling Suppose coffee at a temperature of 170°F is poured into a coffee mug and allowed to cool in a room whose temperature is 70°F. The temperature u (in °F) of the coffee at time t (in minutes) can be modeled by $u(t) = 70 + 100e^{-0.045t}$.

(a) According to the model, what will be the temperature of the coffee in 5 minutes?

(b) According to the model, what will be the temperature of the coffee in 10 minutes?

(c) If the coffee doesn't taste good once its temperature reaches 120°F, will it still be tasty after cooling for 20 minutes?

91. Learning Curve Suppose that a student has 200 vocabulary words to learn. If a student learns 20 words in 30 minutes, the function
$$L(t) = 200(1 - e^{-0.0035t})$$
models the number of words L that the student will learn in t minutes.

(a) How many words will the student learn in 45 minutes?

(b) How many words will the student learn in 60 minutes?

92. Learning Curve Suppose that a student has 50 biology terms to learn. If a student learns 10 terms in 30 minutes, the function
$$L(t) = 50(1 - e^{-0.0223t})$$
models the number of terms L that the student will learn in t minutes.

(a) How many words will the student learn in 45 minutes?

(b) How many words will the student learn in 60 minutes?

93. Current in an RL Circuit The equation governing the amount of current I (in amperes) after time t (in seconds) in a single RL circuit consisting of a resistance R (in ohms), an inductance L (in henrys), and an electromotive force E (in volts) is
$$I = \frac{E}{R}[1 - e^{-(R/L)t}]$$

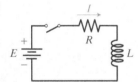

(a) If $E = 120$ volts, $R = 10$ ohms, and $L = 25$ henrys, how much current I is flowing in 0.05 second?

(b) If $E = 240$ volts, $R = 10$ ohms, and $L = 25$ henrys, how much current I is flowing in 0.05 second?

94. Current in an RC Circuit The equation governing the amount of current I (in amperes) after time t (in microseconds) in a single RC circuit consisting of a resistance R (in ohms), a capacitance C (in microfarads), and an electromotive force E (in volts) is
$$I = \frac{E}{R}e^{-t/(RC)}$$

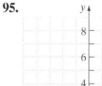

(a) If $E = 120$ volts, $R = 2500$ ohms, and $C = 100$ microfarads, how much current I is flowing initially ($t = 0$)? In 50 microseconds?

(b) If $E = 240$ volts, $R = 2500$ ohms, and $C = 100$ microfarads, how much current I is flowing initially ($t = 0$)? In 50 microseconds?

Extending the Concepts

In Problems 95 and 96, find the exponential function whose graph is given.

95.

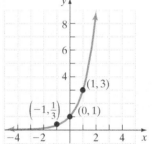

96.

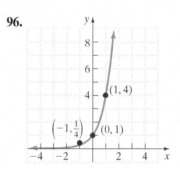

Explaining the Concepts

97. As the base a of an exponential function $f(x) = a^x$ increases (for $a > 1$), what happens to the graph of the exponential function for $x > 0$? What happens to the behavior of the graph for $x < 0$?

98. The graphs of $f(x) = 2^{-x}$ and $g(x) = \left(\dfrac{1}{2}\right)^x$ are identical. Why?

99. Can we solve the equation $2^x = 12$ using the fact that if $a^u = a^v$, then $u = v$? Why or why not?

A.IR6 Logarithmic Functions

Objectives

1. Change Exponential Equations to Logarithmic Equations
2. Change Logarithmic Equations to Exponential Equations
3. Evaluate Logarithmic Functions
4. Determine the Domain of a Logarithmic Function
5. Graph Logarithmic Functions
6. Work with Natural and Common Logarithms
7. Solve Logarithmic Equations
8. Use Logarithmic Models That Describe Our World

Work Smart

It is required that a be positive and not equal to 1 for the same reasons that these restrictions existed for the exponential function.

In Other Words

The logarithm to the base a of x is the number y that we must raise a to in order to obtain x.

Work Smart: Study Skills

In doing problems similar to Examples 1 and 2, it is a good idea to practice saying, "y equals the logarithm to the base a of x is equivalent to a to the y equals x" so that you memorize the definition of a logarithm.

Are You Prepared for This Section?

Before getting started, complete the following problems. If you get a problem wrong, go back to the section cited and review the material.

P1. Solve: $3x + 2 > 0$ [Section 4.IR3, pp. IR-179–IR-183]
P2. What are the square roots of 25? [Section 2.IR2, pp. IR-78–IR-79]
P3. Solve: $19 = 6x + 7$ [Section 2.IR4, pp. IR-93–IR-95]

Consider the "squaring function," $f(x) = x^2$, and the square root function, $g(x) = \sqrt{x}$. How are these two functions related? Well, $f(5) = 25$ and $g(25) = 5$. That is, the input to f is the output of g and the output of f is the input to g, so g "undoes" what f does.

In general, whenever a function is introduced in mathematics, there is a second function that "undoes" it. For example, a square root function undoes the squaring function. A cube root function undoes the cubing function. Is there a function that undoes an exponential function? Yes! This function is the *logarithmic function*.

Definition

The **logarithmic function to the base a**, where $a > 0$ and $a \neq 1$, is denoted by $y = \log_a x$ (read as "y is the logarithm to the base a of x"), and

$$y = \log_a x \quad \text{is equivalent to} \quad x = a^y$$

To evaluate logarithmic functions, convert them into their equivalent exponential form. Therefore, go from logarithmic form to exponential form, and back. For example,

$$0 = \log_3 1 \quad \text{is equivalent to} \quad 3^0 = 1$$
$$2 = \log_5 25 \quad \text{is equivalent to} \quad 5^2 = 25$$
$$-2 = \log_4 \frac{1}{16} \quad \text{is equivalent to} \quad 4^{-2} = \frac{1}{16}$$

Notice that the base of the logarithm is the base of the exponential; the argument of the logarithm is what the exponential equals; and the value of the logarithm is the exponent of the exponential expression. A logarithm is just another way of writing an exponential expression.

To see how the logarithmic function undoes the exponential function, consider the function $y = 2^x$. If the input is $x = 3$, then the output is $y = 2^3 = 8$. To undo this function would require that an input of 8 give an output of 3. If $2^3 = 8$, then $3 = \log_2 8$ using the definition of a logarithm. The input of $\log_2 8$ is 8, and its output is 3.

▶ 1 Change Exponential Equations to Logarithmic Equations

The definition of a logarithm can be used to rewrite exponential equations as logarithmic equations.

EXAMPLE 1 **Changing Exponential Equations to Logarithmic Equations**

Rewrite each exponential equation as an equivalent logarithmic equation.

(a) $6^2 = 36$ (b) $x^2 = 9$ (c) $4^3 = y$

Solution

Use the fact that $y = \log_a x$ is equivalent to $a^y = x$ provided that $a > 0$ and $a \neq 1$.

(a) If $6^2 = 36$, then $2 = \log_6 36$. (b) If $x^2 = 9$, then $2 = \log_x 9$.
(c) If $4^3 = y$, then $3 = \log_4 y$.

Prepared?...Answers

P1. $\left\{x \mid x > -\dfrac{2}{3}\right\}$ or $\left(-\dfrac{2}{3}, \infty\right)$

P2. -5 and 5 **P3.** $\{2\}$

Quick ✓

1. The logarithm to the base a of x, denoted $y = \log_a x$, can be expressed in exponential form as _____, where a __ 0 and a __ 1.

In Problems 2 and 3, rewrite each exponential equation as an equivalent equation involving a logarithm.

2. $4^3 = 64$
3. $p^{-2} = 8$

❷ Change Logarithmic Equations to Exponential Equations

The definition of a logarithm can also be used to rewrite logarithmic equations as exponential equations.

EXAMPLE 2 **Changing Logarithmic Equations to Exponential Equations**

Change each logarithmic equation to an equivalent exponential equation.

(a) $4 = \log_3 81$
(b) $-3 = \log_a \frac{1}{27}$
(c) $2 = \log_4 x$

Solution
In each of these problems, use the fact that $y = \log_a x$ is equivalent to $a^y = x$ provided that $a > 0$ and $a \neq 1$.

(a) If $4 = \log_3 81$, then $3^4 = 81$.
(b) If $-3 = \log_a \frac{1}{27}$, then $a^{-3} = \frac{1}{27}$.
(c) If $2 = \log_4 x$, then $4^2 = x$.

Quick ✓

In Problems 4–6, rewrite each logarithmic equation as an equivalent equation involving an exponent.

4. $4 = \log_2 16$
5. $5 = \log_a 20$
6. $-3 = \log_5 z$

❸ Evaluate Logarithmic Functions

To find the exact value of a logarithm, write the logarithm in exponential notation and use the fact that if $a^u = a^v$, then $u = v$.

EXAMPLE 3 **Finding the Exact Value of a Logarithmic Expression**

Find the exact value of

(a) $\log_2 32$
(b) $\log_4 \frac{1}{16}$

Solution

(a) Let $y = \log_2 32$ and convert this equation into an exponential equation.

$$y = \log_2 32$$
Write the logarithm as an exponent: $2^y = 32$
$32 = 2^5$: $2^y = 2^5$
If $a^u = a^v$, then $u = v$: $y = 5$

Therefore, $\log_2 32 = 5$.

(b) Let $y = \log_4 \dfrac{1}{16}$ and convert this equation into an exponential equation.

$$y = \log_4 \frac{1}{16}$$

Write the logarithm as an exponent: $\quad 4^y = \dfrac{1}{16}$

$\dfrac{1}{16} = 4^{-2}\colon \quad 4^y = 4^{-2}$

If $a^u = a^v$, then $u = v\colon \quad y = -2$

Therefore, $\log_4 \dfrac{1}{16} = -2$.

Quick ✓

In Problems 7 and 8, find the exact value of each logarithmic expression.

7. $\log_5 25$ \qquad\qquad **8.** $\log_2 \dfrac{1}{8}$

The equation $y = \log_a x$ could also be written using function notation as $f(x) = \log_a x$. We use this notation in the next example to evaluate a logarithmic function.

EXAMPLE 4 **Evaluating a Logarithmic Function**

Find the value of each of the following, given that $f(x) = \log_2 x$.

(a) $f(2)$ \qquad\qquad **(b)** $f\left(\dfrac{1}{4}\right)$

Solution

(a) Finding $f(2)$ means evaluating $\log_2 x$ at $x = 2$, or evaluating $\log_2 2$. To determine this value, follow the approach of Example 3 by letting $y = \log_2 2$ and converting the equation into an exponential equation.

$$y = \log_2 2$$

Write the logarithm as an exponent: $\quad 2^y = 2$

$2 = 2^1\colon \quad 2^y = 2^1$

If $a^u = a^v$, then $u = v\colon \quad y = 1$

Therefore, $f(2) = 1$.

(b) $f\left(\dfrac{1}{4}\right)$ means evaluating $\log_2 x$ at $x = \dfrac{1}{4}$ or $\log_2\left(\dfrac{1}{4}\right)$. Let $y = \log_2\left(\dfrac{1}{4}\right)$.

$$y = \log_2\left(\frac{1}{4}\right)$$

Write the logarithm as an exponent: $\quad 2^y = \dfrac{1}{4}$

$\dfrac{1}{4} = 2^{-2}\colon \quad 2^y = 2^{-2}$

If $a^u = a^v$, then $u = v\colon \quad y = -2$

Therefore, $f\left(\dfrac{1}{4}\right) = -2$.

Quick ✓

In Problems 9 and 10, evaluate the function, given that $g(x) = \log_5 x$.

9. $g(25)$ \qquad\qquad **10.** $g\left(\dfrac{1}{5}\right)$

▶ ④ Determine the Domain of a Logarithmic Function

The domain of a function $y = f(x)$ is the set of all x such that the function makes sense, and the range is the set of all outputs of the function. To find the range of the logarithmic function, recognize that $y = f(x) = \log_a x$ is equivalent to $x = a^y$. Because a can be raised to any real number (since $a > 0$ and $a \neq 1$), conclude that y can be any real number in the function $y = f(x) = \log_a x$. In addition, because a^y is positive for any real number, conclude that x must be positive. Since x represents the input of the logarithmic function, the domain of the logarithmic function is the set of all positive real numbers.

Work Smart

Notice that the domain of the logarithmic function is the same as the range of the exponential function. The range of the logarithmic function is the same as the domain of the exponential function.

Domain and Range of The Logarithmic Function

$$\text{Domain of the logarithmic function} = (0, \infty)$$
$$\text{Range of the logarithmic function} = (-\infty, \infty)$$

Because the domain of the logarithmic function is the set of all positive real numbers, the argument of the logarithmic function must be greater than zero. For example, $f(x) = \log_{10} x$ is defined for $x = 2$, but not for $x = -1$, $x = -8$, or any other $x \leq 0$.

EXAMPLE 5   **Finding the Domain of a Logarithmic Function**

Find the domain of each logarithmic function.

(a) $f(x) = \log_6 (x - 5)$ (b) $G(x) = \log_3 (3x + 1)$

Solution

(a) The argument of the function $f(x) = \log_6 (x - 5)$ is $x - 5$. The domain of f is the set of all real numbers x such that $x - 5 > 0$. Solve this inequality:

$$x - 5 > 0$$
Add 5 to both sides: $x > 5$

The domain of f is $\{x | x > 5\}$, or $(5, \infty)$ in interval notation.

(b) The argument of the function $G(x) = \log_3 (3x + 1)$ is $3x + 1$. The domain of G is the set of all real numbers x such that $3x + 1 > 0$.

$$3x + 1 > 0$$
Subtract 1 from both sides: $3x > -1$
Divide both sides by 3: $x > -\dfrac{1}{3}$

The domain of G is $\left\{x \middle| x > -\dfrac{1}{3}\right\}$, or $\left(-\dfrac{1}{3}, \infty\right)$ in interval notation.   ●

Quick ✓

In Problems 11 and 12, find the domain of each logarithmic function.

11. $g(x) = \log_8 (x + 3)$ **12.** $F(x) = \log_2 (5 - 2x)$

▶ ⑤ Graph Logarithmic Functions

To graph a logarithmic function $y = \log_a x$, it helps to rewrite the function in exponential form: $x = a^y$. Then choose "nice" values of y and use the expression $x = a^y$ to find the corresponding values of x.

EXAMPLE 6 Graphing a Logarithmic Function

Graph $f(x) = \log_2 x$ using point plotting. From the graph, state the domain and the range of the function.

Solution

Rewrite $y = f(x) = \log_2 x$ as $x = 2^y$. Table 12 shows various values of y, the corresponding values of x, and points on the graph of $y = f(x) = \log_2 x$. Plot these ordered pairs, connect them in a smooth curve, and obtain the graph of $f(x) = \log_2 x$ in Figure 31. The domain of f is $\{x | x > 0\}$, or $(0, \infty)$ in interval notation. The range of f is the set of all real numbers, or $(-\infty, \infty)$ in interval notation.

Table 12

y	$x = 2^y$	(x, y)
-2	$\frac{1}{4}$	$\left(\frac{1}{4}, -2\right)$
-1	$\frac{1}{2}$	$\left(\frac{1}{2}, -1\right)$
0	1	$(1, 0)$
1	2	$(2, 1)$
2	4	$(4, 2)$

Figure 31

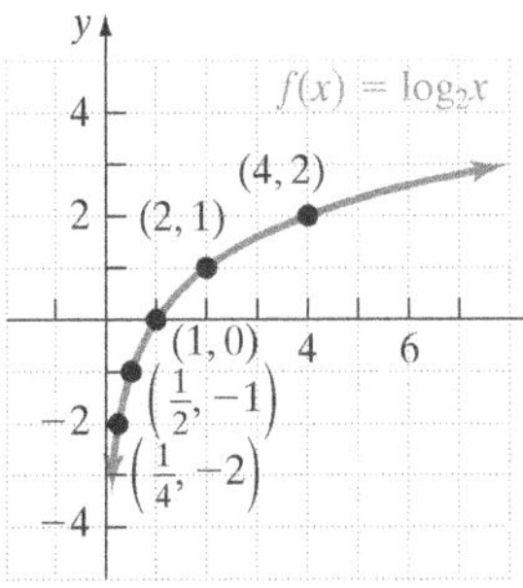

The graph of $f(x) = \log_2 x$ in Figure 31 is typical of all logarithmic functions that have a base larger than 1. Figure 32 shows the graphs of two other logarithmic functions with bases larger than 1, $y = \log_3 x$ and $y = \log_6 x$.

Figure 32

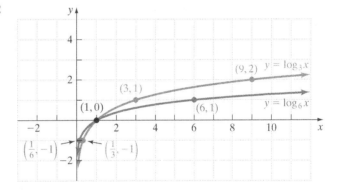

The information about $f(x) = \log_a x$, where $a > 1$, is summarized below.

Properties of the Graph of a Logarithmic Function $f(x) = \log_a x$, $a > 1$

1. The domain is the set of all positive real numbers. The range is the set of all real numbers.
2. There is no y-intercept; the x-intercept is $(1, 0)$.
3. The graph of f contains the points $\left(\frac{1}{a}, -1\right)$, $(1, 0)$, and $(a, 1)$. See Figure 33.

Figure 33
$f(x) = \log_a x$, $a > 1$

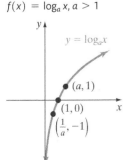

Quick ✓

13. Graph the logarithmic function $f(x) = \log_4 x$ using point plotting. From the graph, state the domain and the range of the function.

Now we consider $f(x) = \log_a x, 0 < a < 1$.

EXAMPLE 7 **Graphing a Logarithmic Function**

Graph $f(x) = \log_{1/2} x$ using point plotting. From the graph, state the domain and the range of the function.

Solution
Rewrite $y = f(x) = \log_{1/2} x$ as $x = \left(\dfrac{1}{2}\right)^y$. Table 13 shows various values of y, the corresponding values of x, and points on the graph of $y = f(x) = \log_{1/2} x$. Plot these points, connect them in a smooth curve, and obtain the graph of $f(x) = \log_{1/2} x$ in Figure 34. The domain of f is $\{x \mid x > 0\}$, or $(0, \infty)$ in interval notation. The range of f is the set of all real numbers, or $(-\infty, \infty)$ in interval notation.

Table 13

y	$x = \left(\dfrac{1}{2}\right)^y$	(x, y)
-2	4	$(4, -2)$
-1	2	$(2, -1)$
0	1	$(1, 0)$
1	$\dfrac{1}{2}$	$\left(\dfrac{1}{2}, 1\right)$
2	$\dfrac{1}{4}$	$\left(\dfrac{1}{4}, 2\right)$

Figure 34

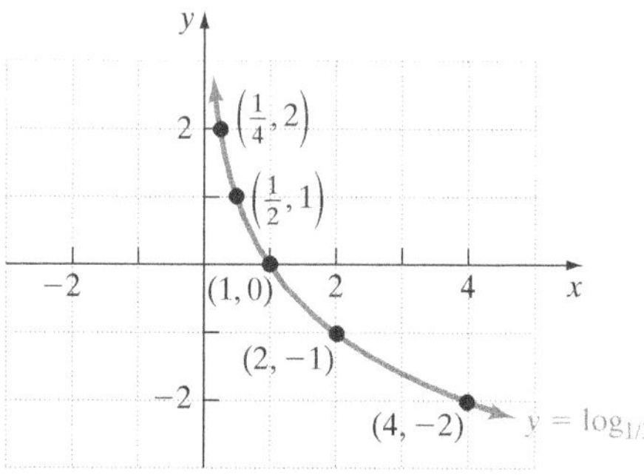

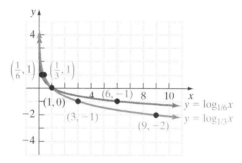

The graph of $f(x) = \log_{1/2} x$ in Figure 34 is typical of all logarithmic functions that have a base between 0 and 1. Figure 35 shows the graph of two more logarithmic functions whose bases are less than 1, $y = \log_{1/3} x$ and $y = \log_{1/6} x$.

Figure 35

The information about $f(x) = \log_a x$, where the base a is between 0 and 1 $(0 < a < 1)$, is summarized below.

Figure 36
$f(x) = \log_a x, 0 < a < 1$

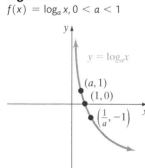

> **Properties of the Graph of a Logarithmic Function $f(x) = \log_a x, 0 < a < 1$**
>
> 1. The domain is the set of all positive real numbers. The range is the set of all real numbers.
> 2. There is no y-intercept; the x-intercept is $(1, 0)$.
> 3. The graph of f contains the points $\left(\dfrac{1}{a}, -1\right)$, $(1, 0)$, and $(a, 1)$. See Figure 36.

Quick ✓

14. Graph the logarithmic function $f(x) = \log_{1/4} x$ using point plotting. From the graph, state the domain and the range of the function.

6 Work with Natural and Common Logarithms

The **natural logarithm function** is a logarithmic function whose base is the number e. This function occurs so frequently in applications that it is given a special symbol, **ln** (from the Latin *logarithmus naturalis*).

In Other Words
A logarithm to the base e is called the natural logarithm because this function can be used to model many things in nature. In addition, $\log_e x$ is written $\ln x$.

Definition

The natural logarithm: $y = \ln x$ if and only if $x = e^y$

Figure 37 shows the graph of $y = \ln x$.

If the base of a logarithmic function is 10, then we have the **common logarithm function.** If the base a of the logarithmic function is not indicated, it is understood to be 10.

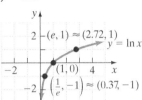

Figure 37
$y = \ln x$

Definition

The common logarithm: $y = \log x$ if and only if $x = 10^y$

Figure 38 shows the graph of $y = \log x$.

Scientific and graphing calculators have both a natural logarithm button, $\boxed{\ln}$, and a common logarithm button, $\boxed{\log}$. This enables us to approximate the values of logarithms to the base e and the base 10, when the results are not exact.

In Other Words
When there is no base on a logarithm, then the base is understood to be 10. That is, $\log_{10} x$ is written $\log x$.

To evaluate logarithmic expressions using a scientific calculator, enter the argument of the logarithm and then press the $\boxed{\ln}$ or $\boxed{\log}$ button, depending on the base of the logarithm. For example, to evaluate log 80, type in 80 and then press the $\boxed{\log}$ button. The display should show 1.90308999. Try it!

To evaluate logarithmic expressions using a graphing calculator, press the $\boxed{\ln}$ or $\boxed{\log}$ button, depending on the base of the logarithm, and then enter the argument of log the logarithm. Finally, press $\boxed{\text{ENTER}}$. For example, to evaluate log 80, press the $\boxed{\log}$ button, type in 80, and then press $\boxed{\text{ENTER}}$. The display should show 1.903089987. Try it!

By the way, it shouldn't be surprising that log 80 is between 1 and 2 since log 10 = 1 (because $10^1 = 10$) and log 100 = 2 (because $10^2 = 100$). It is a good idea to develop number sense for logarithms prior to using your calculator to approximate the value.

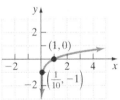

Figure 38
$y = \log x$

EXAMPLE 8 **Evaluating Natural and Common Logarithms on a Calculator**

Using a calculator, evaluate each of the following. Round your answers to three decimal places.

(a) ln 20   (b) log 30   (c) ln 0.5

Solution

(a) $\ln 20 \approx 2.996$ (b) $\log 30 \approx 1.477$ (c) $\ln 0.5 \approx -0.693$ ●

Quick ✓

In Problems 15–17, evaluate each logarithm using a calculator. Round your answers to three decimal places.

15. log 1400   **16.** ln 4.8   **17.** log 0.3

7 Solve Logarithmic Equations

Work Smart
Extraneous solutions can occur while solving logarithmic equations. This is why it is important to always check your solutions!

Equations that contain logarithms are called **logarithmic equations.** Take care when solving logarithmic equations, because *extraneous solutions* (apparent solutions that are not solutions to the original equation) might creep in. To help locate extraneous solutions, remember that in the expression $\log_a M$, both a and M must be positive with $a \neq 1$.

To solve logarithmic equations, rewrite the logarithmic equation as an equivalent exponential equation, using the fact that $y = \log_a x$ is equivalent to $a^y = x$.

EXAMPLE 9   Solving Logarithmic Equations

Solve:

(a) $\log_2(3x + 4) = 4$ (b) $\log_x 25 = 2$

Solution

(a) Find the solution by writing the logarithmic equation as an exponential equation, using the fact that if $y = \log_a x$, then $a^y = x$.

$$\log_2(3x + 4) = 4$$

If $y = \log_a x$, then $a^y = x$: $2^4 = 3x + 4$

$2^4 = 16$: $16 = 3x + 4$

Subtract 4 from both sides: $12 = 3x$

Divide both sides by 3: $4 = x$

Verify this solution by letting $x = 4$ in the original equation:

$$\log_2(3 \cdot 4 + 4) \stackrel{?}{=} 4$$
$$\log_2(16) \stackrel{?}{=} 4$$

Since $2^4 = 16$, $\log_2 16 = 4$. Therefore, the solution set is $\{4\}$.

(b) Change the logarithmic equation to an exponential equation.

$$\log_x 25 = 2$$

If $y = \log_a x$, then $a^y = x$: $x^2 = 25$

$$x = \pm\sqrt{25} = \pm 5$$

Since the base of a logarithm must be positive, $x = -5$ is extraneous. We leave it to you to verify that the solution set is $\{5\}$.

Note
The equation $x^2 = 25$ can be stated in words as, "what number(s), x, when squared, give me 25?". Put another way, the equation is asking for the square roots of 25.

Quick ✓

In Problems 18 and 19, solve each equation. Be sure to verify your solution.

18. $\log_3(5x + 1) = 4$ **19.** $\log_x 16 = 2$

EXAMPLE 10   Solving Logarithmic Equations

Solve each equation and state the exact solution.

(a) $\ln x = 3$ (b) $\log(x + 1) = -2$

Solution
Write each logarithmic equation as an exponential equation.

(a)
$$\ln x = 3$$
Write as an exponent; if $y = \ln x$, then $e^y = x$: $e^3 = x$

Verify this solution by letting $x = e^3$ in the original equation:

$$\ln e^3 \stackrel{?}{=} 3$$

The equation $\ln e^3 = 3$ can be written as $\log_e e^3 = 3$, which is equivalent to $e^3 = e^3$, so this is a true statement. The solution set is $\{e^3\}$.

(b)
$$\log(x + 1) = -2$$
Write as an exponent; if $y = \log x$, then $10^y = x$: $10^{-2} = x + 1$

$10^{-2} = 0.01$: $x + 1 = 0.01$

Subtract 1 from both sides: $x = -0.99$

Work Smart

The solution to Example 10(b) illustrates that negative answers may occur in logarithmic equations. Just because an apparent solution is negative does not automatically mean the solution is extraneous.

Verify this solution by letting $x = -0.99$ in the original equation:
$$\log(-0.99 + 1) \stackrel{?}{=} -2$$
$$\log(0.01) \stackrel{?}{=} -2$$
$$10^{-2} = 0.01 \quad \text{True}$$

The solution set is $\{-0.99\}$.

Quick ✓

In Problems 20 and 21, solve each equation. Be sure to verify your solution.

20. $\ln x = -2$ **21.** $\log(x - 20) = 4$

8 Use Logarithmic Models That Describe Our World

Common logarithms often are used when quantities vary from very large to very small. This is because the common logarithm can "scale down" the measurement. For example, if a certain quantity varied from $0.00000001 = 10^{-8}$ to $100{,}000{,}000 = 10^8$, the common logarithm of the same quantity would vary from $\log 10^{-8} = -8$ to $\log 10^8 = 8$.

Physicists define the **intensity of a sound wave** as the amount of energy the sound wave transmits through a given area. The *loudness L* (measured in **decibels** in honor of Alexander Graham Bell) of a sound of intensity x (measured in watts per square meter) is defined as follows.

Definition

The **loudness** L, measured in decibels, of a sound of intensity x, measured in watts per square meter, is
$$L(x) = 10 \log \frac{x}{10^{-12}}$$

The quantity 10^{-12} watt per square meter in the definition is the least intense sound that a human ear can detect. If $x = 10^{-12}$ watt per square meter, then
$$L(10^{-12}) = 10 \log \frac{10^{-12}}{10^{-12}}$$
$$= 10 \log 1$$
$$= 10(0)$$
$$= 0$$

Thus the loudness of the least intense sound a human ear can detect is 0 decibels.

EXAMPLE 11 **Measuring the Loudness of a Sound**

What is the loudness, in decibels, of normal conversation, which has an intensity level of 10^{-6} watt per square meter?

Solution
Evaluate L at $x = 10^{-6}$.

$$L(10^{-6}) = 10 \log \frac{10^{-6}}{10^{-12}}$$

Laws of exponents; $\frac{a^m}{a^n} = a^{m-n}$: $= 10 \log 10^{-6-(-12)}$

Simplify: $= 10 \log 10^6$

If $y = \log 10^6$, then $10^y = 10^6$, so $y = 6$: $= 10(6)$

$= 60$ decibels

The loudness of normal conversation is 60 decibels.

Quick ✓

22. Music played on an iPhone has an intensity level of 10^{-2} watt per square meter when set at its maximum level. What is the loudness, in decibels, of the iPhone on "full blast"?

A.IR6 Exercises

Underlined exercises have complete video solutions in MyLab.

Problems 1–22 are the Quick ✓s that follow the EXAMPLES.

Building Skills

In Problems 23–30, change each exponential equation to an equivalent equation involving a logarithm. See Objective 1.

<u>23.</u> $64 = 4^3$
24. $16 = 2^4$
25. $\frac{1}{8} = 2^{-3}$
26. $\frac{1}{9} = 3^{-2}$
27. $a^3 = 19$
28. $b^4 = 23$
29. $5^{-6} = c$
30. $10^{-3} = z$

In Problems 31–40, change each logarithmic equation to an equivalent equation involving an exponent. See Objective 2.

31. $\log_2 16 = 4$
32. $\log_3 81 = 4$
33. $\log_3 \frac{1}{9} = -2$
34. $\log_2 \frac{1}{32} = -5$
<u>35.</u> $\log_5 a = -3$
36. $\log_6 x = -4$
37. $\log_a 4 = 2$
38. $\log_a 16 = 2$
39. $\log_{1/2} 12 = y$
40. $\log_{1/2} 18 = z$

In Problems 41–48, find the exact value of each logarithm without using a calculator. See Objective 3.

41. $\log_3 1$
42. $\log_5 5$
43. $\log_2 8$
44. $\log_4 16$
<u>45.</u> $\log_4\left(\frac{1}{16}\right)$
46. $\log_5\left(\frac{1}{125}\right)$
47. $\log_{\sqrt{2}} 4$
48. $\log_{\sqrt{3}} 3$

In Problems 49–52, evaluate each function, given that $f(x) = \log_3 x$ and $g(x) = \log_5 x$. See Objective 3.

49. $f(81)$
50. $f(9)$
51. $g(\sqrt{5})$
52. $g(\sqrt[3]{5})$

In Problems 53–62, find the domain of each function. See Objective 4.

53. $f(x) = \log_2(x - 4)$
54. $f(x) = \log_3(x - 2)$
55. $F(x) = \log_3(2x)$
56. $h(x) = \log_4(5x)$
57. $f(x) = \log_8(3x - 2)$
58. $F(x) = \log_2(4x - 3)$
<u>59.</u> $H(x) = \log_7(2x + 1)$
60. $f(x) = \log_3(5x + 3)$
61. $H(x) = \log_2(1 - 4x)$
62. $G(x) = \log_4(3 - 5x)$

In Problems 63–68, graph each function. From the graph, state the domain and the range of each function. See Objective 5.

<u>63.</u> $f(x) = \log_5 x$
64. $f(x) = \log_7 x$
65. $g(x) = \log_6 x$
66. $G(x) = \log_8 x$
67. $F(x) = \log_{1/5} x$
68. $F(x) = \log_{1/7} x$

In Problems 69–72, change each exponential equation to an equivalent equation involving a logarithm, and change each logarithmic equation to an equivalent equation involving an exponent. See Objective 6.

69. $e^x = 12$
70. $e^4 = M$
71. $\ln x = 4$
72. $\ln(x - 1) = 3$

In Problems 73–76, evaluate each function, given that $H(x) = \log x$ and $P(x) = \ln x$. See Objective 6.

73. $H(0.1)$
74. $H(100{,}000)$
75. $P(e^3)$
76. $P(e^{-3})$

In Problems 77–88, use a calculator to evaluate each expression. Round your answers to three decimal places. See Objective 6.

77. $\log 67$
78. $\log 106$
79. $\ln 5.4$
80. $\ln 10.4$
81. $\log 0.35$
82. $\log 0.78$
83. $\ln 0.2$
84. $\ln 0.4$
85. $\log \frac{5}{4}$
86. $\log \frac{10}{7}$
87. $\ln \frac{3}{8}$
88. $\ln \frac{1}{2}$

In Problems 89–104, solve each logarithmic equation. All answers should be exact. See Objective 7.

<u>89.</u> $\log_3(2x + 1) = 2$
90. $\log_3(5x - 3) = 3$
91. $\log_5(20x - 5) = 3$
92. $\log_4(8x + 10) = 3$
93. $\log_a 36 = 2$
94. $\log_a 81 = 2$
95. $\log_a 1000 = 3$
96. $\log_a 243 = 5$
97. $\ln x = 5$
98. $\ln x = 10$
99. $\log(2x - 1) = -1$
100. $\log(2x + 3) = 1$
101. $\ln e^x = -3$
102. $\ln e^{2x} = 8$
103. $\log_3 81 = x$
104. $\log_4 16 = x + 1$

Mixed Practice

105. Suppose that $f(x) = \log_2 x$.
    (a) What is $f(16)$? What point is on the graph of f?
    (b) If $f(x) = -3$, what is x? What point is on the graph of f?

106. Suppose that $f(x) = \log_5 x$.
    (a) What is $f(5)$? What point is on the graph of f?
    (b) If $f(x) = -2$, what is x? What point is on the graph of f?

107. Suppose that $G(x) = \log_4(x+1)$.
  (a) What is $G(7)$? What point is on the graph of G?
  (b) If $G(x) = 2$, what is x? What point is on the graph of G?

108. Suppose that $F(x) = \log_2 x - 3$.
  (a) What is $F(8)$? What point is on the graph of f?
  (b) If $F(x) = -1$, what is x? What point is on the graph of F?

109. Find a so that the graph of $f(x) = \log_a x$ contains the point $(16, 2)$.

110. Find a so that the graph of $f(x) = \log_a x$ contains the point $\left(\dfrac{1}{4}, -2\right)$.

Applying the Concepts

111. Loudness of a Whisper A whisper has an intensity level of 10^{-10} watt per square meter. How many decibels is a whisper?

112. Loudness of a Concert If you sit in the front row of a rock concert, you will experience an intensity level of 10^{-1} watt per square meter. How many decibels is a rock concert in the front row? If you move back to the 15th row, you will experience an intensity level of 10^{-2} watt per square meter. How many decibels is a rock concert in the 15th row?

113. Threshold of Pain The threshold of pain has an intensity level of 10^1 watt per square meter. How many decibels is the threshold of pain?

114. Exploding Eardrum Instant perforation of the eardrum occurs at an intensity level of 10^4 watts per square meter. At how many decibels does instant perforation of the eardrum occur?

*Problems 115–118 use the following discussion: The **Richter scale** is one way of converting seismographic readings into numbers that provide an easy reference for measuring the **magnitude M** of an earthquake. All earthquakes are compared to a **zero–level earthquake** whose seismographic reading measures 0.001 millimeter at a distance of 100 kilometers from the epicenter. An earthquake whose seismographic reading measures x millimeters has magnitude M given by*

$$M(x) = \log\left(\dfrac{x}{10^{-3}}\right)$$

where 10^{-3} is the reading of a zero–level earthquake 100 kilometers from its epicenter.

115. San Francisco, 1906 According to the United States Geological Survey, the San Francisco earthquake of 1906 resulted in a seismographic reading of 63,096 millimeters 100 kilometers from its epicenter. What was the magnitude of this earthquake?

116. Alaska, 1964 According to the United States Geological Survey, an earthquake on March 28, 1964, in Prince William Sound, Alaska, resulted in a seismographic reading of 1,584,893 millimeters 100 kilometers from its epicenter. What was the magnitude of this earthquake? This earthquake was the second largest ever recorded. The largest was the Great Chilean Earthquake of 1960, whose magnitude was 9.5 on the Richter scale.

117. Japan, 2011 According to the United States Geological Survey, an earthquake on March 11, 2011, off the coast of Japan had a magnitude of 8.9. What was the seismographic reading 100 kilometers from its epicenter?

118. Brazil, 2015 According to the United States Geological Survey, an earthquake on November 26, 2015, in Tarauka, Brazil, had a magnitude of 7.6. What was the seismographic reading 100 kilometers from its epicenter?

119. pH The pH of a chemical solution is given by the formula

$$\text{pH} = -\log[\text{H}^+]$$

where $[\text{H}^+]$ is the concentration of hydrogen ions in moles per liter. Values of pH range from 0 to 14. A solution whose pH is 7 is considered neutral. The pH of pure water at 25 degrees Celsius is 7. A solution whose pH is less than 7 is considered acidic, while a solution whose pH is greater than 7 is considered basic.
  (a) What is the pH of household ammonia for which $[\text{H}^+]$ is 10^{-12}? Is ammonia basic or acidic?
  (b) What is the pH of black coffee for which $[\text{H}^+]$ is 10^{-5}? Is black coffee basic or acidic?
  (c) What is the pH of lemon juice for which $[\text{H}^+]$ is 10^{-2}? Is lemon juice basic or acidic?
  (d) What is the concentration of hydrogen ions in human blood (pH = 7.4)?

120. Energy of an Earthquake The magnitude and the seismic moment are related to the amount of energy that is given off by an earthquake. The relationship between magnitude and energy is

$$\log E_S = 11.8 + 1.5M$$

where E_S is the energy (in ergs) for an earthquake whose magnitude is M. Note that E_S is the amount of energy given off from the earthquake as seismic waves.
  (a) How much energy is given off by an earthquake that measures 5.8 on the Richter scale?
  (b) The earthquake on the Rat Islands in Alaska on February 4, 1965, measured 8.7 on the Richter scale. How much energy was given off by this earthquake?

Explaining the Concepts

121. Explain why the base in the logarithmic function $f(x) = \log_a x$ is not allowed to equal 1.

122. Explain why the base in the logarithmic function $f(x) = \log_a x$ must be greater than 0.

123. Explain why the domain of the function $f(x) = \log_a (x^2 + 1)$ is the set of all real numbers.

124. What are the domain and the range of the exponential function $f(x) = a^x$? What are the domain and the range of the logarithmic function $f(x) = \log_a x$? What is the relation between the domain and the range of each function?

A.IR7 Properties of Logarithms

Objectives

1. Understand the Properties of Logarithms
2. Write a Logarithmic Expression as a Sum or Difference of Logarithms
3. Write a Logarithmic Expression as a Single Logarithm
4. Evaluate a Logarithm Whose Base Is Neither 10 Nor e

Are You Prepared for This Section?

Before getting started, complete the following problems. If you get a problem wrong, go back to the section cited and review the material.

P1. Round 3.03468 to three decimal points.   [Section 1.IR2, pp. IR-13–IR-14]
P2. Evaluate a^0, $a \neq 0$.   [Getting Ready, p. IR-264]

① Understand the Properties of Logarithms

Logarithms have some very useful properties. These properties will be derived directly from the definition of a logarithm and the laws of exponents.

EXAMPLE 1 Deriving Properties of Logarithms

Determine the value of the following logarithmic expressions where a is any positive real number, and $a \neq 1$.

(a) $\log_a 1$ (b) $\log_a a$

Solution

(a) Remember, $y = \log_a x$ is equivalent to $x = a^y$, where $a > 0, a \neq 1$.

$$y = \log_a 1$$
Change to an exponent: $a^y = 1$
Because $a \neq 0, a^0 = 1$: $a^y = a^0$
If $a^u = a^v$, then $u = v$: $y = 0$

Thus $\log_a 1 = 0$.

(b) Again, write the logarithm as an exponent in order to evaluate the logarithm. Let $y = \log_a a$, so that

$$y = \log_a a$$
Change to an exponent: $a^y = a$
$a = a^1$: $a^y = a^1$
If $a^u = a^v$, then $u = v$: $y = 1$

Thus $\log_a a = 1$.

The results of Example 1 are summarized below.

$$\log_a 1 = 0 \qquad \log_a a = 1$$

Prepared?...Answers P1. 3.035   **P2.** 1

Quick ✓

In Problems 1–4, evaluate each logarithm.

1. $\log_5 1$ **2.** $\ln 1$ **3.** $\log_4 4$ **4.** $\log 10$

How can $3^{\log_3 81}$ be evaluated? The exponent, $\log_3 81$, equals 4 (because $y = \log_3 81$ means $3^y = 81$, or $3^y = 3^4$, so $y = 4$). Now $3^{\log_3 81} = 3^4 = 81$, so $3^{\log_3 81} = 81$. This result is true in general.

> **In Other Words**
> When the number a is raised to the power $\log_a M$, the result is M.

An Inverse Property of Logarithms

If a and M are positive real numbers, with $a \ne 1$, then

$$a^{\log_a M} = M$$

This is an Inverse Property of Logarithms because if the logarithm to the base a of a positive number M is computed and then a is raised to this power, the result is M.

EXAMPLE 2 **Using an Inverse Property of Logarithms**

(a) $5^{\log_5 20} = 20$ (b) $0.8^{\log_{0.8} \sqrt{23}} = \sqrt{23}$ ●

Quick ✓

In Problems 5 and 6, evaluate each logarithm.

5. $12^{\log_{12} \sqrt{2}}$ **6.** $10^{\log 0.2}$

How can $\log_5 5^6$ be evaluated? If $y = \log_5 5^6$, then $5^y = 5^6$ (since $y = \log_a x$ means $a^y = x$), so $y = 6$. Thus $\log_5 5^6 = 6$. This result is true in general.

> **In Other Words**
> The logarithm to the base a of a raised to the power of r is r.

An Inverse Property of Logarithms

If a is a positive real number, with $a \ne 1$, and r is any real number, then

$$\log_a a^r = r$$

This is another Inverse Property of Logarithms because if a is raised to some power r and then the logarithm to the base a of a^r is computed, the result is r.

EXAMPLE 3 **Using an Inverse Property of Logarithms**

(a) $\log_4 4^3 = 3$ (b) $\ln e^{-0.5} = -0.5$ ●

Quick ✓

In Problems 7 and 8, evaluate each logarithm.

7. $\log_8 8^{1.2}$ **8.** $\log 10^{-4}$

▶ ❷ Write a Logarithmic Expression as a Sum or Difference of Logarithms

The next two properties deal with logarithms whose arguments are products or quotients.

Notice that $\log_2 8 = 3$ and that $\log_2 2 + \log_2 4 = 1 + 2 = 3$. Therefore, $\log_2 8 = \log_2 2 + \log_2 4$ and $8 = 4 \cdot 2$. This suggests the following result.

The Product Rule of Logarithms

In Other Words
The product rule of logarithms states that the log of a product equals the sum of the logs of the factors.

If M, N, and a are positive real numbers, with $a \neq 1$, then
$$\log_a(MN) = \log_a M + \log_a N$$

EXAMPLE 4 Using the Product Rule of Logarithms

Simplify each of the following logarithms by writing it as the sum of logarithms.

(a) $\log_2(5 \cdot 3)$ (b) $\ln(6z)$

Solution
Do you notice that each argument contains a product? To simplify, use the Product Rule of Logarithms.

(a) $\log_2(5 \cdot 3) = \log_2 5 + \log_2 3$ (b) $\ln(6z) = \ln 6 + \ln z$

Work Smart
$\log_a(M + N)$ does *not* equal $\log_a M + \log_a N$.

Quick ✓

9. True or False $\log(x + 4) = \log x + \log 4$

In Problems 10 and 11, write each logarithm as the sum of logarithms.

10. $\log_4(9 \cdot 5)$ 11. $\log(5w)$

Notice that $\log_2 8 = 3$ and that $\log_2 16 - \log_2 2 = 4 - 1 = 3$. Therefore, $\log_2 8 = \log_2 16 - \log_2 2$ and $8 = \frac{16}{2}$. This suggests the result.

The Quotient Rule of Logarithms

In Other Words
The Quotient Rule states that the log of a quotient equals the difference of the logs of the numerator and denominator.

If M, N, and a are positive real numbers, with $a \neq 1$, then
$$\log_a\left(\frac{M}{N}\right) = \log_a M - \log_a N$$

The relationships in the product and quotient rules for logarithms are not coincidental. After all, when exponential expressions with the same base are multiplied, the exponents are added, and when exponential expressions with the same base are divided, the exponents are subtracted. The same thing is going on here!

EXAMPLE 5 Using the Quotient Rule of Logarithms

Simplify each of the following logarithms by writing it as the difference of logarithms.

(a) $\log_2\left(\frac{5}{3}\right)$ (b) $\log\left(\frac{y}{5}\right)$

Solution
Do you notice that each argument contains a quotient? To simplify, use the Quotient Rule of Logarithms.

(a) $\log_2\left(\frac{5}{3}\right) = \log_2 5 - \log_2 3$ (b) $\log\left(\frac{y}{5}\right) = \log y - \log 5$

Section A.IR7 Properties of Logarithms

Work Smart

$\log_a(M - N)$ does *not* equal $\log_a M - \log_a N$.

Quick ✓

In Problems 12 and 13, write each logarithm as the difference of logarithms.

12. $\log_7\left(\dfrac{9}{5}\right)$ **13.** $\ln\left(\dfrac{p}{3}\right)$

A single logarithm can be written as the sum or difference of logs when the argument of the logarithm contains both products and quotients.

EXAMPLE 6 Writing a Single Logarithm as the Sum or Difference of Logs

Write $\log_3\left(\dfrac{4x}{y}\right)$ as the sum or difference of logarithms.

Solution
Notice that the argument of the logarithm contains a product and a quotient. Thus the Product Rule and the Quotient Rule of Logarithms can be used. When applying both rules, it is typically easier to use the Quotient Rule first.

The Quotient Rule of Logarithms
↓
$$\log_3\left(\dfrac{4x}{y}\right) = \log_3(4x) - \log_3 y$$

The Product Rule of Logarithms: $= \log_3 4 + \log_3 x - \log_3 y$

Work Smart

When you need to use both the Product Rule and the Quotient Rule, use the Quotient Rule first.

Quick ✓

In Problems 14 and 15, write each logarithm as the sum or difference of logarithms.

14. $\log_2\left(\dfrac{3m}{n}\right)$ **15.** $\ln\left(\dfrac{q}{3p}\right)$

Another useful property of logarithms allows us to express powers on the argument of a logarithm as factors.

In Other Words

If the log quantity contains an exponent, it can be "brought down in front" of the log as a factor.

The Power Rule of Logarithms

If M and a are positive real numbers, with $a \neq 1$, and r is any real number, then

$$\log_a M^r = r \log_a M$$

EXAMPLE 7 Using the Power Rule of Logarithms

Use the Power Rule of Logarithms to express all powers as factors.

(a) $\log_8 3^5$ (b) $\ln x^{\sqrt{3}}$

Solution
(a) $\log_8 3^5 = 5 \log_8 3$ (b) $\ln x^{\sqrt{3}} = \sqrt{3} \ln x$

Work Smart

Whenever you see the word "factor," you should think product (or multiplication).

Quick ✓

In Problems 16 and 17, write each logarithm so that all powers are factors.

16. $\log_2 5^{1.6}$ **17.** $\log b^5$

Use the direction *expand the logarithm* to mean "Write a logarithm as a sum or difference with all exponents written as factors."

EXAMPLE 8 **Expanding a Logarithm**

Expand the logarithm.

(a) $\log_2(x^2 y^3)$ (b) $\log\left(\dfrac{100x}{\sqrt{y}}\right)$

Solution

(a) The argument of the logarithm contains a product, so use the Product Rule of Logarithms to write the single log as the sum of two logs.

$$\text{Product Rule} \downarrow$$
$$\log_2(x^2 y^3) = \log_2 x^2 + \log_2 y^3$$
$$\text{Write exponents as factors:} \quad = 2\log_2 x + 3\log_2 y$$

Work Smart

If the argument of the logarithm contains both a quotient and a product, it is easier to write the quotient as the difference of two logs first.

(b) The argument of the logarithm contains a quotient and a product.

$$\text{Quotient Rule} \downarrow$$
$$\log\left(\dfrac{100x}{\sqrt{y}}\right) = \log(100x) - \log\sqrt{y}$$
$$\sqrt{y} = y^{\frac{1}{2}}: \quad = \log(100x) - \log y^{\frac{1}{2}}$$
$$\text{Product Rule:} \quad = \log 100 + \log x - \log y^{\frac{1}{2}}$$
$$\text{Write exponents as factors:} \quad = \log 100 + \log x - \dfrac{1}{2}\log y$$
$$\log 100 = 2: \quad = 2 + \log x - \dfrac{1}{2}\log y$$

Quick ✓

In Problems 18 and 19, expand each logarithm.

18. $\log_4(a^2 b)$ **19.** $\log_3\left(\dfrac{9m^4}{\sqrt[3]{n}}\right)$

Work Smart

To write two logs as a single log, the bases on the logs must be the same.

▶ ❸ Write a Logarithmic Expression as a Single Logarithm

The Product Rule, Quotient Rule, and Power Rule of Logarithms can be used to write the sums and/or differences of logarithms that have the same base as a single logarithm. This skill will be useful when solving certain logarithmic equations in Section A.IR8.

EXAMPLE 9 **Writing Expressions as Single Logarithms**

Write each of the following as a single logarithm.

(a) $\log_6 3 + \log_6 12$ (b) $\log(x-2) - \log x$

Solution

(a) The base of each logarithm is the same, 6. Use the Product Rule to write the sum of the logs as a single log.

$$\text{Product Rule} \downarrow$$
$$\log_6 3 + \log_6 12 = \log_6(3 \cdot 12)$$
$$3 \cdot 12 = 36: \quad = \log_6 36$$
$$6^2 = 36, \text{ so } \log_6 36 = 2: \quad = 2$$

(b) The base of each logarithm is the same, 10. Use the Quotient Rule to write the difference of the logs as a single log.

$$\log(x-2) - \log x = \log\left(\frac{x-2}{x}\right)$$

Quick ✓
In Problems 20 and 21, write each expression as a single logarithm.

20. $\log_8 4 + \log_8 16$ **21.** $\log_3(x+4) - \log_3(x-1)$

Work Smart
If logarithms have coefficients, write them as exponents before using the Product or Quotient Rule.

In order to use the Product Rule or the Quotient Rule to write the sum or difference of logarithms as a single logarithm, the coefficients of the logarithms must be 1. Therefore if logarithms have coefficients, first use the Power Rule to write the coefficient as a power. For example, write $2 \log x$ as $\log x^2$.

EXAMPLE 10 **Writing Expressions as Single Logarithms**

Write each of the following as a single logarithm.

(a) $2 \log_2(x-1) + \frac{1}{2}\log_2 x$ **(b)** $\log(x-1) + \log(x+1) - 3 \log x$

Solution

(a) Each logarithm has the same base. Because the logarithms have coefficients, use the Power Rule to write the coefficients as exponents.

$$2 \log_2(x-1) + \frac{1}{2}\log_2 x = \log_2(x-1)^2 + \log_2 x^{\frac{1}{2}}$$

$x^{\frac{1}{2}} = \sqrt{x}$: $= \log_2(x-1)^2 + \log_2 \sqrt{x}$

Product Rule: $= \log_2\left[(x-1)^2 \sqrt{x}\right]$

(b) Work from left to right. Since the first two logs are being added, use the Product Rule to write these logs as a single log.

Product Rule
↓
$\log(x-1) + \log(x+1) - 3 \log x = \log[(x-1)(x+1)] - 3 \log x$

$(x-1)(x+1) = x^2 - 1$: $= \log(x^2 - 1) - 3 \log x$

Write the coefficient as an exponent: $= \log(x^2 - 1) - \log x^3$

Quotient Rule: $= \log\left(\frac{x^2 - 1}{x^3}\right)$

Quick ✓
In Problems 22 and 23, write each expression as a single logarithm.

22. $\log_5 x - 3 \log_5 2$ **23.** $\log_2(x+1) + \log_2(x+2) - 2 \log_2 x$

▶ ④ Evaluate a Logarithm Whose Base Is Neither 10 Nor e

Section A.IR6 discussed how to use a calculator to approximate logarithms whose base was either 10 (the common logarithm) or e (the natural logarithm). But what if the base of the logarithm is neither 10 nor e? To determine how to evaluate these types of logarithms, the property on the next page is needed. In this property, M, N, and a are positive numbers and $a \neq 1$. The property is a result of the fact that the logarithmic function is a one-to-one function.

One-to-One Property of Logarithms

If $M = N$, then $\log_a M = \log_a N$.

EXAMPLE 11 Approximating a Logarithm Whose Base Is Neither 10 Nor e

Approximate $\log_2 5$. Round the answer to three decimal places.

Solution
Let $y = \log_2 5$ and then convert the logarithmic equation to an equivalent exponential equation using the fact that if $y = \log_a x$, then $x = a^y$.

$$y = \log_2 5$$
$$2^y = 5$$

Now, use the One-to-One Property and "take the logarithm of both sides." So that a calculator can be used, take either the common log or the natural log of both sides (it doesn't matter which). Let's take the common log of both sides.

$$\log 2^y = \log 5$$

Use the Power Rule, $\log a^r = r \log a$: $y \log 2 = \log 5$

Divide both sides by log 2: $y = \dfrac{\log 5}{\log 2}$

Using a calculator, find $\dfrac{\log 5}{\log 2} \approx 2.322$. Thus $\log_2 5 \approx 2.322$.

> **In Other Words**
> "Taking the logarithm" of both sides of an equation is the same type of approach as squaring both sides of an equation.

There is another way to approximate a logarithm whose base is neither 10 nor e. It is called the **Change-of-Base Formula.**

Change-of-Base Formula

If $a \neq 1, b \neq 1$, and M are positive real numbers, then

$$\log_a M = \dfrac{\log_b M}{\log_b a}$$

Because many calculators have keys for only the common logarithm, $\boxed{\log}$, and the natural logarithm, $\boxed{\ln}$, we use the Change-of-Base Formula with either $b = 10$ or $b = e$:

$$\log_a M = \dfrac{\log M}{\log a} \quad \text{or} \quad \log_a M = \dfrac{\ln M}{\ln a}$$

EXAMPLE 12 Using the Change-of-Base Formula

Approximate $\log_4 45$. Round your answer to three decimal places.

Solution
Use the Change-of-Base Formula.

Using common logarithms: $\log_4 45 = \dfrac{\log 45}{\log 4}$

≈ 2.746

Using natural logarithms: $\log_4 45 = \dfrac{\ln 45}{\ln 4}$

≈ 2.746

Thus $\log_4 45 \approx 2.746$.

Quick ✓

24. $\log_3 10 = \dfrac{\log\_\_}{\log\_\_} = \dfrac{\ln\_\_}{\ln\_\_}$

In Problems 25 and 26, approximate each logarithm. Round your answers to three decimal places.

25. $\log_3 32$ 26. $\log_{\sqrt{2}} \sqrt{7}$

Let's review all the properties of logarithms.

Properties of Logarithms

In the following properties, M, N, a, and b are positive real numbers, $a \neq 1$, $b \neq 1$, and r is any real number.

- **Inverse Properties of Logarithms**
  $a^{\log_a M} = M$ and $\log_a a^r = r$
- **The Product Rule of Logarithms**
  $\log_a (MN) = \log_a M + \log_a N$
- **The Quotient Rule of Logarithms**
  $\log_a \left(\dfrac{M}{N}\right) = \log_a M - \log_a N$
- $\log_a 1 = 0$
- **The Power Rule of Logarithms**
  $\log_a M^r = r \log_a M$
- **Change-of-Base Formula**
  $\log_a M = \dfrac{\log_b M}{\log_b a} = \dfrac{\log M}{\log a} = \dfrac{\ln M}{\ln a}$
- **One-to-One Property**
  If $M = N$, then $\log_a M = \log_a N$
- $\log_a a = 1$

A.IR7 Exercises — MyLab Statistics

Underlined exercises have complete video solutions in MyLab.

Problems 1–26 are the Quick ✓s that follow the EXAMPLES.

Building Skills

In Problems 27–38, use properties of logarithms to find the exact value of each expression. Do not use a calculator. See Objective 1.

27. $\log_2 2^3$
28. $\log_5 5^{-3}$
29. $\ln e^{-7}$
30. $\ln e^9$
31. $3^{\log_3 5}$
32. $5^{\log_5 \sqrt{2}}$
33. $e^{\ln 2}$
34. $e^{\ln 10}$
35. $\log_7 7$
36. $\log_5 5$
37. $\log 1$
38. $\ln 1$

In Problems 39–46, suppose that $\ln 2 = a$ and $\ln 3 = b$. Use properties of logarithms to write each logarithm in terms of a and b. See Objective 2.

39. $\ln 6$
40. $\ln \dfrac{3}{2}$
41. $\ln 9$
42. $\ln 4$
43. $\ln 12$
44. $\ln 18$
45. $\ln \sqrt{2}$
46. $\ln \sqrt[4]{3}$

In Problems 47–68, write each expression as a sum or difference of logarithms. Express exponents as factors. See Objective 2.

47. $\log(ab)$
48. $\log_4 \left(\dfrac{a}{b}\right)$
49. $\log_5 x^4$
50. $\log_3 z^{-2}$
51. $\log_2 (xy^2)$
52. $\log_3 (a^3 b)$
53. $\log_5 (25x)$
54. $\log_2 (8z)$
55. $\log_7 \left(\dfrac{49}{y}\right)$
56. $\log_2 \left(\dfrac{16}{p}\right)$
57. $\ln(e^2 x)$
58. $\ln \left(\dfrac{x}{e^3}\right)$
59. $\log_3 (27\sqrt{x})$
60. $\log_2 (32\sqrt[4]{z})$
61. $\log_5 (x^2 \sqrt{x^2 + 1})$
62. $\log_3 (x^3 \sqrt{x^2 - 1})$
63. $\log \left(\dfrac{x^4}{\sqrt[3]{x-1}}\right)$
64. $\ln \left(\dfrac{\sqrt[5]{x}}{(x+2)^2}\right)$
65. $\log_7 \sqrt{\dfrac{x+1}{x}}$
66. $\log_6 \sqrt[3]{\dfrac{x-2}{x+1}}$
67. $\log_2 \left[\dfrac{x(x-1)^2}{\sqrt{x+1}}\right]$
68. $\log_4 \left[\dfrac{x^3(x-3)}{\sqrt[3]{x+1}}\right]$

In Problems 69–92, write each expression as a single logarithm. See Objective 3.

69. $\log 25 + \log 4$
70. $\log_4 32 + \log_4 2$
71. $\log x + \log 3$
72. $\log_2 6 + \log_2 z$
73. $\log_3 36 - \log_3 4$
74. $\log_2 48 - \log_2 3$
75. $10^{\log 8 - \log 2}$
76. $e^{\ln 24 - \ln 3}$
77. $3 \log_3 x$
78. $8 \log_2 z$
79. $\log_4(x + 1) - \log_4 x$
80. $\log_5(2y - 1) - \log_5 y$
81. $2 \ln x + 3 \ln y$
82. $4 \log_2 a + 2 \log_2 b$
83. $\dfrac{1}{2} \log_3 x + 3 \log_3(x - 1)$
84. $\dfrac{1}{3} \log_4 z + 2 \log_4(2z + 1)$
85. $\log x^5 - 3 \log x$
86. $\log_7 x^4 - 2 \log_7 x$
87. $\dfrac{1}{2}[3 \log x + \log y]$
88. $\dfrac{1}{3}[\ln(x - 1) + \ln(x + 1)]$
89. $\log_8(x^2 - 1) - \log_8(x + 1)$
90. $\log_5(x^2 + 3x + 2) - \log_5(x + 2)$
91. $18 \log \sqrt{x} + 9 \log \sqrt[3]{x} - \log 10$
92. $10 \log_4 \sqrt[5]{x} + 4 \log_4 \sqrt{x} - \log_4 16$

In Problems 93–100, use the Change-of-Base Formula and a calculator to evaluate each logarithm. Round your answer to three decimal places. See Objective 4.

93. $\log_2 10$
94. $\log_3 18$
95. $\log_8 3$
96. $\log_7 5$
97. $\log_{1/3} 19$
98. $\log_{1/4} 3$
99. $\log_{\sqrt{2}} 5$
100. $\log_{\sqrt{3}} \sqrt{6}$

Applying the Concepts

101. Find the value of
$\log_2 3 \cdot \log_3 4 \cdot \log_4 5 \cdot \log_5 6 \cdot \log_6 7 \cdot \log_7 8$.

102. Find the value of $\log_2 4 \cdot \log_4 6 \cdot \log_6 8$.

103. Find the value of
$\log_2 3 \cdot \log_3 4 \cdot \,\cdots\, \cdot \log_n(n + 1) \cdot \log_{n+1} 2$.

104. Find the value of
$\log_3 3 \cdot \log_3 9 \cdot \log_3 27 \cdot \,\cdots\, \cdot \log_3 3^n$.

Extending the Concepts

105. Show that
$\log_a\left(x + \sqrt{x^2 - 1}\right) + \log_a\left(x - \sqrt{x^2 - 1}\right) = 0$.

106. Show that
$\log_a\left(\sqrt{x} + \sqrt{x - 1}\right) + \log_a\left(\sqrt{x} - \sqrt{x - 1}\right) = 0$.

107. If $f(x) = \log_a x$, show that $f(AB) = f(A) + f(B)$.

108. Find the domain of $f(x) = \log_a x^2$ and the domain of $g(x) = 2 \log_a x$. Since $\log_a x^2 = 2 \log_a x$, how can it be that the domains are not equal? Write a brief explanation.

Explaining the Concepts

109. State the Product Rule for Logarithms in your own words.

110. State the Quotient Rule for Logarithms in your own words.

111. Write an example to illustrate
$\log_2(x + y) \neq \log_2 x + \log_2 y$.

112. Write an example to illustrate
$(\log_a x)^r \neq r \log_a x$.

A.IR8 Exponential Equations

Objectives

1. Solve Exponential Equations
2. Solve Equations Involving Exponential Models

Are You Prepared for This Section?

Before getting started, complete the following problem. If you get the problem wrong, go back to the section cited and review the material.

P1. Solve: $2x + 5 = 13$ [Section 1.IR4, pp. IR-91–IR-95]

Prepared? . . . Answer **P1.** $\{4\}$

1 Solve Exponential Equations

Section A.IR5 explained how to solve exponential equations using the fact that if $a^u = a^v$, then $u = v$. However, in many situations both sides of the equation cannot be written with the same base.

EXAMPLE 1   Using Logarithms to Solve Exponential Equations

Solve: $3^x = 5$

Solution
Because 5 cannot be written so that it is 3 raised to some integer power, write the equation $3^x = 5$ as a logarithm.

$$3^x = 5$$

If $a^y = x$, then $y = \log_a x$: $x = \log_3 5$ Exact solution

To find a decimal approximation to the solution, use the Change-of-Base Formula.

$$x = \log_3 5 = \frac{\log 5}{\log 3}$$

$$\approx 1.465 \quad \text{Approximate solution}$$

An alternative approach to solving the equation would be to take either the natural logarithm or the common logarithm of both sides of the equation. If the natural logarithm of both sides of the equation is taken, the following results:

$$3^x = 5$$

If $M = N$, then $\ln M = \ln N$: $\ln 3^x = \ln 5$

$\log_a M^r = r \log_a M$: $x \ln 3 = \ln 5$

Divide both sides by $\ln 3$: $x = \dfrac{\ln 5}{\ln 3}$ Exact solution

$$\approx 1.465 \quad \text{Approximate solution}$$

The solution set is $\left\{\dfrac{\ln 5}{\ln 3}\right\}$. If the common logarithm of both sides had been taken, the solution set would have been $\left\{\dfrac{\log 5}{\log 3}\right\}$.

Quick ✓

In Problems 1 and 2, solve each equation. Express answers in exact form and as a decimal rounded to three decimal places.

1. $2^x = 11$ **2.** $5^{2x} = 3$

EXAMPLE 2   Using Logarithms to Solve Exponential Equations

Solve: $4e^{3x} = 10$

Solution
First isolate the exponential expression by dividing both sides of the equation by 4.

$$4e^{3x} = 10$$

$$e^{3x} = \frac{5}{2}$$

Work Smart

You could also take the natural logarithm of both sides:

$$\ln e^{3x} = \ln\left(\frac{5}{2}\right)$$

$$3x = \ln\left(\frac{5}{2}\right)$$

$$x = \frac{\ln\left(\frac{5}{2}\right)}{3}$$

Because $\frac{5}{2}$ cannot be expressed as e raised to an integer power, write the exponential equation as an equivalent logarithmic equation.

$$e^{3x} = \frac{5}{2}$$

If $e^y = x$, then $y = \ln x$: $\ln\left(\frac{5}{2}\right) = 3x$

Divide both sides by 3: $x = \dfrac{\ln\left(\frac{5}{2}\right)}{3}$ Exact solution

$x \approx 0.305$ Approximate solution

The check is left to you. The solution set is $\left\{\dfrac{\ln\left(\frac{5}{2}\right)}{3}\right\}$. •

Quick ✓

In Problems 3 and 4, solve each equation. Express answers in exact form and as a decimal rounded to three decimal places.

3. $e^{2x} = 5$ **4.** $3e^{-4x} = 20$

▶ ❷ Solve Equations Involving Exponential Models

Section A.IR5 examined a variety of models from areas such as statistics, biology, and finance. The following examples, rather than evaluating the models at certain values of the independent variable, solve equations that involve the models.

EXAMPLE 3   **Radioactive Decay**

The half-life of plutonium-239 is 24,360 years. The maximum amount of plutonium-239 that an adult can handle without significant injury is 0.13 microgram ($= 0.000000013$ gram). Suppose a researcher has a 1-gram sample of plutonium-239. The amount A (in grams) of plutonium-239 after t years is given by

$$A(t) = 1\left(\frac{1}{2}\right)^{\frac{t}{24{,}360}}$$

(a) How long will it take until 0.9 gram of plutonium-239 is left in the sample?

(b) How long will it take until the 1-gram sample is safe–that is, until 0.000000013 gram is left?

Solution

(a) To find the time until $A = 0.9$ gram, solve the equation

$$0.9 = 1\left(\frac{1}{2}\right)^{\frac{t}{24{,}360}}$$

for t. How can t be removed from the exponent? Use the fact that $\log_a M^r = r \log_a M$ to move the variable.

$$\log 0.9 = \log\left(\frac{1}{2}\right)^{\frac{t}{24,360}}$$

$$\log 0.9 = \frac{t}{24,360}\log\left(\frac{1}{2}\right)$$

Multiply both sides by 24,360: $\quad 24{,}360 \log 0.9 = t \log\left(\dfrac{1}{2}\right)$

Divide both sides by $\log\left(\dfrac{1}{2}\right)$: $\quad \dfrac{24{,}360 \log 0.9}{\log\left(\dfrac{1}{2}\right)} = t$

Thus $t = \dfrac{24{,}360 \log 0.9}{\log\left(\dfrac{1}{2}\right)} \approx 3702.8$. After approximately 3703 years, there will be 0.9 gram of plutonium-239 left.

(b) To determine the time until $A = 0.000000013$ gram, solve the equation

$$0.000000013 = 1 \cdot \left(\frac{1}{2}\right)^{\frac{t}{24,360}}$$

Take the logarithm of both sides: $\quad \log(0.000000013) = \log\left(\dfrac{1}{2}\right)^{\frac{t}{24,360}}$

$\log_a M^r = r \log_a M$: $\quad \log(0.000000013) = \dfrac{t}{24{,}360}\log\left(\dfrac{1}{2}\right)$

Multiply both sides by 24,360: $\quad 24{,}360 \log(0.000000013) = t \log\left(\dfrac{1}{2}\right)$

Divide both sides by $\log\left(\dfrac{1}{2}\right)$: $\quad \dfrac{24{,}360 \log(0.000000013)}{\log\left(\dfrac{1}{2}\right)} = t$

Thus $t = \dfrac{24{,}360 \log(0.000000013)}{\log\left(\dfrac{1}{2}\right)} \approx 638{,}156.8$. After approximately 638,157 years, the 1-gram sample will be safe!

Quick ✓

5. The half-life of thorium-227 is 18.72 days. Suppose a researcher has a 10-gram sample of thorium-227. The amount A (in grams) of thorium-227 after t days is given by

$$A(t) = 10\left(\frac{1}{2}\right)^{\frac{t}{18.72}}$$

(a) How long will it take until 9 grams of thorium-227 is left in the sample?

(b) How long will it take until 3 grams of thorium-227 is left in the sample?

Now let's look at an example involving compound interest. Remember, the compound interest formula states that the future value of P dollars invested in an account paying an annual interest rate r, compounded n times per year for t years, is given by $A = P\left(1 + \dfrac{r}{n}\right)^{nt}$.

EXAMPLE 4 Future Value of Money

Suppose you deposit $5000 into a Roth IRA today. If the deposit earns 8% interest compounded quarterly, when will it be worth

(a) $7500?

(b) $10,000? That is, when will your money have doubled?

Solution
With $P = 5000$, $r = 0.08$, and $n = 4$ (compounded quarterly),

$$A = 5000\left(1 + \frac{0.08}{4}\right)^{4t} \quad \text{or} \quad A = 5000(1.02)^{4t}$$

(a) Find the time t when $A = 7500$. That is, solve

$$7500 = 5000(1.02)^{4t}$$

Divide both sides by 5000: $\quad 1.5 = (1.02)^{4t}$

Take the logarithm of both sides: $\quad \log 1.5 = \log(1.02)^{4t}$

$\log_a M^r = r \log_a M$: $\quad \log 1.5 = 4t \log(1.02)$

Divide both sides by 4 (1.02): $\quad \dfrac{\log 1.5}{4 \log(1.02)} = t$

So $t = \dfrac{\log 1.5}{4 \log(1.02)} \approx 5.12$. After approximately 5.12 years (5 years, 1.4 months), the account will be worth $7500.

(b) Find the time t when $A = 10,000$. That is, solve

$$10{,}000 = 5000(1.02)^{4t}$$

Divide both sides by 5000: $\quad 2 = (1.02)^{4t}$

Take the logarithm of both sides: $\quad \log 2 = \log(1.02)^{4t}$

$\log_a M^r = r \log_a M$: $\quad \log 2 = 4t \log(1.02)$

Divide both sides by 4 log(1.02): $\quad \dfrac{\log 2}{4 \log(1.02)} = t$

Thus $t = \dfrac{\log 2}{4 \log(1.02)} \approx 8.75$. After approximately 8.75 years (8 years, 9 months), the account will be worth $10,000.

Quick ✓

6. Suppose that you deposit $2000 into a Roth IRA today. If the deposit earns 6% interest compounded monthly, how long will it be before the account is worth
   (a) $3000? (b) $4000? That is, when will your money have doubled?

A.IR8 Exercises — MyLab Statistics
Underlined exercises have complete video solutions in MyLab.

Problems 1–6 are the Quick ✓s that follow the EXAMPLES.

Building Skills

In Problems 7–32, solve each equation. Express irrational solutions in exact form and as a decimal rounded to three decimal places. See Objective 2.

<u>7.</u> $2^x = 10$

8. $3^x = 8$

9. $5^x = 20$

10. $4^x = 20$

11. $\left(\dfrac{1}{2}\right)^x = 7$

12. $\left(\dfrac{1}{2}\right)^x = 10$

13. $e^x = 5$

14. $e^x = 3$

15. $10^x = 5$

16. $10^x = 0.2$

17. $3^{2x} = 13$

18. $2^{2x} = 5$

19. $\left(\dfrac{1}{2}\right)^{4x} = 13$

20. $\left(\dfrac{1}{3}\right)^{2x} = 4$

21. $4 \cdot 2^x + 3 = 8$

22. $3 \cdot 4^x - 5 = 10$

23. $-3e^x = -18$

24. $\dfrac{1}{2}e^x = 4$

25. $0.2^{x+1} = 3^x$

26. $0.4^x = 2^{x-3}$

27. $5^{3x} = 7$

28. $3^{2x} = 4$

29. $\dfrac{1}{3}e^x = 5$

30. $-4e^x = -16$

31. $\left(\dfrac{1}{4}\right)^{x+1} = 8^x$

32. $9^x = 27^{x-4}$

Applying the Concepts

33. **A Population Model** According to the U.S. Census Bureau, the population of the United States in 2016 was 326 million people. In addition, the population of the United States was growing at a rate of 0.99% per year. Assuming that this growth rate continues, the model $P(t) = 326(1.0099)^{t-2016}$ represents the population P (in millions of people) in year t.

    (a) According to this model, when will the population of the United States be 350 million people?

    (b) According to this model, when will the population of the United States be 471 million people?

34. **A Population Model** According to the *United States Census Bureau*, the population of the world in 2016 was 7563 million people. In addition, the population of the world was growing at a rate of 1.02% per year. Assuming that this growth rate continues, the model $P(t) = 7563(1.0102)^{t-2016}$ represents the population P (in millions of people) in year t.

    (a) According to this model, when will the population of the world be 9.84 billion people?

    (b) According to this model, when will the population of the world be 11.58 billion people?

35. **Time Is Money** Suppose that you deposit $5000 in a certificate of deposit (CD) today. If the deposit earns 2% interest compounded monthly, when will the account be worth

    (a) $7000?

    (b) $10,000? That is, when will your money have tripled?

36. **Time Is Money** Suppose that you deposit $8000 in a certificate of deposit (CD) today. If the deposit earns 2% interest compounded quarterly, when will it be worth

    (a) $10,000?

    (b) $24,000? That is, when will your money have tripled?

37. **Depreciation** Based on data obtained from the *Kelley Blue Book*, the value V of a Ford Focus that is t years old can be modeled by $V(t) = 19{,}841(0.88)^t$.

    (a) According to the model, when will the car be worth $15,000?

    (b) According to the model, when will the car be worth $5000?

    (c) According to the model, when will the car be worth $1000?

38. **Depreciation** Based on data obtained from the *Kelley Blue Book*, the value V of a Chevy Malibu that is t years old can be modeled by $V(t) = 25{,}258(0.84)^t$.

    (a) According to the model, when will the car be worth $15,000?

    (b) According to the model, when will the car be worth $5000?

    (c) According to the model, when will the car be worth $1000?

39. **Radioactive Decay** The half-life of beryllium-11 is 13.81 seconds. Suppose that a researcher possesses a 100-gram sample of beryllium-11. The amount A (in grams) of beryllium-11 after t seconds is given by

$$A(t) = 100\left(\dfrac{1}{2}\right)^{\frac{t}{13.81}}$$

    (a) When will there be 90 grams of beryllium-11 left in the sample?

    (b) When will 25 grams be left?

    (c) When will 10 grams of beryllium-11 be left?

40. **Radioactive Decay** The half-life of carbon-10 is 19.255 seconds. Suppose that a researcher possesses a 100-gram sample of carbon-10. The amount A (in grams) of carbon-10 after t seconds is given by

$$A(t) = 100\left(\dfrac{1}{2}\right)^{\frac{t}{19.255}}$$

    (a) When will there be 90 grams of carbon-10 left in the sample?

    (b) When will 25 grams be left?

    (c) When will 10 grams be left?

For Exercises 41 and 42, use Newton's Law of Cooling, which states that the temperature of a heated object decreases exponentially over time toward the temperature of the surrounding medium.

41. Newton's Law of Cooling Suppose that a pizza is removed from a 400°F oven and placed in a room where the temperature is 70°F. The temperature u (in °F) of the pizza at time t (in minutes) can be modeled by $u(t) = 70 + 330e^{-0.072t}$.

(a) According to the model, when will the temperature of the pizza be 300°F?

(b) According to the model, when will the temperature of the pizza be 220°F?

42. Newton's Law of Cooling Suppose that coffee that is 170°F is poured into a coffee mug and allowed to cool in a room where the temperature is 70°F. The temperature u (in °F) of the coffee at time t (in minutes) can be modeled by $u(t) = 70 + 100e^{-0.045t}$.

(a) According to the model, when will the temperature of the coffee be 120°F?

(b) According to the model, when will the temperature of the coffee be 100°F?

43. Learning Curve Suppose that a student has 200 vocabulary words to learn. If a student learns 20 words in 30 minutes, the function

$$L(t) = 200(1 - e^{-0.0035t})$$

models the number of words L that the student will learn in t minutes.

(a) How long will it take the student to learn 50 words?

(b) How long will it take the student to learn 150 words?

44. Learning Curve Suppose that a student has 50 biology terms to learn. If a student learns 10 terms in 30 minutes, the function

$$L(t) = 50(1 - e^{-0.0223t})$$

models the number of terms L that the student will learn in t minutes.

(a) How long will it take the student to learn 10 words?

(b) How long will it take the student to learn 40 words?

45. The Rule of 72 The Rule of 72 states that the time for an investment to double in value is approximately given by 72 divided by the annual interest rate. For example, an investment earning 10% annual interest will double in approximately $\frac{72}{10} = 7.2$ years.

(a) According to the Rule of 72, approximately how long will it take an investment to double if it earns 8% annual interest?

(b) Derive a formula that can be used to find the number of years required for an investment to double. (*Hint*: Let $A = 2P$ in the formula $A = P\left(1 + \frac{r}{n}\right)^{nt}$ and solve for t.)

(c) Use the formula derived in part (b) to determine the exact amount of time it takes an investment to double that earns 8% interest compounded monthly. Compare the result to the results given by the Rule of 72.

46. Critical Thinking Suppose you need to open a certificate of deposit (CD). Bank A offers 2% interest compounded daily, and Bank B offers 2.1% interest compounded quarterly. Which bank offers the better deal? Why?

47. Critical Thinking The bacteria in a 2-liter container double every minute. After 30 minutes the container is full. How long did it take to fill half the container?

Answers to Selected Exercises

Chapter 1 Integrated Review: Getting Ready for Organizing and Summarizing Data

Section 1.IR1 Fundamentals of Fractions 1. False 2. True 3. 9 4. 8; 6 5. proper 6. improper 7. $\frac{3}{4}$ 8. $\frac{9}{30}$ 9. $\frac{4}{6}; \frac{1}{2}$ 10. $\frac{100}{100}, \frac{2}{1}, \frac{16}{3}$
11. $\frac{3}{6}$ 12. $\frac{5}{20}$ 13. $\frac{49}{35}$ 14. $\frac{12}{39}$ 15. $\frac{3}{2}$ 16. $\frac{9}{10}$ 17. $\frac{3}{4}$ 18. $\frac{2}{3}$ 19. already in lowest terms 20. $\frac{2}{5}$ 21. equivalent fractions 22. False
23. equivalent 24. not equivalent 25. $\frac{3}{6}$; numerator: 3; denominator: 6 27. $\frac{12}{16}$; numerator: 12; denominator: 16
29. $\frac{5}{4}$; numerator: 5; denominator: 4 31. $\frac{7}{8}, \frac{3}{4}$ 33. $\frac{9}{8}, \frac{12}{12}, \frac{50}{20}$ 35. $\frac{5}{15}$ 37. $\frac{16}{12}$ 39. $\frac{39}{45}$ 41. $\frac{10}{2}$ 43. $\frac{2}{3}$ 45. already in lowest terms
47. $\frac{6}{7}$ 49. $\frac{14}{11}$ 51. equivalent 53. not equivalent 55. $\frac{1321}{2278}$ 57. $\frac{19}{50}$

Section 1.IR2 Fundamentals of Decimals 1. (a) tenths; $\frac{2}{10}$, or 0.2; (b) thousandths; $\frac{5}{1000}$, or 0.005; (c) tens; (d) hundredths; $\frac{6}{100}$, or 0.06;
(e) ten-thousandths; $\frac{7}{10,000}$, or 0.0007 2. 5 3. 6 4. 4 5. 9 6. tenths; $\frac{3}{10}$ or 0.3 7. hundredths; $\frac{8}{100}$ or 0.08 8. thousandths; $\frac{6}{1000}$ or 0.006
9. hundreds; 4 × 100 or 400 10. and 11. twenty-three hundredths 12. thirty-one and four tenths 13. Eighteen thousandths 14. four thousand, five hundred twenty-one ten-thousandths 15. two hundred and five hundredths 16. two hundred five thousandths 17. ninety-five and $\frac{23}{100}$ dollars 18. $\frac{1}{10}$
19. $\frac{1}{100}$ 20. $\frac{3}{1000}$ 21. $24\frac{1}{2}$ 22. $\frac{5}{8}$ 23. 0.1 24. 0.98 25. 3.68 26. 0.297 27. 24.00 28. 1.9 29. 0.040 30. 74.0 31. $1.91 32. $3.99 33. $0.93 34. $2.95
35. 9 37. 7 39. 9 41. tenths 43. hundredths 45. ones 47. twenty-one hundredths 49. eight hundred forty-one and six tenths 51. three hundred six thousandths 53. six thousandths 55. one hundred twenty-five and $\frac{48}{100}$ dollars 57. eighty-nine and $\frac{99}{100}$ dollars 59. $\frac{3}{10}$ 61. $\frac{1}{4}$ 63. $-\frac{2}{5}$ 65. $\frac{7}{25}$
67. $\frac{1}{20}$ 69. 0.4 71. −0.5 73. 12.8 75. 4.98 77. −19.30 79. 0.78 81. 17.287 83. 0.002 85. 69.000 87. $123.57 89. $17 91. 1.596 minutes
93. 0.91 mm 95. 6.6 cm

Section 1.IR3 The Real Number Line 1. {1, 3, 5, 7} 2. {Alabama, Alaska, Arkansas, Arizona} 3. ∅ or { } 4. True 5. rational 6. 12 7. 12, 0
8. −5, 12, 0 9. $\frac{11}{5}, -5, 12, 2.\overline{76}, 0, \frac{18}{4}$ 10. 2.737737773... 11. All numbers listed 12. origin 13. [number line from −4 to 4 with points at −4, −3, −2, −1, 0, $\frac{1}{2}$, 1, 2, 3, 3.5, 4]
14. inequality 15. < 16. < 17. > 18. > 19. = 20. < 21. > 22. = 23. > 24. < 25. < 26. < 27. Absolute value 28. 15 29. $\frac{3}{4}$
30. −4 31. $A = \{0, 1, 2, 3, 4\}$ 33. $D = \{1, 2, 3, 4\}$ 35. $E = \{\}$ or ∅ 37. 3 39. −4, 3, 0 41. 2.303003000... 43. All numbers listed 45. π 47. $\frac{5}{5} = 1$
49. [number line with points at −1.5, $\frac{3}{3}$, $\frac{4}{3}$] 51. < 53. > 55. > 57. = 59. < 61. > 63. > 65. = 67. 12 69. 4 71. $\frac{3}{8}$ 73. −2.1
75. (a) [number line with points at −4.5, −4, −2, $-\frac{1}{2}$, 0, $\frac{3}{5}$, 2, 3.5, 4, 6, 8] (b) $-4.5, -1, -\frac{1}{2}, \frac{3}{5}, 1, 3, 5, |-7| = 7$ 77. 257, 260, 261, 266, 267, 270, 282
79. 5.02, 5.09, 5.13, 5.15, 5.20, 5.21, 5.23, 5.24, 5.25, 5.26, 5.26, 5.28

Section 1.IR4 Multiplying and Dividing Fractions 1. $\frac{1}{4}$ 2. $\frac{5}{12}$ 3. $\frac{1}{12}$ 4. $\frac{9}{16}$ 5. 60 6. $600 7. $\frac{14}{45}$ 8. $-\frac{5}{14}$ 9. $-\frac{3}{5}$ 10. $-\frac{1}{12}$ 11. $\frac{4}{5}$ 12. −81
13. $-\frac{4}{15}$ 14. 6 15. $\frac{4}{3}$ 16. 10 17. $\frac{24}{5}$ 18. $\frac{1}{8}$ 19. $-\frac{4}{3}$ 20. $\frac{1}{8}$ 21. −10 22. $\frac{1}{2}$ 23. $\frac{24}{5}$ 24. −1 25. $\frac{1}{9}$ 27. $\frac{5}{2}$ 29. $\frac{5}{12}$ 31. $\frac{1}{4}$ 33. $\frac{4}{3}$ 35. $-\frac{1}{15}$ 37. $-\frac{4}{9}$
39. 18 41. −5 43. $\frac{17}{9}$ 45. $-\frac{5}{8}$ 47. $-\frac{1}{3}$ 49. 7 51. $\frac{4}{5}$ 53. 20 55. $\frac{1}{75}$ 57. $\frac{18}{7}$ 59. $\frac{16}{21}$ 61. $-\frac{1}{2}$ 63. $-\frac{1}{49}$ 65. 10 67. $\frac{1}{3}$ 69. $-\frac{13}{10}$ 71. $-\frac{7}{2}$
73. 0 75. $\frac{14}{3}$ 77. 16 79. $\frac{3}{4}$ 81. $\frac{81}{169}$ sq. miles 83. $225 85. 720 87. 285 miles 89. $\frac{1}{2} \cdot \frac{1}{4}$ means one-half of one-fourth. Draw a whole and divide it into 4 equal parts. Shade one part. Divide the shaded region into two equal parts, and shade one of the parts with a different shading. The resulting region represents one-eighth of the whole. Drawings may vary.

Section 1.IR5 Adding and Subtracting Fractions 1. $\frac{5}{9}$ 2. $\frac{3}{16}$ 3. $\frac{2}{3}$ 4. $\frac{3}{2}$ 5. $\frac{10}{11}$ 6. $\frac{1}{4}$ 7. $-\frac{3}{5}$ 8. $\frac{5}{2}$ 9. $-\frac{5}{4}$ 10. $-\frac{3}{10}$ 11. $\frac{1}{3}$ 12. $-\frac{1}{2}$ 13. $-\frac{1}{5}$
14. $-\frac{5}{6}$ 15. $\frac{23}{50}$ 16. $\frac{1}{4}$ of the trip was left; 6 hours 17. 40 18. 12 19. 18 20. 30 21. LCD = 40; $\frac{3}{8} = \frac{15}{40}; \frac{1}{5} = \frac{8}{40}$ 22. LCD = 12; $\frac{5}{12} = \frac{5}{12}; \frac{1}{4} = \frac{3}{12}$
23. LCD = 18; $\frac{8}{9} = \frac{16}{18}; \frac{5}{6} = \frac{15}{18}$ 24. LCD = 30; $\frac{8}{15} = \frac{16}{30}; \frac{7}{10} = \frac{21}{30}$ 25. $\frac{29}{42}$ 26. $\frac{5}{36}$ 27. $\frac{13}{24}$ 28. $\frac{1}{10}$ 29. $-\frac{25}{16}$ 30. $\frac{7}{12}$ 31. $-\frac{19}{20}$ 32. $\frac{1}{45}$ 33. $-\frac{1}{6}$
34. $\frac{9}{5}$ 35. $-\frac{2}{21}$ 36. 45 37. $\frac{13}{40}$ 39. $\frac{1}{3}$ 41. $\frac{5}{21}$ 43. $\frac{5}{7}$ 45. $\frac{6}{7}$ 47. $\frac{1}{2}$ 49. $-\frac{3}{8}$ 51. $-\frac{2}{3}$ 53. 3 55. $-\frac{1}{3}$ 57. $-\frac{1}{5}$ 51. $\frac{1}{2}$ 53. $\frac{1}{15}$ 55. $-\frac{2}{15}$ 57. $\frac{16}{81}$

IR-AN-1

Section 1.IR5 Adding and Subtracting Fractions

59. 12 **61.** 6 **63.** 12 **65.** 216 **67.** 180 **69.** 288 **71.** $\frac{5}{9} = \frac{20}{36}; \frac{3}{4} = \frac{27}{36}$ **73.** $\frac{11}{12} = \frac{22}{24}; \frac{-5}{8} = \frac{-15}{24}$ **75.** $\frac{5}{17} = \frac{15}{51}; \frac{-5}{51} = \frac{-5}{51}$ **77.** $\frac{4}{9} = \frac{32}{72}; \frac{5}{8} = \frac{45}{72}$
79. $-\frac{21}{60} = -\frac{63}{180}; -\frac{55}{90} = -\frac{110}{180}$ **81.** $\frac{4}{21} = \frac{16}{84}; -\frac{3}{28} = -\frac{9}{84}$ **83.** $\frac{23}{24}$ **85.** $\frac{13}{36}$ **87.** $-\frac{1}{6}$ **89.** $\frac{9}{16}$ **91.** $\frac{3}{40}$ **93.** $-\frac{5}{12}$ **95.** $-\frac{17}{9}$ **97.** $\frac{5}{2}$ **99.** $\frac{3}{4}$ **101.** 0 **103.** $\frac{1}{3}$
105. $\frac{9}{16}$ **107.** $-\frac{1}{24}$ **109.** $-\frac{3}{4}$ **111.** $\frac{13}{35}$ **113.** 0 **115.** $\frac{3}{16}$ **117.** $\frac{1}{5}$ **119. (a)** $\frac{8}{15}$ **(b)** $\frac{7}{15}$ **(c)** Armando; $\frac{2}{15}$ **121. (a)** $\frac{41}{100}$ **(b)** $\frac{59}{100}$ **123. (a)** $\frac{1}{6}$ **(b)** 200
125. $\frac{15}{8}$ pounds **127.** $\frac{1}{2}$ of the pizza **129.** The correct answer is $\frac{2+2}{3} = \frac{4}{3}$. The student is confusing the rule for multiplying fractions and the rule for adding fractions with like denominators. When multiplying, multiply the denominators. When adding fractions with like denominators, use the common denominator as the denominator in the answer.
131.

Section 1.IR6 Operations on Decimals
1. 27.49 **2.** 47.64 **3.** 100.67 **4.** 0.48 **5.** −195.305 **6.** −75.37 **7.** −50.0 **8.** 13.21 **9.** 46.3209 **10.** 0.036
11. 9.271 **12.** −17.797 **13.** −13.56 **14.** 502.667 **15.** −11.4 **16.** −8.63 **17.** 65.86 **18.** 0.184 **19.** 24.596 **20.** −0.6848 **21.** 0.00213 **22.** 0.00192
23. 0.000528 **24.** $592.80 **25.** 728.7 **26.** 453 **27.** 207,630 **28.** 3,429,000 **29.** 721¢ **30.** 10,588¢ **31.** 18.598 **32.** 678.095 **33.** 0.00456 **34.** 0.0003
35. dividend; divisor; quotient **36.** divisor; dividend **37.** 0.25 **38.** 24.3 **39.** −84.7 **40.** $17.50 **41.** 14.20 **42.** 81.28 **43.** 2.9 **44.** 49 **45.** 27.5043
46. 0.09106 **47.** 1423.5 **48.** 9830 **49.** 0.6 **50.** 0.375 **51.** 0.833... or 0.8$\overline{3}$ **52.** 0.555... or 0.$\overline{5}$ **53.** > **54.** < **55.** 0.66 **57.** 0.641 **59.** 0.875
61. 0.53 **63.** 0.444 **65.** 0.12 **67.** 0.405 **69.** 0.04674 **71.** 45 **73.** 823 **75.** 45.1 **77.** 0.00357 **79.** 5.87 **81.** 10.23 **83.** 3.244 **85.** 8.27 **87.** 4.78
89. 0.32 **91.** 0.803 **93.** 500 **95.** 4930 **97.** 0.35 **99.** 0.64 **101.** 0.41$\overline{6}$ **103.** > **105.** > **107.** < **109.** > **111.** > **113.** $625.92 **115.** 53.6°F
117. $21.25 **119.** $80 **121.** 39.15 hours; $1409.40 **123.** $14.82 **125.** The 16-ounce can costs $0.0806 per ounce; the 10-ounce can costs $0.089 per ounce. So, the 16-ounce can is the better buy.

Section 1.IR7 Fundamentals of Percent Notation
1. 38% **2.** 25% **3.** 0.9 **4.** 0.02 **5.** 0.045 **6.** 0.0025 **7.** 72% **8.** 180% **9.** 1.2% **10.** 35% **11.** $\frac{3}{5}$
12. $\frac{1}{8}$ **13.** $\frac{5}{4}$ **14.** $\frac{23}{400}$ **15.** 0.05 = 5% **16.** 2.75 = 275% **17.** 0.375 = 37.5% **18.** 1.6 = 160% **19.** 0.417 = 41.7% **20.** 0.692 = 69.2% **21.** 85% **23.** 99%
25. 46% **27.** 84% **29.** 0.45 **31.** 0.6 **33.** 4 **35.** 0.01 **37.** 0.255 **39.** 0.002 **41.** 0.0725 **43.** 96% **45.** 30% **47.** 3.5% **49.** 160% **51.** 0.03%
53. $\frac{2}{5}$ **55.** $\frac{9}{4}$ **57.** $\frac{2}{3}$ **59.** $\frac{5}{8}$ **61.** $\frac{9}{500}$ **63.** $\frac{3}{16}$ **65.** 0.15 = 15% **67.** 0.28 = 28% **69.** 0.38 = 38% **71.** 0.115 = 11.5% **73.** 0.276 = 27.6%
75. 0.5833 = 58.33% **77.** 0.571 = 57.1% **79.** 0.7852 = 78.52% **81.** 0.833 = 83.3% **83.** 0.0053 = 0.53% **85. (a)** 0.06; 6% **(b)** $\frac{1}{4}$; 25% **(c)** $\frac{17}{20}$; 0.85
(d) $\frac{2}{5}$; 40% **(e)** 0.095; 9.5% **(f)** $\frac{1}{8}$; 0.125 **87.** 38% **89.** 63.9%

Section 1.IR8 Language Used In Modeling
1. mathematical modeling **2.** 12 − 8 **3.** 15 + 6 **4.** 80 + 90 **5.** 25 − 10 **6.** 30 − 18 **7.** 100 − 36
8. 85 + 45 **9. (a)** 344 **(b)** 310 **(c)** 1050 **10.** 2 × 18 or 2 · 18 **11.** 0.45 × 18 or 0.45 · 18 **12.** 27 × 100 or 27 · 100 **13.** 168 **14.** 130 ÷ 18 or $\frac{130}{18}$
15. 500 ÷ 12 or $\frac{500}{12}$ **16.** $\frac{100 \text{ miles}}{3 \text{ hours}}$ **17.** $\frac{830}{2400} = \frac{83}{240}$; 0.346 **18.** $24,251.25 **19.** 1932 calories; over by 132 calories **21.** 14,932,000 **23.** 89,812
25. (a) 178 students **(b)** 379 students **27.** 372 drivers **29.** $\frac{338}{2252} = \frac{169}{1126}$; 0.150 **31.** $44.00 **33.** 205 lbs.

Chapter 2 Integrated Review: Getting Ready for Numerically Summarizing Data

Section 2.IR1 Exponents and the Order of Operations
1. base; exponent; power **2.** 11^5 **3.** $(-7)^4$ **4.** 16 **5.** 49 **6.** $-\frac{1}{216}$ **7.** 0.81 **8.** −16
9. 16 **10.** 15 **11.** 24 **12.** 23 **13.** −19 **14.** 40 **15.** −63 **16.** 4 **17.** $-\frac{8}{7}$ **18.** $\frac{4}{7}$ **19.** $\frac{5}{3}$ **20.** 20 **21.** 40 **22.** −9 **23.** 48 **24.** $-\frac{1}{6}$ **25.** 12 **26.** −108
27. 10 **28.** 4 **29.** 5^2 **31.** $\left(-\frac{3}{5}\right)^3$ **33.** 64 **35.** 64 **37.** 1000 **39.** −1000 **41.** −1000 **43.** 2.25 **45.** −64 **47.** −1 **49.** 0 **51.** $\frac{1}{64}$ **53.** $-\frac{1}{27}$
55. 14 **57.** −3 **59.** 2500 **61.** 160 **63.** 20 **65.** 4 **67.** $\frac{3}{5}$ **69.** −1 **71.** 42 **73.** 5 **75.** −4 **77.** 42 **79.** −5 **81.** $\frac{169}{4}$ **83.** −24 **85.** $-\frac{13}{12}$ **87.** 11
89. 12 **91.** 1.7 **93.** 36 **95.** 0.19 **97.** 24 **99.** 5 **101.** 0.0975 **103.** 3 **105.** $\frac{3}{2}$ **107.** 21.5 **109.** $-\frac{1}{6}$ **111.** $2^3 \cdot 3^2$ **113.** $2^4 \cdot 3$ **115.** $(4 \cdot 3 + 6) \cdot 2$
117. $(4 + 3) \cdot (4 + 2)$ **119.** $(6 − 4) + (3 − 1)$ **121.** $514.93 **123.** 603.19 in.^{2} **125.** $1060.90 **127.** 115.75° **129.** The expression -3^2 means "take the opposite of three squared": $-(3 \cdot 3) = -9$. The base is the number 3. The expression $(-3)^2$ means use −3 as a base twice: $-3 \cdot -3 = 9$.

Section 2.IR2 Square Roots
1. b^2; a **2.** −4 and 4 **3.** $-\frac{3}{10}$ and $\frac{3}{10}$ **4.** −0.02 and 0.02 **5.** principal square root **6.** radicand **7.** 10 **8.** −3
9. $\frac{5}{7}$ **10.** 0.6 **11.** 12 **12.** 13 **13.** 17 **14.** 7 **15.** ≈5.92 **16.** ≈−2.45 **17.** rational; 7 **18.** irrational; ≈8.43 **19.** not a real number
20. rational; −4 **21.** −1, 1 **23.** $-\frac{1}{3}, \frac{1}{3}$ **25.** −13, 13 **27.** −0.5, 0.5 **29.** 12 **31.** −3 **33.** 15 **35.** $\frac{1}{11}$ **37.** 0.2 **39.** −18 **41.** $\frac{4}{9}$ **43.** 35 **45.** 10
47. 14 **49.** 8 **51.** 6 **53.** 2.828 **55.** 5.48 **57.** 7.5 **59.** not a real number **61.** rational; 20 **63.** rational; $\frac{1}{2}$ **65.** irrational; 7.35 **67.** irrational; 7.07
69. 1.73 **71.** 0.63 **73.** not a real number **75.** 6 **77.** −8 **79.** 5 **81.** 6.71 **83.** 2.24 **85.** 25 ft **87.** 16 km **89.** 7 m **91.** 14 in. **93.** 398 m

95. The number $-\sqrt{9}$ is a real number because it denotes the opposite, or negative, of the principal square root of 9, namely, 3. The number $\sqrt{-9}$ is not a real number because there is no real number whose square is -9.

Section 2.IR3 Simplifying Algebraic Expressions; Summation Notation **1.** variable **2.** evaluate **3.** -7 **4.** 2 **5.** $216 **6.** True **7.** $5x^2$; $3xy$
8. $9ab$; $-3bc$; $5ac$; $-ac^2$ **9.** $\dfrac{2mn}{5}$; $-\dfrac{3n}{7}$ **10.** 2 **11.** 1 **12.** -1 **13.** 5 **14.** $-\dfrac{2}{3}$ **15.** False **16.** like **17.** like **18.** unlike **19.** unlike **20.** like **21.** b; c
22. $6x + 12$ **23.** $-5x - 10$ **24.** $-2k + 14$ **25.** $6x + 9$ **26.** 4; 9 **27.** $-5x$ **28.** $-4x^2$ **29.** $-8x + 3$ **30.** $-2a + 9b - 4$ **31.** $12ac - 5a + b$
32. $8ab^2 - a^2b$ **33.** $2rs - \dfrac{3}{2}r^2 - 5$ **34.** remove all parentheses and combine like terms. **35.** $-2x - 1$ **36.** $-2m - n - 7$ **37.** $a - 11b$ **38.** $6x - 2$
39. 72 **40.** 75 **41.** 320 **42.** 20 **43.** 13 **45.** 17 **47.** -21 **49.** $\dfrac{1}{4}$ **51.** 81 **53.** 4 **55.** $2x^3$, $3x^2$; $-x, 6; 2, 3, -1, 6$ **57.** z^2, $\dfrac{2y}{3}$; $1, \dfrac{2}{3}$ **59.** unlike **61.** like
63. like **65.** unlike **67.** $3m + 6$ **69.** $18n^2 + 12n - 6$ **71.** $-x + y$ **73.** $-4x + 3y$ **75.** $3x$ **77.** $6z$ **79.** $10m + 10n$ **81.** $2.2x^7$ **83.** $10y^6$
85. $-6w - 12y + 13z$ **87.** $-3k + 15$ **89.** $4n - 8$ **91.** $-3x + 3$ **93.** $4n - 2$ **95.** $-4n + 20$ **97.** $\dfrac{5}{6}x$ **99.** $-\dfrac{11}{2}$ **101.** $-3.5x - 6$ **103.** 49 **105.** 11
107. 40 **109.** 32 **111.** 27 **113.** 0 **115.** -13 **117.** -7 **119.** -12 **121.** 44 **123.** $-\dfrac{3}{2}$ **125.** 36 **127.** $\dfrac{4}{3} \approx 1.33$ **129.** 0.1536 **131.** 14 **133.** $78.70

135. $4819 **137. (a)** $8w - 8$ **(b)** 32 yards **139.** $228.88 **141.** $-3x^2 + 7x - 3$ **143.** The sum $2x^2 + 4x^2$ is not equal to $6x^4$ because when we combine like terms, we add the coefficients of the like terms and keep the variables and exponents the same. Put another way, $2x^2 + 4x^2 = (2 + 4)x^2 = 6x^2$.

Section 2.IR4 Solving Linear Equations **1.** linear; sides **2.** solution **3.** $x = 1$ **4.** $x = -7$ **5.** $z = -1$ **6.** True **7.** isolate **8.** $\{3\}$ **9.** $\{-2\}$
10. $\left\{\dfrac{1}{5}\right\}$ **11.** $\{2\}$ **12.** $\{-1\}$ **13.** $\left\{\dfrac{3}{2}\right\}$ **14.** $\{4\}$ **15.** $\{-4\}$ **16.** $\left\{\dfrac{3}{4}\right\}$ **17.** $\left\{\dfrac{1}{2}\right\}$ **18.** least common denominator **19.** $\{2\}$ **20.** $\{-5\}$ **21.** $\{-4\}$
22. $\left\{-\dfrac{3}{5}\right\}$ **23.** $\{10\}$ **24.** $\{32\}$ **25.** conditional equation **26.** contradiction; identity **27.** $\varnothing$ or $\{\ \}$; contradiction **28.** $\{x | x$ is any real number$\}$; identity **29.** $\{0\}$; conditional **30.** $\{z | z$ is any real number$\}$; identity **31.** $x = 2$ **33.** $m = 1$ **35.** $x = 5$ **37.** $\{2\}$ **39.** $\left\{-\dfrac{1}{4}\right\}$ **41.** $\{9\}$
43. $\left\{-\dfrac{2}{3}\right\}$ **45.** $\{-4\}$ **47.** $\{2\}$ **49.** $\left\{-\dfrac{3}{5}\right\}$ **51.** $\{-40\}$ **53.** $\{3\}$ **55.** $\{-5\}$ **57.** $\{\ \}$ or $\varnothing$; contradiction **59.** $\{\ \}$ or $\varnothing$; contradiction
61. $\{y | y$ is any real number$\}$ or $\mathbb{R}$; identity **63.** $\{-3\}$; conditional **65.** $\{\ \}$ or $\varnothing$; contradiction **67.** $\left\{\dfrac{5}{2}\right\}$; conditional
69. $\{z | z$ is any real number$\}$ or $\mathbb{R}$; identity **71.** $\left\{-\dfrac{7}{2}\right\}$; conditional **73.** $\left\{\dfrac{1}{7}\right\}$; conditional **75.** $\left\{\dfrac{7}{2}\right\}$; conditional **77.** $\{p | p$ is any real number$\}$ or $\mathbb{R}$; identity **79.** $\{\ \}$ or $\varnothing$; contradiction **81.** $\left\{-\dfrac{4}{3}\right\}$; conditional **83.** $\{-14\}$; conditional **85.** $\{-3\}$; conditional **87.** $\{-1.6\}$; conditional
89. $\{2\}$; conditional **91.** $a = -4$ **93.** $a = 3$ **95.** The card's annual interest rate is 0.15 or 15% **97.** Your adjusted income for 2015 was $32,492.
99. $4(x + 1) - 2$ is an algebraic expression and $4(x + 1) = 2$ is an equation. An algebraic expression is any combination of variables, grouping symbols, and mathematical operations but does not contain an equal sign. An equation is a statement made up of two algebraic expressions that are equal. **101.** Answers will vary.

Section 2.IR5 Using Linear Equations to Solve Problems **1.** equations **2.** $x + 7 = 12$ **3.** $3y = 21$ **4.** $2(n + 3) = 5$ **5.** $x - 10 = \dfrac{x}{2}$
6. $2n + 3 = 5$ **7.** False **8.** 18, 20, 22 **9.** 25, 26, 27 **10.** $15 per hour **11.** $12 per hour **12.** 150 miles **13.** 925 minutes **14.** 100 **15.** 40
16. 160 **17.** 75% **18.** $30 **19.** $1.20 **20.** Interest; principal **21.** $32.50 **22.** $10.50; $1410.50 **23. (a)** $h = \dfrac{2A}{b}$ **(b)** 5 inches
24. (a) $b = \dfrac{P - 2a}{2}$ **(b)** 10 cm **25.** $P = \dfrac{I}{rt}$ **26.** $y = \dfrac{C - Ax}{B}$ **27.** $h = \dfrac{4x - 3}{2x - 3}$ **28.** $n = \dfrac{S + d}{a + d}$ **29.** 10 **31.** 40 **33.** 37.5% **35.** $x + 12 = 20$; 8
37. $2(y + 3) = 16$; 5 **39.** $w - 22 = 3w$; -11 **41.** $4x = 2x + 14$; 7 **43.** $0.8x = x + 5$; -25 **45.** $r = \dfrac{d}{t}$ **47.** $m = \dfrac{y - y_1}{x - x_1}$ **49.** $x = \mu + \sigma Z$
51. $m_1 = \dfrac{r^2 F}{Gm_2}$ **53.** $P = \dfrac{A}{1 + rt}$ **55.** $F = \dfrac{9}{5}C + 32$ **57.** $y = -2x + 13$ **59.** $y = 3x - 5$ **61.** $y = -\dfrac{4}{3}x + \dfrac{13}{3}$ **63.** $y = -3x + 12$ **65.** 13 and 26
67. 24, 25, and 26 **69.** Kendra needs an 83 on her final exam to have an average of 80. **71.** Jacob would need to print 2500 pages for the cost to be the same for the two printers. **73.** Connor: $400,000; Olivia: $300,000; Avery: $100,000. **75.** The final bill will be $637.04. **77.** The dealer's cost is about $20,826.09. **79.** The flash drives originally cost $68.88. **81.** The Nissan Altima weighs 3193 pounds, the Mazda 6s weighs 3329 pounds, and the Honda Accord EX weighs 3312 pounds. **83.** Adam will get $8500 and Krissy will get $11,500. **85.** You should invest $15,000 in stocks and $9000 in bonds. **87.** The interest charge after one month will be $29.17. **89. (a)** $h = \dfrac{V}{\pi r^2}$ **(b)** The height of the cylinder is 8 inches.
91. (a) $A = \dfrac{206.3 - M}{0.711}$ **(b)** An individual whose maximum heart rate is 160 should be about 65 years old. **93. (a)** $P = \dfrac{A}{(1 + r)^t}$
(b) Approximately $4109.64 should be deposited today to have $5000 in 5 years in an account that pays 4% annual interest. **95.** Mathematical modeling is the process of developing an equation or inequality to find a solution to a problem. Just as there is more than one way to solve a problem, there is typically more than one way to develop a mathematical model.

Chapter 3 Integrated Review: Getting Ready for Least-Squares Regression

Section 3.IR1 The Rectangular Coordinate System and Equations in Two Variables   **1.** x-axis; y-axis; origin   **2.** x-coordinate; y-coordinate
3. False   **4.** False   **5.**   (a) I   (b) III   **6.**   (a) II   (b) I
   (c) IV   (d) x-axis   (c) III   (d) x-axis
   (e) y-axis   (f) II   (e) y-axis   (f) IV

7. (a) $(2, 3)$ (b) $(1, -3)$ (c) $(-3, 0)$ (d) $(-2, -1)$ (e) $(0, 2)$ **8.** True   **9.** (a) Yes   (b) No   (c) No
10. (a) No   (b) Yes   (c) Yes   **11.** $(3, 4)$ **12.** $(-3, 1)$ **13.** $\left(\dfrac{1}{2}, -\dfrac{2}{3}\right)$

14.

x	y	(x, y)
-2	-12	$(-2, -12)$
0	-2	$(0, -2)$
1	3	$(1, 3)$

15.

x	y	(x, y)
-1	7	$(-1, 7)$
2	-2	$(2, -2)$
5	-11	$(5, -11)$

16.

x	y	(x, y)
-5	2	$(-5, 2)$
-2	-4	$(-2, -4)$
2	-12	$(2, -12)$

17.

x	y	(x, y)
-6	-6	$(-6, -6)$
-1	-4	$(-1, -4)$
2	$-\dfrac{14}{5}$	$\left(2, -\dfrac{14}{5}\right)$

18. (a)

x therms	50 therms	100 therms	150 therms
C($)	$54.34	$96.68	$140.01

$(50, 53.34); (100, 96.68); (150, 140.01)$

(b)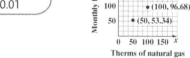

19.   **21.**   **23.**   **25.** $A(4, 0)$: x-axis; $B(-3, 2)$: quadrant II; $C(1, -4)$:
   quadrant IV; $D(-2, -4)$: quadrant III; $E(3, 5)$:
   quadrant I; $F(0, -3)$: y-axis

27. A No	**29.** A Yes	**31.** A Yes
B Yes	B No	B No
C Yes	C Yes	C Yes

Quadrant I: B Quadrant I: C, E Positive x-axis: A **33.** $(4, 1)$ **35.** $(5, -1)$ **37.** $(-3, 3)$
Quadrant II: A, E Quadrant III: F Negative x-axis: D
Quadrant III: C Quadrant IV: B Positive y-axis: C
Quadrant IV: D, F x-axis: A, G;   Negative y-axis: B
   y-axis: D, G

39.

x	y	(x, y)
-3	3	$(-3, 3)$
0	0	$(0, 0)$
1	-1	$(1, -1)$

41.

x	y	(x, y)
-2	7	$(-2, 7)$
-1	4	$(-1, 4)$
4	-11	$(4, -11)$

43.

x	y	(x, y)
-1	8	$(-1, 8)$
2	2	$(2, 2)$
3	0	$(3, 0)$

45.

x	y	(x, y)
-4	6	$(-4, 6)$
1	6	$(1, 6)$
12	6	$(12, 6)$

47.

x	y	(x, y)
1	$\dfrac{7}{2}$	$\left(1, \dfrac{7}{2}\right)$
-4	1	$(-4, 1)$
-2	2	$(-2, 2)$

49.

x	y	(x, y)
4	7	$(4, 7)$
-4	3	$(-4, 3)$
-6	2	$(-6, 2)$

51.

x	y	(x, y)
0	-3	$(0, -3)$
-2	0	$(-2, 0)$
2	-6	$(2, -6)$

53. $A\ (2, -16)$ **55.** $A\ (2, -6)$
     $B\ (-3, -1)$ $B\ (0, 0)$
     $C\left(-\dfrac{1}{3}, -9\right)$ $C\left(\dfrac{1}{6}, -\dfrac{1}{2}\right)$

57. $A\ (4, -8)$ **59.** $A\ (3, 4)$ **61.** $A\left(-4, -\dfrac{4}{3}\right)$ **63.** $A\ (20, 23)$ **65.** (a) $265,000   (b) $235,000   (c) After 8 years
     $B\ (4, -19)$ $B\ (-6, -2)$ $B\ (-2, -1)$ $B\ (-4, -17)$ (d) After 3 years, the book value is $255,000.
     $C\ (4, 5)$ $C\left(\dfrac{1}{2}, \dfrac{7}{3}\right)$ $C\left(-\dfrac{2}{3}, -\dfrac{7}{9}\right)$ $C\ (2.6, -6)$

67. (a) 30.5%   (b) 32.78%   (c) 34.68%   (d) 2060
(e) Answers may vary. The result is not reasonable since it is unlikely the trend will be linear over this time frame.

69.

a	b	(a, b)
2	−8	(2, −8)
0	−4	(0, −4)
−5	6	(−5, 6)

71.

p	q	(p, q)
0	10/3	(0, 10/3)
5/2	0	(5/2, 0)
−10	50/3	(−10, 50/3)

73. $k = 4$ **75.** $k = 2$

77. $k = \dfrac{1}{2}$

79. Points may vary; line

81.

x	y	(x, y)
−2	0	(−2, 0)
−1	−3	(−1, −3)
0	−4	(0, −4)
1	−3	(1, −3)
2	0	(2, 0)

83.

x	y	(x, y)
−2	10	(−2, 10)
−1	3	(−1, 3)
0	2	(0, 2)
1	1	(1, 1)
2	−6	(2, −6)

85. The first quadrant is the upper right-hand corner of the rectangular coordinate system. The quadrants are then II, III, and IV going in a counterclockwise direction. Points in quadrant I have both the x- and y-coordinates positive; points in quadrant II have a negative x-coordinate and a positive y-coordinate; points in quadrant III have both the x- and y-coordinates negative; points in quadrant IV have a positive x-coordinate and a negative y-coordinate. A point on the x-axis has y-coordinate equal to zero. A point on the y-axis has x-coordinate that is equal to zero.

87.   **89.**   **91.** 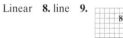 **93.**

Section 3.IR2 Graphing Equations in Two Variables

1. graph   **2.**   **3.**   **4.** linear; standard form   **5.** Linear   **6.** Not linear   **7.** Linear   **8.** line   **9.**

10.   **11. (a)** $(0, 3000), (10{,}000, 3800), (25{,}000, 5000)$ **(b)** 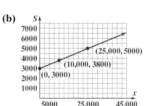   **12.** intercepts

13. Intercepts: $(0, 3), (4, 0)$; x-intercept: $(4, 0)$; y-intercept: $(0, 3)$ **14.** Intercept: $(0, −2)$; y-intercept: $(0, −2)$; no x-intercept   **15.** False

16.   **17.**   **18.**   **19.**   **20.**   **21.** vertical; $(a, 0)$ **22.** horizontal; $(0, b)$ **23.**

24.   **25.** Linear   **27.** Not linear   **29.** Not linear   **31.** Linear   **33.** $y = 2x$ **35.** $y = 4x − 2$

37. $y = −2x + 5$ **39.** $x + y = 5$ **41.** $−2x + y = 6$ **43.** $4x − 2y = −8$ **45.** $x = −4y$ **47.** $y + 7 = 0$

IR-AN-6 Section 3.IR2 Graphing Equations in Two Variables

49. $y - 2 = 3(x + 1)$ **51.** $(0, -5), (5, 0)$ **53.** $(0, 4), (2, 0)$ **55.** $(0, -3)$ **57.** $(-5, 0)$ **59.** $(0, -4), (-6, 0)$

61. $(0, 0)$ **63.** $(0, -5), (5, 0)$ **65.** $(0, 8), (6, 0)$ **67.** $(4, 0)$ **69.** $(0, -2)$

71. $3x + 6y = 18$ **73.** $-x + 5y = 15$ **75.** $\frac{1}{2}x = y + 3$ **77.** $9x - 2y = 0$ **79.** $y = -\frac{1}{2}x + 3$ **81.** $\frac{1}{3}y + 2 = 2x$

83. $\frac{x}{2} + \frac{y}{3} = 1$ **85.** $4y - 2x + 1 = 0$ **87.** $x = 5$ **89.** $y = -6$ **91.** $y - 12 = 0$ **93.** $3x - 5 = 0$

95. $y = 2x - 5$ **97.** $y = -5$ **99.** $2x + 5y = -20$ **101.** $2x = -6y + 4$ **103.** $x - 3 = 0$ **105.** $3y - 12 = 0$

107. $y = 2$ **109.** $x = 7$ **111.** $y = 5$ **113.** $x = -2$ **115.** $y = 4$ **123. (a)** $(0, 500), (4, 900), (10, 1500)$

117. $x = -9$ **(b)**

119. $x = 2y$

121. $y = x + 2$

  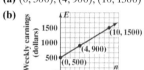

125. The "steepness" of the lines is the same.   **127.** The lines get more steep as the coefficient of x gets larger.   **(c)** If she sells 0 cars, her weekly earnings are $500.

129. $(0, -6), (-2, 0), (3, 0)$

131. $(0, 14), (-3, 0), (2, 0), (5, 0)$

133. The graph of an equation is the set of all ordered pairs (x, y) that make the equation a true statement.

135. Two points are needed to graph a line. A third point is used to verify your results.

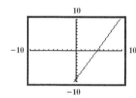 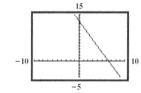

137. $y = 2x - 9$ **139.** $y + 2x = 13$ or $y = -2x + 13$ **141.** $y = -6x^2 + 1$

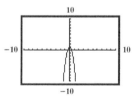

Section 3.IR3 Slope

1. $\frac{3}{5}$

2. False

3. True

4. positive

5.

$m = 2$; y increases by 2 when x increases by 1

6.

$m = -\frac{9}{5}$; y decreases by 9 when x increases by 5, or y increases by 9 when x decreases by 5.

7. 0; undefined

8. Slope undefined; when y increases by 1, there is no change in x.

9. $m = 0$; there is no change in y when x increases by 1 unit.

10. (a) **(b)** **(c)**

11. 8%   **12.** $m = 0.12$; between 10,000 and 14,000 miles driven, the average annual cost of operating a Chevy Cobalt is $0.12 per mile.

13. $-\dfrac{3}{2}$ **15.** $\dfrac{1}{2}$ **17.** $-\dfrac{2}{3}$

19. (a), (b)   **(c)** $m = \dfrac{1}{2}$; for every 2-unit increase in x, there is a 1-unit increase in y.   **21. (a), (b)**   **(c)** $m = -2$; the value of y decreases by 2 when x increases by 1.

23. $m = -2$; y decreases by 2 when x increases by 1.   **25.** $m = -1$; y decreases by 1 when x increases by 1.   **27.** $m = -\dfrac{5}{3}$; y decreases by 5 when x increases by 3.   **29.** $m = \dfrac{2}{3}$; y increases by 2 when x increases by 3.   **31.** $m = 2$; y increases by 2 when x increases by 1.   **33.** $m = \dfrac{1}{3}$; y increases by 1 when x increases by 3.   **35.** $m = -4.95$; y decreases by 4.95 when x increases by 1.   **37.** m is undefined.   **39.** $m = 0$ **41.** $m = 0$; the line is horizontal, so there is no change in y when x increases by 1.   **43.** slope is undefined; the line is vertical, so there is no change in x when y increases by 1.

45.   **47.**   **49.** 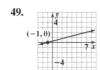   **51.**   **53.**   **55.**

57.   **59.**   **61.**   **63.** 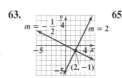   **65.**   **67.** $\dfrac{1}{3}$ **69.** 12 in. or 1 ft

71. 16%   **73.** $m = 2.325$ million; the population was increasing at an average rate of about 2.325 million people per year.

75. Points may vary. $(-2, 1)$, $(0, -5)$; $m = -3$

77. Points may vary. $(-2, -2)$, $(0, 4)$; $m = 3$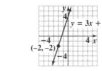

79. $m = -2$ **81.** $m = \dfrac{q}{p}$ **83.** $m = \dfrac{6}{a - 6}$ **85.** $MR = 2$; For every hot dog sold, revenue increases by $2.   **87.** The line is a vertical line. Answers will vary, but the points should be of the form (a, y_1) and (a, y_2), where a is a specific value and y_1, y_2 are any two different values. Because the line is a vertical line, the slope is undefined.

Section 3.IR4 Slope-Intercept Form of a Line

1. $y = mx + b$ **2.** slope: 4; y-intercept: $(0, -3)$ **3.** slope: -3; y-intercept: $(0, 7)$
4. slope: $-\dfrac{2}{5}$; y-intercept: $(0, 3)$ **5.** slope: 0; y-intercept: $(0, 8)$ **6.** slope: undefined; y-intercept: none

7.   **8.**   **9.**   **10.**   **11.**   **12.**   **13.**

14. point-plotting, using intercepts, using slope and a point

15. $y = 3x - 2$ **16.** $y = -\dfrac{1}{4}x + 3$ **17.** $y = -1$ **18.** $y = x$ **19. (a)** 2075 grams   **(b)** 2933 grams   **(c)** slope $= 143$. Birth weight increases by 143 grams for each additional week of pregnancy.   **(d)** A gestation period of 0 weeks does not make sense.   **(e)**

     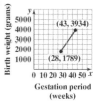

20. (a) $y = 0.38x + 50$ **(b)** $78.50 **(c)** 90 miles **(d)**

21. $m = 5$; y-intercept $= (0, 2)$ **23.** $m = 1$; y-intercept $= (0, -9)$

25. $m = -10$; y-intercept $= (0, 7)$ **27.** $m = -1$; y-intercept $= (0, -9)$

29. $m = -2$; y-intercept $= (0, 4)$ **31.** $m = -\dfrac{2}{3}$; y-intercept $= (0, 8)$

33. $m = \dfrac{5}{3}$; y-intercept $= (0, -3)$ **35.** $m = \dfrac{1}{2}$; y-intercept $= \left(0, -\dfrac{5}{2}\right)$ **37.** $m = 0$; y-intercept $= (0, -5)$ **39.** m is undefined; no y-intercept

41. **43.** **45.** **47.** **49.** **51.** **53.**

55. **57.** $y = -x + 8$ **59.** $y = \dfrac{6}{7}x - 6$ **61.** $y = -\dfrac{1}{3}x + \dfrac{2}{3}$ **63.** $x = -5$ **65.** $y = 3$ **67.** $y = 5x$

69. **71.** **73.** **75.** **77.** **79.**

81. **83.** **85.** **87.** **89.** **91.**

93. (a) 62 minutes **(b)** 2012 **(c)** Each year, Americans spent 21.7 additional minutes on their mobile devices. **(d)** No, because the number of daily minutes isn't infinite. **(e)**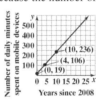

95. (a) $y = 0.08x + 400$ **(b)** $496 **(c)**

97. $B = -4$ **99.** $A = -4$ **101.** $B = -3$

103. (a) $y = 40x + 4000$ **(b)** $24,000 **(c)** 375 calculators **(d)**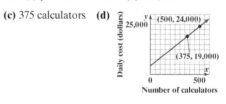

105. The line has a positive slope with a negative y-intercept. Possible equations are (a) and (e).

Section 3.IR5 Point-Slope Form of a Line

1. $y - y_1 = m(x - x_1)$ **2.** True

3. $y = 3x - 5$ **4.** $y = \dfrac{1}{3}x - 5$ **5.** $y = -4x - 3$ **6.** $y = -\dfrac{5}{2}x - 5$ **7.** $y = 3$ **8.** $y = x + 2$ **9.** $y = -3x + 1$

10. $x = 3$

11. Horizontal line: $y = b$;
Vertical line: $x = a$;
Point-slope: $y - y_1 = m(x - x_1)$;
Slope-intercept: $y = mx + b$;
Standard form: $Ax + By = C$

12. (a)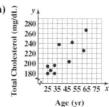

(b) As age increase, total cholesterol also increases.

13. (a) nonlinear (b) linear, positive slope
14. (a) Answers will vary. Using (25, 180) and (65, 269): $y = 2.225x + 124.375$ (b) 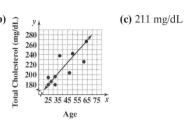 (c) 211 mg/dL
(d) Each year, a male's total cholesterol increases by 2.225 mg/dL; no

15. $y = 3x - 1$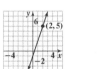
17. $y = -2x$
19. $y = \dfrac{1}{4}x - 3$
21. $y = -6x + 13$
23. $y = -7$
25. $x = -4$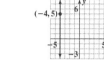

27. $y = \dfrac{2}{3}x + 2$
29. $y = -\dfrac{3}{4}x$
31. $x = -3$
33. $y = -5$
35. $y = -4.3$
37. $y = 2x + 4$
39. $y = -4x + 6$
41. $y = -\dfrac{3}{2}x - \dfrac{5}{2}$
43. $y = 2x - 5$
45. $y = -3$
47. $x = 2$
49. $y = 0.25x + 0.575$
51. $y = x - \dfrac{11}{4}$

53. $y = 5x - 22$
55. $y = 5$
57. $y = x + 2$
59. $y = \dfrac{1}{2}x + 4$
61. $x = 5$
63. $y = -7x + 2$

65. $y = -\dfrac{2}{3}x + 7$
67. $y = -\dfrac{3}{2}x$
69. $y = \dfrac{2}{5}x - 2$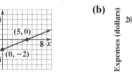
71. (a) When 60 packages are shipped, the expenses are $1635.
(b)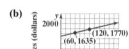
(c) $y = \dfrac{9}{4}x + 1500$
(d) $1950
(e) Expenses increase by $2.25 for each additional package.

73. (a) (600, 15); (750, 6) (b) 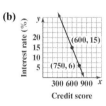 (c) $y = -0.06x + 51$ (d) 9% (e) For each 1-point increase in credit score, interest rate goes down 0.06%.
75. (a) $y = 7250x - 1735$ (b) $3847.50
(c) If the weight of a diamond increases by 1 carat, the cost increases by $7250.
(d) 0.97 carat

77. (a) $y = 0.7737x + 1544.4115$ (b) $12,547.2 billion
(c) If personal disposable income increases by $1, personal consumption increase by $0.77. (d) $14,957 billion

79. (a) 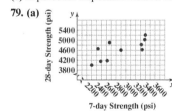 (b) Linear (c) Answers will vary. Using the points (2300, 4070) and (3390, 5220), the equation is $y = 1.06x + 1632$ (d) 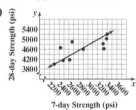 (e) 4812 psi (f) If the 7-day strength is increased by 1 psi, then the 28-day strength will increase by 1.06 psi.

81. (a) 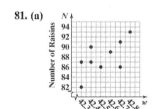 (b) Answers will vary. Using the points (42.3, 82) and (42.8, 93), the equation is $y = 22x - 848.6$. (c) 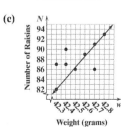 (d) approximately 86 raisins (e) If the weight increases by 1 gram, then the number of raisins increases by 22 raisins.

83. No. The data do not follow a linear pattern.

Chapter 4 Integrated Review

Section 4.IR1 Scientific Notation 1. scientific 2. positive 3. false 4. 4.32×10^2 5. 1.0302×10^4 6. 5.432×10^6 7. 9.3×10^{-2} 8. 4.59×10^{-5} 9. 8×10^{-8} 10. True 11. True 12. 310 13. 0.901 14. 170,000 15. 7 16. 0.00089 17. 6×10^7 18. 8×10^{-3} 19. 1.5×10^4 20. 2.8×10^{-5} 21. 4×10^5 22. 2×10^{-4} 23. 5×10^3 24. 6.25×10^{-5} 25. $1.047 \times 10^{10} = 10{,}470{,}000{,}000$ gallons 26. $1.314 \times 10^9 = 1{,}314{,}000{,}000$ customers 27. 3×10^5 29. 6.4×10^7 31. 5.1×10^{-4} 33. 1×10^{-9} 35. 8.007×10^9 37. 3.09×10^{-5} 39. 6.2×10^2 41. 4×10^0 43. 7.303×10^9 45. $\$1.89 \times 10^{13}$ 47. 9.3×10^7 miles 49. 3×10^{-5} mm 51. 2.5×10^{-7} m 53. 3.11×10^{-3} kg 55. 420,000 57. 100,000,000 59. 0.0039 61. 0.4 63. 3760 65. 0.0082 67. 0.00006 69. 7,050,000 71. 0.000000000000001 73. 0.00225 75. 500,000 77. 3×10^9 79. 8.4×10^{-3} 81. 3×10^8 83. 2.5×10^1 85. 8×10^8 87. 9×10^{15} 89. 1.116×10^7 miles 91. 0.000001 mile; 0.00000168 mile 93. (a) 4.1×10^9 pounds (b) 3.19×10^8 (c) 12.9 pounds 95. 2.5×10^{-4} m 97. 8×10^{-10} m 99. 7.15×10^{-8} m 101. $1.2348\pi \times 10^{-14}$ m^3 103. $2.88\pi \times 10^{-25}$ m^3 105. To convert a number written in decimal notation to scientific notation, count the number of a decimal places, N, that the decimal point must be moved to arrive at a number x such that $1 \leq x < 10$. If the original number is greater than or equal to 1, move the decimal point to the left that many places and write the number in the form $x \times 10^N$. If the original number is between 0 and 1, move the decimal point to the right that many places and write the number in the form $x \times 10^{-N}$. 107. The number 34.5×10^4 is incorrect because the number 34.5 is not a number between 1 (inclusive) and 10. The correct answer is 3.45×10^5.

Section 4.IR2 Linear Inequalities in One Variable 1. closed interval 2. left endpoint; right endpoint 3. False 4. $[-3, 2]$ 5. $[3, 6)$ 6. $(-\infty, 3]$ 7. $\left(\frac{1}{2}, \frac{7}{2}\right)$ 8. $0 < x \leq 5$ 9. $-6 < x < 0$ 10. $x > 5$ 11. $x \leq \frac{8}{3}$ 12. intersection 13. and; or 14. True 15. $\{1, 3, 5\}$ 16. $\{2, 4, 6\}$ 17. $\{1, 2, 3, 4, 5, 6, 7\}$ 18. $\{1, 2, 3, 4, 5, 6, 8\}$ 19. $\varnothing$ or $\{\ \}$ 20. $\{1, 2, 3, 4, 5, 6, 7, 8\}$ 21. $\{x | 2 < x < 7\}; (2, 7)$ 22. $\{x | x \leq -3 \text{ or } x > 2\}; (-\infty, -3] \cup (2, \infty)$ 23. $[2, 10]$; 25. $[-4, 0)$; 27. $[6, \infty)$; 29. $\left(-\infty, \frac{3}{2}\right)$; 31. $1 < x < 8$; 33. $-5 < x \leq 1$; 35. $x < 5$; 37. $x \leq 3$; 39. $\{1, 4, 5, 6, 7, 8, 9\}$ 41. $\{5, 7, 9\}$ 43. $\varnothing$ or $\{\ \}$ 45. (a) $A \cap B = \{x | -2 < x \leq 5\}; (-2, 5]$  (b) $A \cup B = \{x | x \text{ is any real number}\}; (-\infty, \infty)$ 47. (a) $E \cap F = \varnothing$ or $\{\ \}$ (b) $E \cup F = \{x | x < -1 \text{ or } x > 3\}; (-\infty, -1) \cup (3, \infty)$

Section 4.IR3 Solving Linear Inequalities in One Variable 1. solve 2. $10 < 15$; Addition Property 3. $\{n | n > 3\}; (3, \infty)$ 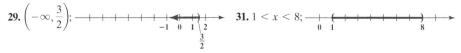 4. $\{x | x < 4\}; (-\infty, 4)$ 5. $\{n | n \leq -4\}; (-\infty, -4]$ 6. $\{x | x > -1\}; (-1, \infty)$ 7. $1 < 4$; Multiplication Property 8. $-3 > -5$; Multiplication Property 9. $\{k | k < 12\}; (-\infty, 12)$ 10. $\{n | n \geq -3\}; [-3, \infty)$ 11. $\{k | k < -8\}; (-\infty, -8)$ 12. $\left\{p | p \geq \frac{3}{5}\right\}; \left[\frac{3}{5}, \infty\right)$ 13. $\{x | x > 7\}; (7, \infty)$ 14. $\{n | n > -3\}; (-3, \infty)$ 15. $\{x | x > -4\}; (-4, \infty)$

16. $\{x | x \leq -6\}; (-\infty, -6]$
17. $\{x | x > 8\}; (8, \infty)$
18. $\left\{x | x \leq \dfrac{19}{8}\right\}; \left(-\infty, \dfrac{19}{8}\right]$
19. (a) True (b) True
20. $\{x | x \geq 3\}; [3, \infty)$
21. $\varnothing$ or $\{\ \}$
22. $\{x | x > 4\}; (4, \infty)$
23. $\{x | x \text{ is any real number}\}; (-\infty, \infty)$
24. at most 20 boxes   25. <; Addition Property of Inequality   27. >; Multiplication Property of Inequality   29. ≤; Addition Property of Inequality   31. ≤; Multiplication Property of Inequality
33. $\{x | x < 4\}; (-\infty, 4)$ 35. $\{x | x \geq 2\}; [2, \infty)$ 37. $\{x | x \leq 5\}; (-\infty, 5]$
39. $\{x | x > -7\}; (-7, \infty)$ 41. $\{x | x > 3\}; (3, \infty)$ 43. $\{x | x \geq 2\}; [2, \infty)$
45. $\{x | x \geq -1\}; [-1, \infty)$ 47. $\{x | x > -7\}; (-7, \infty)$ 49. $\{x | x \leq 0\}; (-\infty, 0]$
51. $\{x | x < -20\}; (-\infty, -20)$ 53. $\varnothing$ or $\{\ \}$
55. $\{n | n \text{ is any real number}\}; (-\infty, \infty)$ 57. $\{n | n > 5\}; (5, \infty)$
59. $\{w | w \text{ is any real number}\}; (-\infty, \infty)$ 61. $\left\{y | y < -\dfrac{3}{2}\right\}; \left(-\infty, -\dfrac{3}{2}\right)$
63. $x \geq 16{,}000$ 65. $x \leq 20{,}000$ 67. $x > 12{,}000$ 69. $x > 0$ 71. $x \leq 0$ 73. $\{x | x > 4\}; (4, \infty)$
75. $\left\{x | x < \dfrac{3}{4}\right\}; \left(-\infty, \dfrac{3}{4}\right)$ 77. $\{x | x \text{ is any real number}\}; (-\infty, \infty)$
79. $\{a | a < -1\}; (-\infty, -1)$ 81. $\{n | n \text{ is any real number}\}; (-\infty, \infty)$
83. $\left\{x | x \geq \dfrac{4}{3}\right\}; \left[\dfrac{4}{3}, \infty\right)$ 85. $\{x | x < 25\}; (-\infty, 25)$ 87. $\varnothing$ or $\{\ \}$
89. $\{x | x > 5.9375\}; (5.9375, \infty)$ 91. at most 1250 miles   93. at least 32   95. more than 400 minutes
97. greater than $50,361.11   99. at least 74   101. A left parenthesis is used to indicate that the solution is greater than a number. A left bracket is used to show that the solution is greater than or equal to a given number.   103. In solving an inequality, when the variables are eliminated and a true statement results, the solution is all real numbers. In solving an inequality, when the variables are eliminated and a false statement results, the solution is the empty set.

Chapter 6 Integrated Review: Getting Ready for the Normal Probability Distribution

Section 6.IR1 Perimeter and Area of Polygons and Circles
1. perimeter   2. area   3. Perimeter: 22 feet; Area: 24 square feet
4. Perimeter: 26 m; Area: 30 square m   5. (a) 30 feet   (b) $29.85   6. (a) 108 ft²   (b) $565.92   7. (a) 180 ft²   (b) 25,920 in²   (c) 240   8. False
9. Perimeter: 16 cm; Area: 16 square cm   10. Perimeter: 6 yards; Area: 2.25 square yards   11. (a) 12.25 ft²   (b) $91.02   12. Perimeter: 130 yards; Area: 650 square yards   13. (a) 378 ft²   (b) 2   14. $\dfrac{1}{2}h(b + B); h; b; B$ 15. Perimeter: 36 m; Area: 70 square m   16. Perimeter: 33 yards; Area: 51 square yards
17. $11.25   18. True   19. Perimeter: 19 mm; Area: 12 square mm   20. Perimeter: 30 feet; Area: 30 square feet   21. 13 square feet   22. radius   23. True
24. 12 feet   25. $\dfrac{15}{2}$ in., or 7.5 in.   26. 18 cm   27. 7.2 yards   28. circumference   29. False   30. Circumference: 8π feet. ≈ 25.12 feet; Area: 16π square feet ≈ 50.24 square feet   31. Circumference: 24π cm ≈ 75.36 cm; Area: 144π square cm ≈ 452.16 square cm   32. 180,864 square miles
33. Perimeter: 28 feet; Area: 40 square feet   35. Perimeter: 40 m; Area: 75 square m   37. Perimeter: 72 feet; Area: 218 square feet   39. Perimeter: 54 m; Area: 62 square m   41. Perimeter: 30 feet; Area: 45 square feet   43. Perimeter: 28 mm; Area: 36 square mm   45. Perimeter: 40 in.; Area: 84 square in.
47. Perimeter: 45 cm; Area: 94.5 square cm   49. Perimeter: 32 m; Area: 42 square m   51. Perimeter: 32 feet; Area: 24 square feet   53. 10 in.   55. 5 cm
57. 7 cm   59. $\dfrac{11}{2}$ yards or 5.5 yards   61. Circumference: 32π in. ≈ 100.48 in.; Area: 256π square in. ≈ 803.84 square in.   63. Circumference: 20π cm ≈ 62.82 cm; Area: 100π square cm ≈ 314 square cm   65. $44.73   67. $8.38   69. 26 in.   71. 6 in.   73. 7918 mi   75. 696,342 km   77. $\dfrac{3}{8}$ inch   79. 282,600 square miles   81. (a) 94.2 feet;   (b) 706.5 square feet   83. 4,906,250 square feet   85. (a) 20.38 miles;   (b) 326.05 square miles   87. (a) 62 tiles;

IR-AN-12 Section 6.IR1 Perimeter and Area of Polygons and Circles

(b) $372; (c) Yes 89. (a) 4948 ft^2 (b) $1237 91. 1 square units 93. about 26.17 feet 95. 30.28 square feet 97. 0.8413
99. Ted incorrectly multiplied $\frac{1}{2}$ by both 12 ft. and 4 ft. The correct solution is

$$A = \frac{1}{2}bh$$
$$= \frac{1}{2}(12\,\text{ft.})(4\,\text{ft.})$$
$$= \frac{1}{2}\left(\overset{6}{\cancel{12}}\,\text{ft.}\right)(4\,\text{ft.})$$
$$= 6\,\text{ft.} \cdot 4\,\text{ft.}$$
$$= 24\,\text{ft.}^2$$

Chapter 8 Integrated Review: Getting Ready for Confidence Intervals

Section 8.IR1 Compound Inequalities 1. $\{x|x \geq 2\}; [2, \infty)$
2. $\{x|-3 < x < 3\}; (-3, 3)$ 3. $\{x|1 < x < 3\}; (1, 3)$
4. $\{\ \}$ or $\varnothing$ 5. $\{1\}$ 6. $\{x|-1 < x < 3\}; (-1, 3)$
7. $\left\{x\left|\frac{5}{4} < x \leq 2\right.\right\}; \left(\frac{5}{4}, 2\right]$ 8. $\{x|-6 \leq x \leq -2\}; [-6, -2]$
9. $\{x|x < -2 \text{ or } x > 5\}; (-\infty, -2) \cup (5, \infty)$
10. $\{x|x \leq 2 \text{ or } x > 6\}; (-\infty, 2] \cup (6, \infty)$ 11. $\{x|x \geq 1\}; [1, \infty)$
12. $\{x|x < 4 \text{ or } x > 9\}; (-\infty, 4) \cup (9, \infty)$
13. $\{x|x \text{ is any real number}\}; (-\infty, \infty)$ 14. $\{x|x \geq -1\}; [-1, \infty)$
15. Income between $37,450 and $90,750. 16. Total usage was between 45 and 245 minutes.
17. (a) $\{x|-2 \leq x \leq 2\}; [-2, 2]$ (b) $(-\infty, -2) \cup (2, \infty)$
19. (a) $\{x|-3 < x < 3\}; (-3, 3)$ (b) $(-\infty, -3] \cup [3, \infty)$
21. $\{x|-2 \leq x < 3\}; [-2, 3)$ 23. $\varnothing$ or $\{\ \}$ 25. $\{x|x < -2\}; (-\infty, -2)$
27. $\{1\}$ 29. $\{x|-1 \leq x < 3\}; [-1, 3)$
31. $\left\{x\left|-\frac{2}{3} \leq x \leq \frac{3}{2}\right.\right\}; \left[-\frac{2}{3}, \frac{3}{2}\right]$ 33. $\left\{x\left|-1 < x \leq \frac{4}{5}\right.\right\}; \left(-1, \frac{4}{5}\right]$
35. $\{x|0 \leq x \leq 8\}; [0, 8]$ 37. $\{x|-6 \leq x \leq -2\}; [-6, -2]$
39. $\varnothing$ or $\{\ \}$ 41. $\left\{x\left|-\frac{5}{3} < x \leq 5\right.\right\}; \left(-\frac{5}{3}, 5\right]$ 43. $\{x|-4 < x \leq 3\}; (-4, 3]$
45. $\{x|x < -2 \text{ or } x > 3\}; (-\infty, -2) \cup (3, \infty)$
47. $\{x|x < -2 \text{ or } x > 5\}; (-\infty, -2) \cup (5, \infty)$
49. $\{x|x \text{ is any real number}\}; (-\infty, \infty)$
51. $\{x|x < -1 \text{ or } x > 4\}; (-\infty, -1) \cup (4, \infty)$
53. $\{x|x \leq -3 \text{ or } x > 6\}; (-\infty, -3] \cup (6, \infty)$
55. $\{x|x \text{ is any real number}\}; (-\infty, \infty)$

57. $\{x | x < 0 \text{ or } x > \frac{5}{2}\}; (-\infty, 0) \cup \left(\frac{5}{2}, \infty\right)$ **59.** $\{a | -3 \le a < 0\}; [-3, 0)$

61. $\{x | x \text{ is any real number}\}; (-\infty, \infty)$ **63.** $\{x | -2 \le x \le \frac{8}{3}\}; \left[-2, \frac{8}{3}\right]$

65. $\{x | x < -10 \text{ or } x > 2\}; (-\infty, -10) \cup (2, \infty)$ **67.** $\{x | 2 < x < 5\}; (2, 5)$

69. $\{x | x \le -3 \text{ or } x > \frac{15}{4}\}; (-\infty, -3] \cup \left(\frac{15}{4}, \infty\right)$

71. $\{x | -5 < x \le \frac{1}{2}\}; \left(-5, \frac{1}{2}\right]$ **73.** $a = 1$ and $b = 8$ **75.** $a = 12$ and $b = 30$ **77.** $a = -1$ and $b = 23$

79. $90 < x < 140$ **81.** Joanna needs to score at least a 77 on the final. That is, $77 \le x \le 100$ (assuming 100 is the max score; otherwise, $77 \le x \le 104$). **83.** The amount withheld ranges between \$89.50 and \$104.50, inclusive. **85.** Total sales between \$50,000 and \$250,000 **87.** The electrical usage ranged from 850 kwh to 1370 kwh.

89. Step 1:

$a < b$
$a + a < a + b$
$2a < a + b$
$\frac{2a}{2} < \frac{a+b}{2}$
$a < \frac{a+b}{2}$

Step 2:

$a < b$
$a + b < b + b$
$a + b < 2b$
$\frac{a+b}{2} < \frac{2b}{2}$
$\frac{a+b}{2} < b$

Step 3: Since $a < \frac{a+b}{2}$ and $\frac{a+b}{2} < b$, it follows that $a < \frac{a+b}{2} < b$.

91. $\{\ \}$ or $\varnothing$ **93.** This is a contradiction. There is no solution. If, during simplification, the variable terms all cancel out and a contradiction results, then there is no solution to the inequality. **95.** If $x < 2$ then $x - 2 < 2 - 2 \Rightarrow x - 2 < 0$. When multiplying both sides of the inequality by $x - 2$ in the second step, the direction of the inequality must switch.

Section 8.IR2 Absolute Value Equations and Inequalities **1.** $\{-7, 7\}$ **2.** $\{-1, 1\}$ **3.** $a; -a$ **4.** $2x + 3 = -5$ **5.** $\{-2, 5\}$ **6.** $\{-\frac{5}{3}, 3\}$

7. $\{-1, \frac{9}{5}\}$ **8.** $\{-5, 1\}$ **9.** True **10.** $\{\ \}$ or $\varnothing$ **11.** $\{\ \}$ or $\varnothing$ **12.** $\{-1\}$ **13.** $u; v; u; -v$ **14.** $\{-8, -\frac{2}{3}\}$ **15.** $\{-2, 3\}$ **16.** $\{-3, 0\}$

17. $\{2\}$ **18.** $-a < u < a$ **19.** $-10; 10$ **20.** $\{x | -5 \le x \le 5\}; [-5, 5]$

21. $\{x | -\frac{3}{2} < x < \frac{3}{2}\}; \left(-\frac{3}{2}, \frac{3}{2}\right)$ **22.** $\{x | -8 < x < 2\}; (-8, 2)$

23. $\{x | -2 \le x \le 5\}; [-2, 5]$ **24.** $\{\ \}$ or $\varnothing$

25. False **26.** $\{x | -2 < x < 2\}; (-2, 2)$ **27.** $\{x | -1 \le x \le 7\}; [-1, 7]$

28. $\{x | -2 \le x \le 1\}; [-2, 1]$ **29.** $\{x | -\frac{7}{3} < x < 3\}; \left(-\frac{7}{3}, 3\right)$

30. $u < -a; u > a$ **31.** $-7; 7$ **32.** $\{x | x \le -6 \text{ or } x \ge 6\}; (-\infty, -6] \cup [6, \infty)$

33. $\{x | x < -\frac{5}{2} \text{ or } x > \frac{5}{2}\}; \left(-\infty, -\frac{5}{2}\right) \cup \left(\frac{5}{2}, \infty\right)$ **34.** $<$

35. $\{x | x < -7 \text{ or } x > 1\}; (-\infty, -7) \cup (1, \infty)$

36. $\{x | x \le -\frac{1}{2} \text{ or } x \ge 2\}; \left(-\infty, -\frac{1}{2}\right] \cup [2, \infty)$

37. $\{x | x < -\frac{5}{3} \text{ or } x > 3\}; \left(-\infty, -\frac{5}{3}\right) \cup (3, \infty)$

38. $\{x | x \ne -\frac{5}{2}\}; \left(-\infty, -\frac{5}{2}\right) \cup \left(-\frac{5}{2}, \infty\right)$

Section 8.IR2 Absolute Value Equations and Inequalities

39. $\{x \mid x \text{ is any real number}\}$; $(-\infty, \infty)$

40. $\{x \mid x \text{ is any real number}\}$; $(-\infty, \infty)$

41. The acceptable belt width is between $\dfrac{127}{32}$ inches and $\dfrac{129}{32}$ inches.

42. The percentage of Americans that have been shot at is between 7.3% and 10.7%, inclusive. **43.** $\{-10, 10\}$ **45.** $\{-1, 7\}$

47. $\left\{-1, \dfrac{13}{3}\right\}$ **49.** $\{-5, 5\}$ **51.** $\left\{-\dfrac{11}{2}, \dfrac{5}{2}\right\}$ **53.** $\{-4, 10\}$ **55.** $\{0\}$ **57.** $\left\{-\dfrac{7}{3}, 3\right\}$ **59.** $\left\{-7, \dfrac{3}{5}\right\}$ **61.** $\{1, 3\}$ **63.** $\{2\}$

65. $\{x \mid -9 < x < 9\}$; $(-9, 9)$ **67.** $\{x \mid -3 \leq x \leq 11\}$; $[-3, 11]$

69. $\left\{x \mid -3 < x < \dfrac{7}{3}\right\}$; $\left(-3, \dfrac{7}{3}\right)$ **71.** $\varnothing$ or $\{\ \}$

73. $\{x \mid 0 < x < 6\}$; $(0, 6)$ **75.** $\left\{x \mid -1 < x < \dfrac{9}{5}\right\}$; $\left(-1, \dfrac{9}{5}\right)$

77. $\{x \mid 1.995 < x < 2.005\}$; $(1.995, 2.005)$

79. $\{y \mid y < 3 \text{ or } y > 7\}$; $(-\infty, 3) \cup (7, \infty)$

81. $\left\{x \mid x \leq -2 \text{ or } x \geq \dfrac{1}{2}\right\}$; $(-\infty, -2] \cup \left[\dfrac{1}{2}, \infty\right)$

83. $\{y \mid y \text{ is any real number}\}$; $(-\infty, \infty)$

85. $\left\{x \mid x < -2 \text{ or } x > \dfrac{4}{5}\right\}$; $(-\infty, -2) \cup \left(\dfrac{4}{5}, \infty\right)$

87. $\{x \mid x < 0 \text{ or } x > 1\}$; $(-\infty, 0) \cup (1, \infty)$

89. $\{x \mid x \leq -2 \text{ or } x \geq 3\}$; $(-\infty, -2] \cup [3, \infty)$

91. (a) $\{-5, 5\}$ (b) $\{x \mid -5 \leq x \leq 5\}$; $[-5, 5]$ (c) $\{x \mid x < -5 \text{ or } x > 5\}$; $(-\infty, -5) \cup (5, \infty)$

93. (a) $\{-5, 1\}$ (b) $\{x \mid -5 < x < 1\}$; $(-5, 1)$ (c) $\{x \mid x \leq -5 \text{ or } x \geq 1\}$; $(-\infty, -5] \cup [1, \infty)$

95. $\{x \mid x < -5 \text{ or } x > 5\}$; $(-\infty, -5) \cup (5, \infty)$

97. $\{-4, -1\}$ **99.** $\{-5, 5\}$ **101.** $\left\{x \mid -2 \leq x \leq \dfrac{6}{5}\right\}$; $\left[-2, \dfrac{6}{5}\right]$ **103.** $\varnothing$ or $\{\ \}$

105. $\left\{x \mid x \leq -\dfrac{7}{3} \text{ or } x \geq 1\right\}$; $\left(-\infty, -\dfrac{7}{3}\right] \cup [1, \infty)$

107. $\left\{x \mid x < 0 \text{ or } x > \dfrac{4}{3}\right\}$; $(-\infty, 0) \cup \left(\dfrac{4}{3}, \infty\right)$ **109.** $\{-1, 1\}$ **111.** $\varnothing$ or $\{\ \}$ **113.** $\left\{-8, \dfrac{4}{7}\right\}$

115. $|x - 5| < 3$; $\{x \mid 2 < x < 8\}$; $(2, 8)$ **117.** $|2x - (-6)| > 3$; $\left\{x \mid x < -\dfrac{9}{2} \text{ or } x > -\dfrac{3}{2}\right\}$; $\left(-\infty, -\dfrac{9}{2}\right) \cup \left(-\dfrac{3}{2}, \infty\right)$

119. The acceptable rod lengths are between 5.6995 inches and 5.7005 inches, inclusive. **121.** An unusual IQ score would be less than 70.6 or greater than 129.4. **123.** $\left\{-\dfrac{5}{2}\right\}$ **125.** $\{2\}$ **127.** $\varnothing$ or $\{\ \}$ **129.** $\{x \mid x \leq -5\}$; $(-\infty, -5]$ **131.** The absolute value, when isolated, is equal to a negative number, which is not possible. **133.** The absolute value, when isolated, is less than -3. Since absolute values are always nonnegative, this is not possible.

Appendix A Functions, Exponential Functions, Logarithmic Functions

Section A.IR1 Relations **1.** corresponds; depends **2.** {(Max, November 8), (Alesia, January 20), (Trent, March 3), (Yolanda, November 8), (Wanda, July 6), (Elvis, January 8)} **3.** Domain   Range **4.** domain; range **5.** Domain: {Max, Alesia, Trent, Yolanda, Wanda, Elvis}; Range: {January 20, March 3, July 6, November 8, January 8} **6.** Domain: {1, 5, 8, 10}; Range: {3, 4, 13} **7.** Domain: {-2, -1, 2, 3, 4}; Range: {-3, -2, 0, 2, 3} **8.** True **9.** False **10.** Domain: $\{x \mid -2 \leq x \leq 4\}$ or $[-2, 4]$; Range: $\{y \mid -2 \leq y \leq 2\}$ or $[-2, 2]$ **11.** Domain: $\{x \mid x \text{ is a real number}\}$ or $(-\infty, \infty)$; Range: $\{y \mid y \text{ is a real number}\}$ or $(-\infty, \infty)$

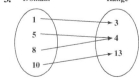

12.
Domain: $\{x | x$ is a real number$\}$ or $(-\infty, \infty)$;
Range: $\{y | y$ is a real number$\}$ or $(-\infty, \infty)$

13.
Domain: $\{x | x$ is a real number$\}$ or $(-\infty, \infty)$;
Range: $\{y | y \geq -8\}$ or $[-8, \infty)$

14.
Domain: $\{x | x \geq 1\}$ or $[1, \infty)$;
Range: $\{y | y$ is a real number$\}$ or $(-\infty, \infty)$

15. $\{$ (USA Today, 2.28.), (Wall Street Journal, 2.06), (New York Times, 1.1), (Los Angeles Times, 0.82), (Washington Post, 0.70) $\}$; Domain: $\{$ USA Today, Wall Street Journal, New York Times, Los Angeles Times, Washington Post $\}$; Range: $\{2.28, 2.06, 1.1, 0.82, 0.70\}$ **17.** $\{$(Less than 9th Grade, $20,791), (9th–12th Grade – no diploma, $23,234), (High School Graduate, $32,456), (Associate's Degree, $42,508), (Bachelor's Degree, $62,240) $\}$; Domain: $\{$ Less than 9th Grade, 9th–12th Grade – no diploma, High School Graduate, Associate's Degree, Bachelor's Degree $\}$; Range: $\{ \$20,791, \$23,234, \$32,456, \$42,508, \$62,240 \}$

19.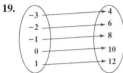
Domain: $\{-3, -2, -1, 0, 1\}$;
Range: $\{4, 6, 8, 10, 12\}$

21.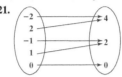
Domain: $\{-2, -1, 0, 1, 2\}$;
Range: $\{0, 2, 4\}$

23.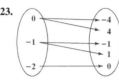
Domain: $\{-2, -1, 0\}$;
Range: $\{-4, -1, 0, 1, 4\}$

25. Domain: $\{-3, -2, 0, 2, 3\}$; Range: $\{-3, -1, 2, 3\}$
27. Domain: $\{x | -4 \leq x \leq 4\}$ or $[-4, 4]$;
Range: $\{y | -2 \leq y \leq 2\}$ or $[-2, 2]$
29. Domain: $\{x | -1 \leq x \leq 3\}$ or $[-1, 3]$;
Range: $\{y | 0 \leq y \leq 4\}$ or $[0, 4]$
31. Domain: $\{x | x$ is a real number$\}$ or $(-\infty, \infty)$;
Range: $\{y | y \geq -3\}$ or $[-3, \infty)$

33. Domain: $\{x | x$ is a real number$\}$ or $(-\infty, \infty)$; Range: $\{y | y$ is a real number$\}$ or $(-\infty, \infty)$ **35.** Domain: $\{x | x$ is a real number$\}$ or $(-\infty, \infty)$; Range: $\{y | y$ is a real number$\}$ or $(-\infty, \infty)$ **37.** Domain: $\{x | x$ is a real number$\}$ or $(-\infty, \infty)$; Range: $\{y | y$ is a real number$\}$ or $(-\infty, \infty)$ **39.** Domain: $\{x | x$ is a real number$\}$ or $(-\infty, \infty)$; Range: $\{y | y \leq 0\}$ or $(-\infty, 0]$ **41.** Domain: $\{x | x$ is a real number$\}$ or $(-\infty, \infty)$; Range: $\{y | y \geq -8\}$ or $[-8, \infty)$ **43.** Domain: $\{x | x$ is a real number$\}$ or $(-\infty, \infty)$; Range: $\{y | y \geq 0\}$ or $[0, \infty)$ **45.** Domain: $\{x | x$ is a real number$\}$ or $(-\infty, \infty)$; Range: $\{y | y$ is a real number$\}$ or $(-\infty, \infty)$ **47.** Domain: $\{x | x$ is a real number$\}$ or $(-\infty, \infty)$; Range: $\{y | y$ is a real number$\}$ or $(-\infty, \infty)$ **49.** Domain: $\{x | x$ is a real number$\}$ or $(-\infty, \infty)$; Range: $\{y | y$ is a real number$\}$ or $(-\infty, \infty)$ **51.** Domain: $\{x | x$ is a real number$\}$ or $(-\infty, \infty)$; Range: $\{y | y \geq -4\}$ or $[-4, \infty)$ **53.** Domain: $\{x | x \geq -1\}$ or $[-1, \infty)$; Range: $\{y | y$ is a real number$\}$ or $(-\infty, \infty)$

55. (a) Domain: $\{x | 0 < x < 50\}$ or $(0, 50)$; Range: $\{y | 0 < y \leq 625\}$ or $(0, 625]$ **(b)** Width must be less than $\dfrac{1}{2}$ the perimeter.

57. (a) Domain: $\{m | 0 \leq m \leq 15,120\}$ or $[0, 15,120]$; Range: $\{C | 100 \leq C \leq 3380\}$ or $[100, 3380]$ **(b)** $21 \cdot 12 \cdot 60 = 15,120$ minutes

59. Actual graphs will vary but all should be horizontal lines.   **61.** A relation is a correspondence between two sets called the domain and range. The domain is the set of all inputs, and the range is the set of all outputs.

Section A.IR2 An Introduction to Functions

1. function   **2.** false   **3.** Function; Domain: $\{$ Max, Alesia, Trent, Yolanda, Wanda, Elvis $\}$; Range: $\{$ January 20, March 3, July 6, November 8, January 8 $\}$ **4.** Not a function   **5.** Function; Domain: $\{-3, -2, -1, 0, 1\}$; Range: $\{0, 1, 2, 3\}$ **6.** Not a function   **7.** Function   **8.** Not a function   **9.** Function   **10.** True   **11.** Function   **12.** Not a function   **13.** 14   **14.** -4 **15.** -18 **16.** 2   **17.** dependent; independent; argument   **18.** $2x - 9$ **19.** $2x - 4$ **20.** domain   **21.** $(-\infty, \infty)$ **22.** $\{x | x \neq 3\}$ **23.** $\{r | r > 0\}$ or $(0, \infty)$ **24. (a)** Independent variable: t; dependent variable: A **(b)** $A(30) \approx 706.86$ square miles. After 30 days, the area contaminated with oil will be a circle covering about 706.86 square miles.   **25.** Function; Domain: $\{$ Virginia, Nevada, Arkansas, Tennessee, Texas $\}$; Range: $\{4, 9, 11, 36\}$ **27.** Not a function; Domain: $\{150, 174, 180\}$; Range: $\{118, 130, 140\}$ **29.** Function; Domain: $\{0, 1, 2, 3\}$; Range: $\{3, 4, 5, 6\}$ **31.** Function; Domain: $\{-3, 1, 4, 7\}$; Range: $\{5\}$ **33.** Not a function; Domain: $\{-10, -5, 0\}$; Range: $\{1, 2, 3, 4\}$ **35.** Function   **37.** Function   **39.** Not a function   **41.** Function   **43.** Not a function   **45.** Function   **47.** Not a function   **49.** Function   **51.** Function   **53. (a)** $f(0) = 3$;   **(b)** $f(3) = 9$ **(c)** $f(-2) = -1$ **55. (a)** $f(0) = 2$ **(b)** $f(3) = -13$ **(c)** $f(-2) = 12$ **57. (a)** $f(0) = 0$ **(b)** $f(3) = 0$ **(c)** $f(-2) = 10$ **59. (a)** $f(0) = 3$ **(b)** $f(3) = -3$ **(c)** $f(-2) = -3$ **61. (a)** $f(-x) = -2x - 5$ **(b)** $f(x + 2) = 2x - 1$ **(c)** $f(2x) = 4x - 5$ **(d)** $-f(x) = -2x + 5$ **(e)** $f(x + h) = 2x + 2h - 5$ **63. (a)** $f(-x) = 7 + 5x$ **(b)** $f(x + 2) = -3 - 5x$ **(c)** $f(2x) = 7 - 10x$ **(d)** $-f(x) = -7 + 5x$ **(e)** $f(x + h) = 7 - 5x - 5h$ **65.** $f(2) = 7$ **67.** $s(-2) = 16$ **69.** $F(-3) = 5$ **71.** $F(4) = -6$ **73.** $\{x | x$ is any real number$\}$ or $(-\infty, \infty)$ **75.** $\{z | z \neq 5\}$ **77.** $\{x | x$ is any real number$\}$ or $(-\infty, \infty)$ **79.** $\left\{ x \middle| x \neq -\dfrac{1}{3} \right\}$ **81.** $C = -6$ **83.** $A = 5$ **85.** $A(r) = \pi r^2$; 50.27 in.^{2} **87.** $G(h) = 15h$; $375   **89. (a)** The dependent variable is the population, P, and the independent variable is the age, a.   **(b)** $P(20) = 4433.3$ thousand; The population of U.S. residents who were 20 years of age in 2014 was roughly 4,433,000.   **(c)** $P(0) = 3694.1$ thousand; $P(0)$ represents residents of the U.S. that are 0 years of age in 2014 was roughly 3,694,000.   **91. (a)** The dependent variable is revenue, R, and the independent variable is price, p.   **(b)** $R(50) = 7500$; Selling MP3 players for $50 will yield a daily revenue of $7500 for the company.   **(c)** $R(120) = 9600$; Selling MP3 players for $120 will yield a daily revenue of $9600 for the company.   **93.** $\{r | r > 0\}$ or $(0, \infty)$ **95.** $\{h | 0 \leq h \leq 60\}$ or $[0, 60]$ **97.** $\{p | 0 \leq p \leq 120\}$ or $[0, 120]$ **99. (a) (i)** -5 **(ii)** 1   **(iii)** 1   **(b) (i)** 13   **(ii)** 4   **(iii)** 4   **101.** Answers will vary.   **103.** A function is a relation between two sets, the domain and the range. The domain is the set of all inputs to the function, and the range is the set of all outputs. In a function, each input in the domain corresponds to exactly one output in the range.   **105.** The four forms of a function are map, ordered pairs, equation, and graph.   **107.** $f(2) = 7$ **109.** $F(-3) = 5$ **111.** $H(7) = 5$ **113.** $F(4) = -0.65$ or $-\dfrac{13}{20}$

Section A.IR3 Functions and Their Graphs

1. graph; function   2. 4; −7

3.   4.   5.   6. (a) Domain: $\{x \mid x \text{ is a real number}\}$; $(-\infty, \infty)$; Range: $\{y \mid y \leq 1\}$; $(-\infty, 1]$
(b) x-intercepts: $(-2, 0)$ and $(2, 0)$; y-intercept: $(0, 1)$ 7. $f(3) = 8$; $(-2, 4)$
8. (a) $f(-3) = -15; f(1) = -3$ (b) Domain: $\{x \mid x \text{ is a real number}\}$ or $(-\infty, \infty)$
(c) Range: $\{y \mid y \text{ is a real number}\}$ or $(-\infty, \infty)$ (d) x-intercepts: $(-2, 0)$, $(0, 0)$ and $(2, 0)$; y-intercept: $(0, 0)$ (e) $\{3\}$

9. (a) No   (b) $f(3) = -2$; $(3, -2)$ is on the graph   (c) $x = 5$; $(5, -8)$ is on the graph   10. yes   11. no   12. yes   13. −2 and 2

14. 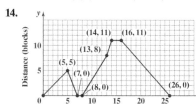   15. $f(x) = 4x - 6$ 17. $h(x) = x^2 - 2$ 19. $G(x) = |x - 1|$ 21. $g(x) = x^3$

23. (a) Domain: $\{x \mid x \text{ is a real number}\}$ or $(-\infty, \infty)$; Range: $\{y \mid y \text{ is a real number}\}$ or $(-\infty, \infty)$ (b) $(0, 2)$ and $(1, 0)$ (c) 1
25. (a) Domain: $\{x \mid x \text{ is a real number}\}$ or $(-\infty, \infty)$; Range: $\{y \mid y \geq -2.25\}$ or $[-2.25, \infty)$ (b) $(-2, 0)$, $(4, 0)$, and $(0, -2)$ (c) −2, 4
27. (a) Domain: $\{x \mid x \text{ is a real number}\}$ or $(-\infty, \infty)$; Range: $\{y \mid y \text{ is a real number}\}$ or $(-\infty, \infty)$ (b) $(-3, 0)$, $(-1, 0)$, $(2, 0)$, and $(0, -3)$
(c) −3, −1, 2   29. (a) Domain: $\{x \mid x \text{ is a real number}\}$ or $(-\infty, \infty)$; Range: $\{y \mid y \geq 0\}$ or $[0, \infty)$ (b) $(-3, 0)$, $(3, 0)$, and $(0, 9)$ (c) −3, 3
31. (a) Domain: $\{x \mid x \leq 4\}$ or $(-\infty, 4]$; Range: $\{y \mid y \leq 3\}$ or $(-\infty, 3]$ (b) $(-2, 0)$ and $(0, 2)$ (c) −2   33. (a) $f(-7) = -2$ (b) $f(-3) = 3$
(c) $f(6) = 2$ (d) negative   (e) $\{-6, -1, 4\}$ (f) $\{x \mid -7 \leq x \leq 6\}$ or $[-7, 6]$ (g) $\{y \mid -2 \leq y \leq 3\}$ or $[-2, 3]$ (h) $(-6, 0)$, $(-1, 0)$ and $(4, 0)$
(i) $(0, -1)$ (j) $\{-7, 2\}$ (k) $x = -3$ (l) −6, −1, 4   35. (a) $F(-2) = 3$ (b) $F(3) = -6$ (c) $x = -1$ (d) $(-4, 0)$ (e) $(0, 2)$ 37. (a) no
(b) $f(3) = 3$; $(3, 3)$ (c) 4; $(4, 7)$ (d) no   39. (a) yes   (b) $g(6) = 1$; $(6, 1)$ (c) −12; $(-12, 10)$ (d) yes   41. (c)   43. (e)   45. (d)

47. 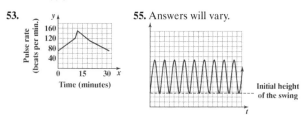   49.   51. (a) III   (b) I   (c) IV   (d) V   (e) II

53.   55. Answers will vary.   57. The person's weight increases until age 30, then oscillates back and forth between 158 pounds and 178 pounds, then slowly levels off at about 150 pounds.
59. One possibility: Answers will vary.
61. A function cannot have more than one output for a given input. So, there cannot be two outputs for the input 0.
63. The range of a function is the set of all outputs of the elements in the domain.   65. The zeros of a function are the input values that make the function equal to zero.

Section A.IR4 Linear Functions and Models

1. slope; y-intercept   2. line   3. False   4. −2; $(0, 3)$

5.   6.   7.   8. 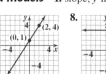   9. 5   10. −8   11. 12

12. (a) $\{x \mid x \geq 0\}$; $[0, \infty)$ 13. (a) $C(x) = 81x + 2000$ 14. (a) $C(x) = 0.18x + 250$
(b) $40   (c) $68   (d) 130 miles   (b) $2405   (c) 10 bicycles   (b) $[0, \infty)$ (c) $307.60
(e) 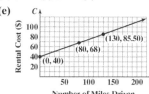   (d)   (d) 180 miles
(f) You may drive between 0 and 250 miles.   (e)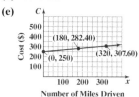

15. $F(x) = 5x - 2$ 17. $G(x) = -3x + 7$ 19. $H(x) = -2$ 21. $f(x) = \frac{1}{2}x - 4$ 23. $F(x) = -\frac{5}{2}x + 5$ 25. $G(x) = -\frac{3}{2}x$

   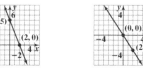

27. -5 **29.** 8 **31.** 6 **33.** 9 **35.** nonlinear **37.** linear; positive slope

39. (a) 3 **(b)** $(0,2)$ **(c)** $-\dfrac{2}{3}$ **(d)** 1; $(1,5)$ **41. (a)** $\{3\}; -2; (3,-2); (3,-2)$ **43.** $f(x)=2x+2; f(-2)=-2$
(e) $\{x|x\le -1\}; (-\infty,-1]$ **(b)** $\{x|x>3\}; (3,\infty)$ **45.** $h(x)=-\dfrac{7}{4}x+\dfrac{49}{4}; h\left(\dfrac{1}{2}\right)=\dfrac{91}{8}$
(f) 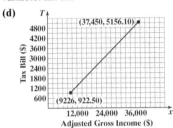 **(c)** 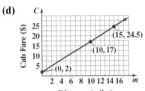 **47. (a)** 3 **(b)** -1 **(c)** 2
(d) x-intercept: $(2,0)$; y-intercept: $(0,-2)$
(e) $f(x)=x-2$

49. (a) $\{x|9926\le x\le 37{,}450\}; [9926, 37{,}450]$ **(b)** $\$2538.60$ **51. (a)** $\{m|m\ge 0\}$ or $[0,\infty)$
(c) Independent variable: adjusted gross income; dependent variable: tax bill   **(b)** 2; The base fare is \$2.00 before any distance is driven.
(c) \$9.50
(d) 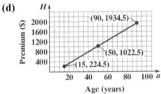 **(d)**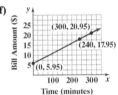

(e) A person can travel 7.5 miles in a cab for \$13.25.
(f) A person can ride between 0 miles and 25 miles.

(e) \$23,051

53. (a) The independent variable is age; the dependent variable is insurance cost. **55. (a)** $B(m)=0.05m+5.95$ **57. (a)** $V(x)=-900x+2700$
(b) $\{a|15\le a\le 90\}$ or $[15,90]$ **(b)** The independent variable is minutes; the dependent variable is bill amount.   **(b)** $\{x|0\le x\le 3\}$ or $[0,3]$
(c) \$566.50   **(c)** \$1800
(d) 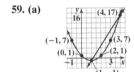   **(c)** $\{m|m\ge 0\}$ or $[0,\infty)$ **(d)** \$20.95   **(d)** The V-intercept is $(0,2700)$, and the x-intercept is $(3,0)$.
(e) 240 minutes
(f)   **(e)** After two years
(f)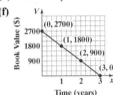

(e) 48 years

(g) You can talk between 0 minutes and 250 minutes.

59. (a) [graph] **(b)** 6 **(c)** $y=6x-7$ **(d)** [graph] $x=3$: slope $=4$; $y=4x-5$; $x=2$: slope $=2$; $y=2x-3$; $x=1.5$: slope $=1$; $y=x-2$; $x=1.1$: slope $=0.2$; $y=0.2x-1.2$ **(e)** As x approaches 1, the slope decreases and gets closer to 0.

Getting Ready for Exponential and Logarithmic Functions

1. base; power; exponent **2.** a^{m+n} **3.** 125 **4.** -243 **5.** y^7 **6.** $-10x^7$ **7.** $-6y^5$ **8.** True **9.** 25 **10.** y^2 **11.** $\dfrac{8}{5}a$ **12.** $-\dfrac{3}{2}b^2$ **13.** $1; 0$ **14.** $\dfrac{1}{a^n}; 0$

15. $\dfrac{1}{125}$ **16.** $\dfrac{5}{z^7}$ **17.** x^4 **18.** $5y^3$ **19.** -1 **20.** 1 **21.** $\dfrac{9}{16}$ **22.** -64 **23.** $\dfrac{x^2}{9}$ **24.** 20 **25.** $\dfrac{1}{36}$ **26.** 100 **27.** $\dfrac{20x^3}{y}$ **28.** $\dfrac{2}{3}ab^4$ **29.** $-\dfrac{3}{2}b^8$ **30.** $\dfrac{10t^5}{3s^3}$

31. 64 **32.** 1 **33.** 4096 **34.** a^{15} **35.** $\dfrac{1}{z^{18}}$ **36.** s^{21} **37.** $125y^3$ **38.** 1 **39.** $81x^8$ **40.** $\dfrac{1}{16a^6}$ **41.** $\dfrac{z^4}{81}$ **42.** $\dfrac{32}{x^5}$ **43.** $\dfrac{x^8}{y^{12}}$ **44.** $\dfrac{27}{a^6b^{12}}$ **45.** -25 **47.** $-\dfrac{1}{25}$

49. -1 **51.** $\dfrac{81}{16}$ **53.** $-\dfrac{1}{27}$ **55.** -64 **57.** 288 **59.** 1 (assuming $x\ne 0$) **61.** $-10t^5$ **63.** $\dfrac{5x^2}{y}$ **65.** $4x^2y$ **67.** $\dfrac{3b^3}{2a}$ **69.** $\dfrac{1}{x^8}$ **71.** $27x^6y^3$ **73.** $\dfrac{64}{z^3}$ **75.** $\dfrac{a^6}{9}$

77. x^6 cubic units **79.** 2^{x+1} **81.** 3^{10x+3} **83.** 625 **85.** $\dfrac{1}{2401}$ **87.** No, your friend is incorrect. He added exponents instead of multiplying.
89. Explanations will vary. Essentially, if a were to equal zero and m, n, or $m+n$ were negative, we would end up with division by 0, which is not defined. For example, $0^{-2}=\dfrac{1}{0^2}=\dfrac{1}{0}$, which is undefined. If m, n or $m+n$ were to equal 0, we would end up with 0^0, which is indeterminate.
91. Explanations will vary. If a or b were to equal 0 when n was negative, we would end up with division by 0, which is not defined. If a or b were to equal 0 when n was zero, we end up with 0^0, which is indeterminate. **93.** Explanations will vary. One possibility: $\dfrac{x^5}{x^2}=\dfrac{x\cdot x\cdot x\cdot x\cdot x}{x\cdot x}=x^3=x^{5-2}$
95. Explanations will vary. One possibility: $(xy)^3=(xy)(xy)(xy)=x\cdot x\cdot x\cdot y\cdot y\cdot y=x^3y^3$

Section A.IR5 Exponential Functions

1. $>$; $\neq$ **2. (a)** 3.249009585   **(b)** 3.317278183   **(c)** 3.321880096   **(d)** 3.32211036   **(e)** 3.321997085

3. The domain of f is all real numbers or, in interval notation, $(-\infty, \infty)$. The range of f is $\{y|y > 0\}$ or, in interval notation, $(0, \infty)$.   **4.** $\left(-1, \dfrac{1}{a}\right); (0, 1); (1, a)$ **5.** True   **6.** False   **7.** The domain of f is all real numbers or, in interval notation, $(-\infty, \infty)$. The range of f is $\{y|y > 0\}$ or, in interval notation, $(0, \infty)$.

8. The domain of f is all real numbers or, in interval notation, $(-\infty, \infty)$. The range of f is $\{y|y > 0\}$ or, in interval notation, $(0, \infty)$.

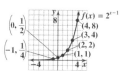

9. The domain of f is all real numbers or, in interval notation, $(-\infty, \infty)$. The range of f is $\{y|y > 1\}$ or, in interval notation, $(1, \infty)$.   **10.** 2.71828   **11. (a)** 54.598   **(b)** 0.018   **12.** $\{3\}$ **13.** $\{2\}$ **14. (a)** 0.918 or 91.8%   **(b)** 0.998 or 99.8%   **15. (a)** approximately 6.91 grams   **(b)** 5 grams   **(c)** 0.625 gram   **(d)** approximately 0.247 gram   **16. (a)** \$2102.32   **(b)** \$4227.41   **(c)** \$8935.49   **17. (a)** 11.212   **(b)** 11.587   **(c)** 11.664   **(d)** 11.665   **(e)** 11.665   **19. (a)** 73.517   **(b)** 77.708   **(c)** 77.924   **(d)** 77.881   **(e)** 77.880   **21.** g   **23.** e   **25.** f   **27.** h

29. Domain: all real numbers or $(-\infty, \infty)$; Range: $\{y|y > 0\}$ or $(0, \infty)$

31. 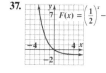 Domain: all real numbers or $(-\infty, \infty)$; Range: $\{y|y > 0\}$ or $(0, \infty)$

33. Domain: all real numbers or $(-\infty, \infty)$; Range: $\{y|y > 0\}$ or $(0, \infty)$

35. Domain: all real numbers or $(-\infty, \infty)$; Range: $\{y|y > 3\}$ or $(3, \infty)$

37. 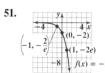 Domain: all real numbers or $(-\infty, \infty)$; Range: $\{y|y > -1\}$ or $(-1, \infty)$

39. 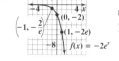 Domain: all real numbers or $(-\infty, \infty)$; Range: $\{y|y > 0\}$ or $(0, \infty)$

41. (a) 21.217   **(b)** 22.472   **(c)** 22.460   **(d)** 22.460   **(e)** 22.459   **43.** 7.389   **45.** 0.135   **47.** 9.974   **49.** 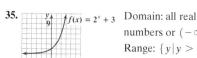 Domain: all real numbers or $(-\infty, \infty)$; Range: $\{y|y > 0\}$ or $(0, \infty)$

51. 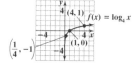 Domain: all real numbers or $(-\infty, \infty)$; Range: $\{y|y < 0\}$ or $(-\infty, 0)$ **53.** $\{5\}$ **55.** $\{-4\}$ **57.** $\{5\}$ **59.** $\{5\}$ **61.** $\left\{\dfrac{3}{2}\right\}$ **63.** $\{1\}$ **65.** $\{9\}$ **67.** $\{-2\}$ **69.** $\{-2\}$ **71.** $\{2\}$ **73. (a)** $f(3) = 8; (3, 8)$ **(b)** $x = -3; \left(-3, \dfrac{1}{8}\right)$ **75. (a)** $g(-1) = -\dfrac{3}{4}; \left(-1, -\dfrac{3}{4}\right)$ **(b)** $x = 2; (2, 15)$ **77. (a)** $H(-3) = 24; (-3, 24)$ **(b)** $x = 2; \left(2, \dfrac{3}{4}\right)$

79. (a) approximately 342.5 million people   **(b)** approximately 455.7 million people   **(c)** Answers may vary. One possibility is that the population is not growing exponentially.   **81. (a)** \$5100.92   **(b)** \$5308.92   **(c)** \$5525.29   **83. (a)** \$2318.55   **(b)** \$2322.37   **(c)** \$2323.23   **(d)** \$2323.65   **(e)** The future value is higher with more compounding periods.   **85. (a)** \$19,841   **(b)** \$15,365   **(c)** \$10,471   **87. (a)** approximately 95.105 grams   **(b)** 50 grams   **(c)** 25 grams   **(d)** approximately 0.661 gram   **89. (a)** approximately 300.233°F   **(b)** approximately 230.628°F   **(c)** yes   **91. (a)** approximately 29 words   **(b)** approximately 38 words   **93. (a)** approximately 0.238 ampere   **(b)** approximately 0.475 ampere   **95.** $y = 3^x$ **97.** As x increases, the graph increases very rapidly. As x decreases, the graph approaches the x-axis.   **99.** Because 12 cannot be written as 2 raised to an integer (or rational) power.

Section A.IR6 Logarithmic Functions

1. $x = a^y$; $>$; $\neq$ **2.** $3 = \log_4 64$ **3.** $-2 = \log_p 8$ **4.** $2^4 = 16$ **5.** $a^5 = 20$ **6.** $5^{-3} = z$ **7.** 2   **8.** -3 **9.** 2   **10.** -1 **11.** $\{x|x > -3\}$ or $(-3, \infty)$ **12.** $\left\{x \middle| x < \dfrac{5}{2}\right\}$ or $\left(-\infty, \dfrac{5}{2}\right)$

13. The domain of f is $\{x|x > 0\}$ or, in interval notation, $(0, \infty)$. The range of f is all real numbers or, in interval notation, $(-\infty, \infty)$.

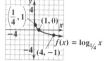

14. The domain of f is $\{x|x > 0\}$ or, in interval notation, $(0, \infty)$. The range of f is all real numbers or, in interval notation, $(-\infty, \infty)$.

15. 3.146 **16.** 1.569 **17.** −0.523 **18.** {16} **19.** {4} **20.** {e^{-2}} **21.** {10,020} **22.** 100 decibels **23.** $3 = \log_4 64$ **25.** $-3 = \log_2\left(\frac{1}{8}\right)$ **27.** $\log_a 19 = 3$ **29.** $\log_5 c = -6$ **31.** $2^4 = 16$ **33.** $3^{-2} = \frac{1}{9}$ **35.** $5^{-3} = a$ **37.** $a^2 = 4$ **39.** $\left(\frac{1}{2}\right)^y = 12$ **41.** 0 **43.** 3 **45.** −2 **47.** 4 **49.** 4 **51.** $\frac{1}{2}$ **53.** $\{x|x > 4\}$ or $(4, \infty)$ **55.** $\{x|x > 0\}$ or $(0, \infty)$ **57.** $\left\{x\mid x > \frac{2}{3}\right\}$ or $\left(\frac{2}{3}, \infty\right)$ **59.** $\left\{x\mid x > -\frac{1}{2}\right\}$ or $\left(-\frac{1}{2}, \infty\right)$ **61.** $\left\{x\mid x < \frac{1}{4}\right\}$ or $\left(-\infty, \frac{1}{4}\right)$

63. 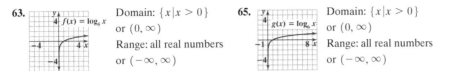 Domain: $\{x|x > 0\}$ or $(0, \infty)$; Range: all real numbers or $(-\infty, \infty)$ **65.** Domain: $\{x|x > 0\}$ or $(0, \infty)$; Range: all real numbers or $(-\infty, \infty)$ **67.** Domain: $\{x|x > 0\}$ or $(0, \infty)$; Range: all real numbers or $(-\infty, \infty)$

69. $\ln 12 = x$ **71.** $e^4 = x$ **73.** -1 **75.** 3 **77.** 1.826 **79.** 1.686 **81.** −0.456 **83.** −1.609 **85.** 0.097 **87.** −0.981 **89.** {4} **91.** $\left\{\frac{13}{2}\right\}$ **93.** {6} **95.** {10} **97.** {e^5} **99.** $\left\{\frac{11}{20}\right\}$ **101.** {−3} **103.** {4} **105.** (a) $f(16) = 4; 116, 42$ (b) $x = \frac{1}{8}; \left(\frac{1}{8}, -3\right)$ **107.** (a) $G(7) = \frac{3}{2}; \left(7, \frac{3}{2}\right)$ (b) $x = 15; 115, 22$ **109.** $a = 4$ **111.** 20 decibels **113.** 130 decibels **115.** approximately 7.8 on the Richter scale **117.** approximately 794,328 **119.** (a) 12; basic (b) 5; acidic (c) 2; acidic (d) $10^{-7.4}$ mole per liter **121.** The base of $f(x) = \log_a x$ cannot equal 1 because $y = \log_a x$ is equivalent to $x = a^y$ and a does not equal 1 in the exponential function. In addition, the graph would be a vertical line ($x = 1$), which is not a function. **123.** The domain of $f(x) = \log_a(x^2 + 1)$ is the set of all real numbers because $x^2 + 1 > 0$ for all x.

Section A.IR7 Properties of Logarithms **1.** 0 **2.** 0 **3.** 1 **4.** 1 **5.** $\sqrt{2}$ **6.** 0.2 **7.** 1.2 **8.** −4 **9.** False **10.** $\log_4 9 + \log_4 5$ **11.** $\log 5 + \log w$ **12.** $\log_7 9 - \log_7 5$ **13.** $\ln p - \ln 3$ **14.** $\log_2 3 + \log_2 m - \log_2 n$ **15.** $\ln q - \ln 3 - \ln p$ **16.** $1.6 \log_2 5$ **17.** $5 \log b$ **18.** $2 \log_4 a + \log_4 b$ **19.** $2 + 4 \log_3 m - \frac{1}{3} \log_3 n$ **20.** 2 **21.** $\log_3\left(\frac{x+4}{x-1}\right)$ **22.** $\log_5 \frac{x}{8}$ **23.** $\log_2 \frac{x^2 + 3x + 2}{x^2}$ **24.** 10; 3; 10; 3 **25.** 3.155 **26.** 2.807 **27.** 3 **29.** −7 **31.** 5 **33.** 2 **35.** 1 **37.** 0 **39.** $a + b$ **41.** $2b$ **43.** $2a + b$ **45.** $\frac{1}{2}a$ **47.** $\log a + \log b$ **49.** $4 \log_5 x$ **51.** $\log_2 x + 2 \log_2 y$ **53.** $2 + \log_5 x$ **55.** $2 - \log_7 y$ **57.** $2 + \ln x$ **59.** $3 + \frac{1}{2} \log_3 x$ **61.** $2 \log_5 x + \frac{1}{2} \log_5(x^2 + 1)$ **63.** $4 \log x - \frac{1}{3} \log(x - 1)$ **65.** $\frac{1}{2} \log_7(x + 1) - \frac{1}{2} \log_7 x$ **67.** $\log_2 x + 2 \log_2(x - 1) - \frac{1}{2} \log_2(x + 1)$ **69.** 2 **71.** $\log(3x)$ **73.** 2 **75.** 4 **77.** $\log_3 x^3$ **79.** $\log_4\left(\frac{x+1}{x}\right)$ **81.** $\ln(x^2 y^3)$ **83.** $\log_3[\sqrt{x}(x-1)^3]$ **85.** $\log(x^2)$ **87.** $\log(x\sqrt{xy})$ **89.** $\log_8(x-1)$ **91.** $\log\left(\frac{x^{12}}{10}\right)$ **93.** 3.322 **95.** 0.528 **97.** −2.680 **99.** 4.644 **101.** 3 **103.** 1

105. $\log_a(x + \sqrt{x^2 - 1}) + \log_a(x - \sqrt{x^2 - 1})$
$= \log_a[(x + \sqrt{x^2 - 1})(x - \sqrt{x^2 - 1})]$
$= \log_a[x^2 - x\sqrt{x^2 - 1} + x\sqrt{x^2 - 1} - (x^2 - 1)]$
$= \log_a(x^2 - x^2 + 1)$
$= \log_a 1$
$= 0$

107. If $f(x) = \log_a x$, then
$f(AB) = \log_a(AB)$
$= \log_a A + \log_a B$
$= f(A) + f(B)$

109. Answers will vary. One possibility: The logarithm of the product of two expressions equals the sum of the logarithms of the two expressions. **111.** Answers will vary. One possibility: $\log_2(2 + 4) \neq \log_2 2 + \log_2 4$

Section A.IR8 Exponential and Logarithmic Equations **1.** $\left\{\frac{\ln 11}{\ln 2}\right\}$ or $\left\{\frac{\log 11}{\log 2}\right\}$; {3.459} **2.** $\left\{\frac{\ln 3}{2\ln 5}\right\}$ or $\left\{\frac{\log 3}{2\log 5}\right\}$; {0.341} **3.** $\left\{\frac{\ln 5}{2}\right\}$; {0.805} **4.** $\left\{-\frac{\ln\left(\frac{20}{3}\right)}{4}\right\}$; {−0.474} **5.** (a) approximately 2.85 days (b) approximately 32.52 days **6.** (a) approximately 6.77 years (b) approximately 11.58 years **7.** $\left\{\frac{1}{\log 2}\right\} \approx \{3.322\}$ or $\left\{\frac{\ln 10}{\ln 2}\right\} \approx \{3.322\}$ **9.** $\left\{\frac{\log 20}{\log 5}\right\} \approx \{1.861\}$ or $\left\{\frac{\ln 20}{\ln 5}\right\} \approx \{1.861\}$ **11.** $\left\{\frac{\log 7}{\log\left(\frac{1}{2}\right)}\right\} \approx \{-2.807\}$ or $\left\{\frac{\ln 7}{\ln\left(\frac{1}{2}\right)}\right\} \approx \{-2.807\}$ **13.** $\{\ln 5\} \approx \{1.609\}$ **15.** $\{\log 5\} \approx \{0.699\}$ **17.** $\left\{\frac{\log 13}{2\log 3}\right\} \approx \{1.167\}$ or $\left\{\frac{\ln 13}{2\ln 3}\right\} \approx \{1.167\}$ **19.** $\left\{\frac{\log 3}{4\log\left(\frac{1}{2}\right)}\right\} \approx \{-0.396\}$ or $\left\{\frac{\ln 3}{4\ln\left(\frac{1}{2}\right)}\right\} \approx \{-0.396\}$ **21.** $\left\{\frac{\log\left(\frac{5}{4}\right)}{\log 2}\right\} \approx \{0.322\}$ or $\left\{\frac{\ln\left(\frac{5}{4}\right)}{\ln 2}\right\} \approx \{0.322\}$ **23.** $\{\ln 6\} \approx \{1.792\}$ **25.** $\left\{\frac{\log 0.2}{\log 3 - \log 0.2}\right\} \approx \{-0.594\}$ or $\left\{\frac{\ln 0.2}{\ln 3 - \ln 0.2}\right\} \approx \{-0.594\}$ **27.** $\left\{\frac{\log 7}{3\log 5}\right\} \approx \{0.403\}$ or $\left\{\frac{\ln 7}{3\ln 5}\right\} \approx \{0.403\}$ **29.** $\{\ln 15\} \approx \{2.708\}$ **31.** $\left\{-\frac{2}{5}\right\}$ **33.** (a) 2023 (b) 2053 **35.** (a) approximately 16.8 years (b) approximately 34.7 years **37.** (a) approximately 2.188 years (b) approximately 10.782 years (c) approximately 23.372 years **39.** (a) approximately 2.099 seconds (b) 27.62 seconds (c) approximately 45.876 seconds **41.** (a) approximately 5 minutes (b) approximately 11 minutes **43.** (a) approximately 82 minutes (b) approximately 396 minutes (or 6.6 hours) **45.** (a) approximately 9 years (b) $t = \dfrac{\log 2}{n \log\left(1 + \dfrac{r}{n}\right)}$ (c) approximately 8.693 years, which is about the same as the result from the Rule of 72 **47.** 29 minutes